12·95

Higher Electrical Technolo

PROPERTY OF

allan Barclay

173 KIRKLAND WALK

METHIL FIFE

KY8 2HZ.

Higher Electrical Technology

J. O. BIRD

BSc(Hons), CEng, MIEE, FIMA,
TEng, MIElecIE, FCollP

Heinemann Newnes

Heinemann Newnes
An imprint of Heinemann Professional Publishing Ltd
Halley Court, Jordan Hill, Oxford OX2 8EJ

OXFORD LONDON MELBOURNE AUCKLAND SINGAPORE
IBADAN NAIROBI GABORONE KINGSTON

First published 1987
Reprinted 1988, 1990

© J.O. Bird 1987

British Library Cataloguing in Publication Data
Bird, J.O.
   Higher electrical technology.
   1. Electric engineering
   I. Title
   621.3   TK145

ISBN 0 434 90144 X

Printed in Great Britain by
Courier International Ltd, Tiptree, Essex

To Elizabeth, Amanda and Yvonne

Contents

Preface

The study of electrical principles and circuit theory is essential to the BTEC Higher National Certificate/Diploma courses in electrical and electronic engineering and telecommunications engineering and in the earlier years of degree courses in electrical and electronic engineering. Syllabuses currently in use for BTEC Higher National Certificate/Diploma studies involving electrical principles and circuit theory vary from college to college, the material being drawn from the extensive BTEC suggested material or being 'college devised'.

I am very grateful for the response given by colleges to a feasibility study on actual material being used; the choice of the contents of this book is the result of the conclusions drawn from that study. It was felt that there was sufficient common ground upon which a useful textbook could be based. All of the topic areas included in this book are likely to be studied on most BTEC Higher National Certificate/Diploma courses in electrical and electronic engineering and in telecommunications engineering and are also basic to undergraduate studies.

It is not really possible to acquire a thorough understanding of electrical engineering principles and circuit theory without working through a large number of numerical problems. It is for this reason that 210 detailed worked problems are included in this book, together with over 400 further problems with answers. Understanding is reinforced by the 500 diagrams that are used throughout to illustrate both theory and problems.

The competent application of complex numbers is essential to higher electrical engineering and circuit theory. Applications of complex numbers that are explained and exemplified include those in series and paralled a.c. circuits, in the calculation of power in a.c. circuits, in a.c. bridges, in the use of Kirchhoff's laws with a.c. circuits, with mesh and nodal analysis, in Thévenin's and Norton's theorems, and in delta–star and star–delta transformations. The text also includes studies of complex waveforms, the magnetic and dielectric materials used in electrical engineering, field theory and attenuators, and provides an introduction to transmission line theory.

The book has no pretensions to be exhaustive; an attempt has been made to deal with the major topic areas likely to be covered by the majority of colleges, polytechnics and universities. Each topic considered is presented in a way that assumes in the reader only the knowledge required for the BTEC National Certificate in Electrical and Electronic Principles, or its equivalent.

The graphical symbols used throughout the text are those set out in BS3939: Parts 2–13: 1985.

I would like to thank Mr David Stokes, Mr Bob Cotton and Mr Anthony May for their helpful advice, and also to express my appreciation of the friendly cooperation and helpful advice given by the publishers.

Thanks are also due to Mrs Elaine Woolley for the excellent typing of the manuscript.

Finally, a words of thanks is due to my wife Elizabeth for her patience, help and encouragement during the preparation of this book.

John O. Bird

Highbury College of Technology
Portsmouth
January 1987

1 Complex numbers

1. Introduction

The solutions of the quadratic equation $ax^2 + bx + c = 0$ may be obtained by using the quadratic formula:

$$x = \frac{-b \pm \sqrt{(b^2 - 4ac)}}{2a}$$

Hence the solutions of $2x^2 + x - 3 = 0$ are

$$x = \frac{-1 \pm \sqrt{[1^2 - 4(2)(-3)]}}{2(2)}$$

$$= \frac{-1 \pm \sqrt{25}}{4} = -\tfrac{1}{4} \pm \tfrac{5}{4}$$

i.e., $x = 1$ or $x = -1\tfrac{1}{2}$.

However, a problem exists if $(b^2 - 4ac)$ in the quadratic formula results in a negative number, since we cannot obtain in real terms the square root of a negative number. The only numbers we have met to date have been real numbers. These are either integers (such as $+1$, $+5$, -7, etc.) or rational numbers (such as $\tfrac{4}{1}$ or 4.000, $-\tfrac{7}{9}$ or $-0.7\dot{7}$, $\tfrac{1}{2}$ or 0.5000, etc.) or irrational numbers (such as $\pi = 3.141\,592\ldots$, $\sqrt{3} = 1.732\ldots$, etc).

If $x^2 - 2x + 5 = 0$, then

$$x = \frac{2 \pm \sqrt{[(-2)^2 - 4(1)(5)]}}{2(1)} = \frac{2 \pm \sqrt{-16}}{2}$$

In order to deal with such a problem as determining $\sqrt{-16}$, the concept of complex numbers was developed.

2. Definition of a complex number

(a) Imaginary numbers

Let b and c be real numbers. An **imaginary number** is written in the form jb or jc (i.e. $j \times b$ or $j \times c$) where operator j is defined by the following two rules:

(i) For addition, $\qquad\qquad jb + jc = j(b + c)$ $\qquad\qquad\qquad$ (1)

(ii) For multiplication, $\qquad jb \times jc = -bc$ $\qquad\qquad\qquad\quad$ (2)

i.e. $j^2 bc = -bc$

Thus $j^2 = -1$ or $j = \sqrt{-1}$.

It is immaterial whether the operator j is placed in front of or following the number, i.e. $j4 = 4j$, and so on. An imaginary number $j3$ means $(\sqrt{-1}) \times 3$ and $-j2$ means $-(\sqrt{-1}) \times 2$ or $(-2) \times (\sqrt{-1})$.

Similarly, from equation (1), $j2 + j5 = j7$ and $j5 - j2 = j3$.

(b) Complex numbers

From section 1, $x^2 - 2x + 5 = 0$, from which

$$x = \frac{2 \pm \sqrt{-16}}{2}$$

$\sqrt{-16}$ can be split into $(\sqrt{-1}) \times \sqrt{16}$, i.e. $j\sqrt{16}$ or $\pm j4$. Hence

$$x = \frac{2 \pm j4}{2} = 1 \pm j2$$

i.e. $x = 1 + j2$ or $x = 1 - j2$.

The solutions of the quadratic equation $x^2 - 2x + 5 = 0$ are of the form $a + jb$, where a is a real number and jb is an imaginary number. Numbers in the form $a + jb$ are called **complex numbers**. Hence $1 + j2$, $5 - j7$, $-2 + j3$ and $-\pi + j\sqrt{2}$ are all examples of complex numbers.

In algebra, if a quantity x is added to a quantity $3y$ the result is written as $x + 3y$ since x and y are separate quantities. Similarly, it is important to appreciate that real and imaginary numbers are different types of numbers and must be kept separate.

When imaginary numbers were first introduced, the symbol i (i.e. the first letter of the word imaginary) was used to indicate $\sqrt{-1}$ and this symbol is still used in pure mathematics. However, in engineering, the symbol i indicates electric current, and to avoid any possible confusion the next letter in the alphabet, i.e. j, is used to represent $\sqrt{-1}$.

Complex numbers are widely used in the analysis of series, parallel and series-parallel electrical circuits supplied by alternating voltages (see chapters 2 to 4), in deriving balance equations with a.c. bridges (see chapter 5), in analysing a.c. circuits using Kirchhoff's laws, mesh and nodal analysis and the superposition theorem (see chapter 7) and with Thévenin's and Norton's theorems (see chapter 8) and also in many other aspects of higher electrical engineering. The advantage of the use of complex numbers is that the manipulative processes become simply algebraic processes.

Worked problems on the introduction to complex numbers

Problem 1. Solve the following quadratic equations:

(a) $x^2 + 9 = 0$.
(b) $4y^2 - 3y + 5 = 0$.

(a) $x^2 + 9 = 0$.
    Thus $x^2 = -9$
    and $x = \sqrt{-9} = (\sqrt{-1})(\sqrt{9}) = (\sqrt{-1})(\pm 3) = (j)(\pm 3) = \pm j3$.

(b) $4y^2 - 3y + 5 = 0$. Using the quadratic formula,

$$y = \frac{-(-3) \pm \sqrt{[(-3)^2 - 4(4)(5)]}}{2(4)} = \frac{3 \pm \sqrt{(9 - 80)}}{8}$$

$$= \frac{3 \pm \sqrt{-71}}{8} = \frac{3 \pm (\sqrt{-1})(\sqrt{71})}{8} = \frac{3 \pm j\sqrt{71}}{8}$$

$$= \frac{3}{8} \pm j\frac{\sqrt{71}}{8} = \mathbf{0.375 \pm j1.053}, \text{ correct to three decimal places}$$

Problem 2. Evaluate: (a) j^3; (b) j^4; (c) j^5; (d) j^6; (e) j^7; (f) j^{19}.

(a) $j = \sqrt{-1}$ and $j^2 = -1$. $j^3 = j \times j^2 = (j)(-1) = -\mathbf{j}$.
(b) $j^4 = j \times j^3 = (j)(-j) = -j^2 = +\mathbf{1}$. Alternatively, $j^4 = (j^2)^2 = (-1)^2 = +\mathbf{1}$.
(c) $j^5 = j \times j^4 = j \times (j^2)^2 = j \times (+1) = \mathbf{j}$.
(d) $j^6 = (j^2)^3 = (-1)^3 = -\mathbf{1}$.
(e) $j^7 = j \times j^6 = j \times (j^2)^3 = j \times (-1) = -\mathbf{j}$.
(f) $j^{19} = j \times j^{18} = j \times (j^2)^9 = j \times (-1)^9 = -\mathbf{j}$.

Further problems on the introduction to complex numbers may be found in section 9, problems 1 to 6, page 18.

3. The Argand diagram

A complex number can be represented pictorially on rectangular or cartesian axes. The horizontal axis is used to represent the real axis and the vertical axis is used to represent the imaginary axis. Such a diagram is called an **Argand diagram** (named after the eighteenth-century French mathematician), and is shown in Fig. 1(a).

In Fig. 1(b) the point A represents the complex number $2 + j3$ and is obtained by plotting the coordinates $(2, j3)$ as in graphical work. Often an

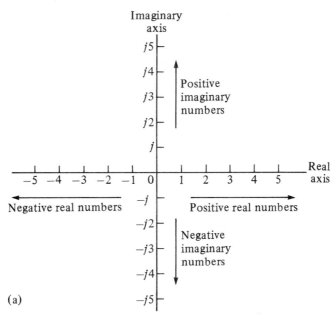

(a)

Figure 1 The Argand diagram

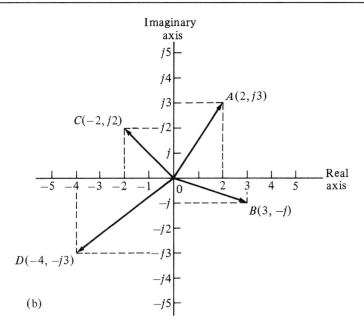

Figure 1 (*continued*)

(b)

arrow is drawn from the origin to the point as shown. Figure 1(b) also shows the Argand points B, C and D representing the complex numbers $3 - j$, $-2 + j2$ and $-4 - j3$ respectively.

A complex number of the form $a + jb$ is called a **cartesian or rectangular complex number**. The significance of the j operator is shown in Fig. 2. In Fig. 2(a) the number 4 (i.e., $4 + j0$) is shown drawn as a phasor horizontally to the right of the origin on the real axis. (Such a phasor could represent an alternating current, $i = 4 \sin \omega t$ amperes, at time t is zero.)

The number $j4$ (i.e., $0 + j4$) is shown in Fig. 2(b) drawn vertically upwards from the origin on the imaginary axis. Hence multiplying the number 4 by the operator j results in an anticlockwise phase-shift of $90°$ without altering its magnitude.

Multiplying $j4$ by j gives $j^2 4$, i.e. -4, and is shown in Fig. 2(c) as a phasor four units long on the horizontal real axis to the left of the origin—an anticlockwise phase-shift of $90°$ compared with the position shown in Fig. 2(b). Thus multiplying by j^2 reverses the original direction of a phasor.

Multiplying $j^2 4$ by j gives $j^3 4$, i.e. $-j4$, and is shown in Fig. 2(d) as a phasor four units long on the vertical, imaginary axis downward from the origin—an anticlockwise phase-shift of $90°$ compared with the position shown in Fig. 2(c).

Multiplying $j^3 4$ by j gives $j^4 4$, i.e. 4, which is the original position of the phasor shown in Fig. 2(a).

Summarising, application of the operator j to any number rotates it $90°$ anticlockwise on the Argand diagram, multiplying a number by j^2 rotates it $180°$ anticlockwise, multiplying a number by j^3 rotates it $270°$ anticlockwise and multiplication by j^4 rotates it $360°$ anticlockwise, i.e., back to its original position. In each case the phasor is unchanged in its magnitude.

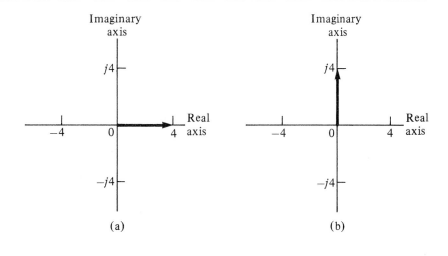

(a)   (b)

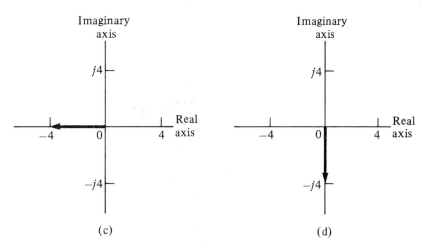

Figure 2   (c)   (d)

By similar reasoning, if a phasor is operated on by $-j$ then a phasor shift of $-90°$ (i.e., clockwise direction) occurs, again without change of magnitude.

4. Operations involving cartesian complex numbers

(a) Addition and subtraction

Let two complex numbers be represented by $Z_1 = a + jb$ and $Z_2 = c + jd$. Two complex numbers are added/subtracted by adding/subtracting separately the two real parts and the two imaginary parts. Hence

$$Z_1 + Z_2 = (a + jb) + (c + jd)$$
$$= (a + c) + j(b + d)$$

and

$$Z_1 - Z_2 = (a + jb) - (c + jd)$$
$$= (a - c) + j(b - d)$$

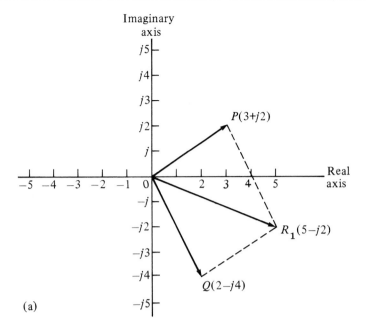

(a)

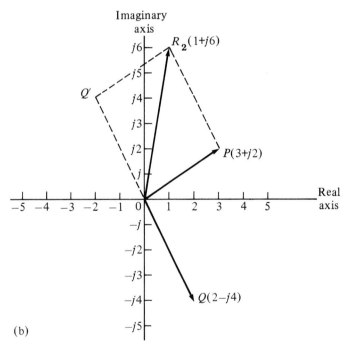

Figure 3 (a) $0P + 0Q = 0R_1$
(b) $0P - 0Q = 0R_2$

(b)

Thus, if $Z_1 = 3 + j2$ and $Z_2 = 2 - j4$, then

$$Z_1 + Z_2 = (3 + j2) + (2 - j4) = 3 + j2 + 2 - j4 = \mathbf{5 - j2}$$

and

$$Z_1 - Z_2 = (3 + j2) - (2 - j4) = 3 + j2 - 2 + j4 = \mathbf{1 + j6}$$

The addition and subtraction of complex numbers may be achieved graphically as shown in the Argand diagram in Fig. 3.

In Fig. 3(a), by complex addition, $\mathbf{0P} + \mathbf{0Q} = \mathbf{0R_1}$. R_1 is found to be the Argand point $(5, -j2)$, i.e. $(3 + j2) + (2 - j4) = \mathbf{5 - j2}$ (as above).

In Fig. 3(b), to determine $\mathbf{0P} - \mathbf{0Q}$, complexor $\mathbf{0Q}$ is reversed (shown as $\mathbf{0Q'}$) since it is being subtracted. (Note that $\mathbf{0Q} = 2 - j4$. Thus $-\mathbf{0Q} = -(2 - j4) = -2 + j4$, shown as $\mathbf{0Q'}$.)

Thus $\mathbf{0P} - \mathbf{0Q} = \mathbf{0P} + \mathbf{0Q'} = \mathbf{0R_2}$. R_2 is found to be the Argand point $(1, j6)$, i.e. $(3 + j2) - (2 - j4) = \mathbf{1 + j6}$ (as above).

(b) Multiplication

Two complex numbers are multiplied by assuming that all quantities involved are real and then, by using $j^2 = -1$, expressing the product in the form $a + jb$. Hence

$$(a + jb)(c + jd) = ac + a(jd) + (jb)c + (jb)(jd)$$
$$= ac + jad + jbc + j^2bd.$$

But $j^2 = -1$, thus

$$(a + jb)(c + jd) = (ac - bd) + j(ad + bc).$$

For example, if $Z_1 = 3 + j2$ and $Z_2 = 2 - j4$ then

$$Z_1 Z_2 = (3 + j2)(2 - j4) = 6 - j12 + j4 - j^2 8$$
$$= (6 - (-1)8) + j(-12 + 4)$$
$$= 14 + j(-8) = \mathbf{14 - j8}$$

(c) Complex conjugate

The complex conjugate of a complex number is obtained by changing the sign of the imaginary part. Hence $a + jb$ is the complex conjugate of $a - jb$ and $-2 + j$ is the complex conjugate of $-2 - j$. (Note that only the sign of the imaginary part is changed.)

The product of a complex number Z and its complex conjugate Z^* is always a real number, and this is an important property used when dividing complex numbers. Thus if

$$Z = a + jb \quad \text{then} \quad Z^* = a - jb$$

and

$$ZZ^* = (a + jb)(a - jb)$$
$$= a^2 - jab + jab - j^2 b^2$$
$$= a^2 - (-b^2) = a^2 + b^2 \text{ (i.e. a real number)}.$$

Similarly, if
$$Z = 1 + j2 \quad \text{then} \quad Z^* = 1 - j2$$
and
$$ZZ^* = (1 + j2)(1 - j2) = 1^2 + 2^2 = 5.$$

(d) Division

The expression of one complex number divided by another, in the form $a + jb$, is accomplished by multiplying the numerator and denominator by the complex conjugate of the denominator. This has the effect of making the denominator a real number. Hence, for example,

$$\frac{2 + j4}{3 - j4} = \frac{2 + j4}{3 - j4} \times \frac{3 + j4}{3 + j4}$$

$$= \frac{6 + j8 + j12 + j^2 16}{3^2 + 4^2}$$

$$= \frac{-10 + j20}{25} = \frac{-10}{25} + j\frac{20}{25} \quad \text{or} \quad -0.4 + j0.8$$

The elimination of the imaginary part of the denominator by multiplying both the numerator and denominator by the conjugate of the denominator is often termed **"rationalising"**.

Worked problems on operations involving cartesian complex numbers

Problem 1. If $Z_1 = 1 + j2$, $Z_2 = 2 - j3$, $Z_3 = -4 + j$ and $Z_4 = -3 - j2$, evaluate in $a + jb$ form the following:

(a) $Z_1 + Z_2 - Z_3$ (b) $Z_1 Z_3$ (c) $Z_2 Z_3 Z_4$

(d) $\dfrac{Z_2}{Z_4}$ (e) $\dfrac{Z_1 - Z_4}{Z_2 + Z_3}$

(a) $Z_1 + Z_2 - Z_3 = (1 + j2) + (2 - j3) - (-4 + j)$
$$= 1 + j2 + 2 - j3 + 4 - j = \mathbf{7 - j2}$$
(b) $Z_1 Z_3 = (1 + j2)(-4 + j)$
$$= -4 + j - j8 + j^2 2 = \mathbf{-6 - j7}$$
(c) $Z_2 Z_3 Z_4 = (2 - j3)(-4 + j)(-3 - j2)$
$$= (-8 + j2 + j12 - j^2 3)(-3 - j2) = (-5 + j14)(-3 - j2)$$
$$= 15 + j10 - j42 - j^2 28 = \mathbf{43 - j32}$$
(d) $\dfrac{Z_2}{Z_4} = \dfrac{2 - j3}{-3 - j2} = \dfrac{2 - j3}{-3 - j2} \times \dfrac{-3 + j2}{-3 + j2}$

$$= \frac{-6 + j4 + j9 - j^2 6}{3^2 + 2^2} = \frac{0 + j13}{13} = \mathbf{0 + j1} \text{ or } \boldsymbol{j}$$
(e) $\dfrac{Z_1 - Z_4}{Z_2 + Z_3} = \dfrac{(1 + j2) - (-3 - j2)}{(2 - j3) + (-4 + j)}$

$$= \frac{4 + j4}{-2 - j2} = \frac{4(1 + j)}{-2(1 + j)} = \mathbf{-2 + j0} \text{ or } \mathbf{-2}$$

Problem 2. In an electrical circuit the total impedance Z_T is given by

$$Z_T = \frac{Z_1 Z_2}{Z_1 + Z_2} + Z_3$$

Determine Z_T in $(a + jb)$ form, correct to two decimal places, when $Z_1 = 5 - j3$, $Z_2 = 4 + j7$ and $Z_3 = 3.9 - j6.7$.

$$Z_1 Z_2 = (5 - j3)(4 + j7) = 20 + j35 - j12 - j^2 21$$
$$= 20 + j35 - j12 + 21 = 41 + j23$$

$$Z_1 + Z_2 = (5 - j3) + (4 + j7) = 9 + j4$$

Hence

$$\frac{Z_1 Z_2}{Z_1 + Z_2} = \frac{41 + j23}{9 + j4} = \frac{(41 + j23)(9 - j4)}{(9 + j4)(9 - j4)}$$

$$= \frac{369 - j164 + j207 - j^2 92}{9^2 + 4^2} = \frac{369 - j164 + j207 + 92}{97}$$

$$= \frac{461 + j43}{97} = 4.753 + j0.443$$

Thus

$$\frac{Z_1 Z_2}{Z_1 + Z_2} + Z_3 = (4.753 + j0.443) + (3.9 - j6.7)$$

$$= \mathbf{8.65 - j6.26}, \text{ correct to two decimal places.}$$

Problem 3. Given $Z_1 = 3 + j4$ and $Z_2 = 2 - j5$ determine in cartesian form correct to three decimal places:

(a) $\dfrac{1}{Z_1}$ (b) $\dfrac{1}{Z_2}$ (c) $\dfrac{1}{Z_1} + \dfrac{1}{Z_2}$ (d) $\dfrac{1}{(1/Z_1) + (1/Z_2)}$

(a) $\dfrac{1}{Z_1} = \dfrac{1}{3 + j4} = \dfrac{3 - j4}{(3 + j4)(3 - j4)} = \dfrac{3 - j4}{3^2 + 4^2} = \dfrac{3 - j4}{25}$

$$= \frac{3}{25} - j\frac{4}{25} = \mathbf{0.120 - j0.160}$$

(b) $\dfrac{1}{Z_2} = \dfrac{1}{2 - j5} = \dfrac{2 + j5}{(2 - j5)(2 + j5)} = \dfrac{2 + j5}{2^2 + 5^2} = \dfrac{2 + j5}{29}$

$$= \tfrac{2}{29} + j\tfrac{5}{29} = \mathbf{0.069 + j0.172}$$

(c) $\dfrac{1}{Z_1} + \dfrac{1}{Z_2} = (0.120 - j0.160) + (0.069 + j0.172)$

$$= \mathbf{0.189 + j0.012}$$

(d) $\dfrac{1}{(1/Z_1) + (1/Z_2)} = \dfrac{1}{0.189 + j0.012} = \dfrac{0.189 - j0.012}{(0.189 + j0.012)(0.189 - j0.012)}$

$$= \frac{0.189 - j0.012}{0.189^2 + 0.012^2} = \frac{0.189 - j0.012}{0.03587}$$

$$= \frac{0.189}{0.03587} - \frac{j0.012}{0.03587} = \mathbf{5.269 - j0.335}$$

Further problems on operations involving cartesian complex numbers may be found in section 9, problems 10 to 23, page 18.

5. Complex equations

If two complex numbers are equal, then their real parts are equal and their imaginary parts are equal. Hence, if $a + jb = c + jd$ then $a = c$ and $b = d$. This is a useful property, since equations having two unknown quantities can be solved from one equation. Complex equation are used when deriving balance equations with a.c. bridges (see chapter 5).

Worked problems on complex equations

Problem 1. Solve the following complex equations:

(a) $3(a + jb) = 9 - j2$
(b) $(2 + j)(-2 + j) = x + jy$
(c) $(a - j2b) + (b - j3a) = 5 + j2$

(a) $3(a + jb) = 9 - j2$. Thus

$$3a + j3b = 9 - j2$$

Equating real parts gives $3a = 9$, i.e. $a = 3$
Equating imaginary parts gives $3b = -2$, i.e., $b = -\frac{2}{3}$
(b) $(2 + j)(-2 + j) = x + jy$. Thus

$$-4 + j2 - j2 + j^2 = x + jy$$
$$-5 + j0 = x + jy$$

Equating real and imaginary parts gives $x = -5, y = 0$.
(c) $(a - j2b) + (b - j3a) = 5 + j2$. Thus

$$(a + b) + j(-2b - 3a) = 5 + j2$$

Hence

$$a + b = 5 \qquad (1)$$

and

$$-2b - 3a = 2 \qquad (2)$$

We have two simultaneous equations to solve. Multiplying equation (1) by 2 gives

$$2a + 2b = 10 \qquad (3)$$

Adding equations (2) and (3) gives

$$-a = 12, \text{ i.e. } a = -12$$

From equation (1), $b = 17$.

Problem 2. An equation derived from an a.c. bridge network is given by

$$R_1 R_3 = (R_2 + j\omega L_2)\left(\frac{1}{(1/R_4) + j\omega C_4}\right)$$

R_1, R_3, R_4 and C_4 are known values. Determine expressions for R_2 and L_2 in terms of the known components.

Multiplying both sides of the equation by $(1/R_4 + j\omega C_4)$ gives

$$(R_1 R_3)\left(\frac{1}{R_4} + j\omega C_4\right) = R_2 + j\omega L_2$$

i.e.

$$\frac{R_1 R_3}{R_4} + jR_1 R_3 \omega C_4 = R_2 + j\omega L_2$$

Equating the real parts gives $\boldsymbol{R_2 = R_1 R_3/R_4}$.
Equating the imaginary parts gives $\omega L_2 = R_1 R_3 \omega C_4$ from which $\boldsymbol{L_2 = R_1 R_3 C_4}$.

Further problems on complex equations may be found in section 9, problems 24 to 29, page 19.

6. The polar form of a complex number

Let a complex number Z in cartesian form be $x + jy$. This is shown in the Argand diagram of Fig. 4(a). Let r be the distance $0Z$ and let θ be the angle that $0Z$ makes with the positive real axis. From trigonometry,

$$\cos \theta = \frac{x}{r}, \quad \text{i.e.} \quad x = r \cos \theta$$

and

$$\sin \theta = \frac{y}{r}, \quad \text{i.e.} \quad y = r \sin \theta$$

$$Z = x + jy = r \cos \theta + jr \sin \theta$$
$$= r(\cos \theta + j \sin \theta)$$

This latter form is usually abbreviated to $Z = r \underline{/\theta}$, and is called the **polar form** of a complex number. The complex number is now specified in terms of r and θ instead of x and y.

r is called the **modulus** (or magnitude of Z) and is written as mod Z or $|Z|$. r is determined from Pythagoras's theorem on triangle $0AZ$, i.e. $\boldsymbol{|Z| = r = \sqrt{(x^2 + y^2)}}$.

The modulus is represented on the Argand diagram by the distance $0Z$. θ is called the **argument** (or amplitude) of Z and is written as arg Z. θ is also deduced from triangle $0AZ$, giving $\boldsymbol{\arg Z = \theta = \arctan y/x}$.

In Fig. 4(a) the Argand point Z is shown in the first quadrant. However, the above results apply to any point in the Argand diagram.

By convention, the principal value of θ is used, i.e. the numerically least value such that $-\pi \leqslant \theta \leqslant \pi$. For example, in Fig. 4(b), $\theta' = \arctan(4/3) = 53°8'$. Hence $\theta = 180° - 53°8' = 126°52'$. Therefore $-3 + j4 = 5 \underline{/126°52'}$. Similarly, in Fig. 4(c), θ is $180° - 45°$, i.e. $135°$ measured in the negative direction. Hence $-2 - j2 = \sqrt{8} \underline{/-135°}$. (This is the same as $\sqrt{8} \underline{/225°}$: however, the principal value, $\sqrt{8} \underline{/-135°}$, is normally used.) In Fig. 4(d), $\theta = \arctan(5/12) = 22°37'$. Hence $12 - j5 = 13 \underline{/-22°37'}$.

Whenever changing from a cartesian form of complex number to a polar form, or *vice versa*, a sketch is invaluable for deciding the quadrant in which the complex number occurs. There are always two possible values of $\theta = \arctan(y/x)$, only one of which is correct for a particular complex number.

Figure 4

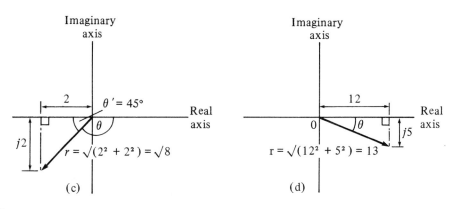

(a)   (b)   (c)   (d)

7. Multiplication and division using complex numbers in polar form

An important use of the polar form of a complex number is in multiplication and division which is achieved more easily than with cartesian form.

(a) Multiplication

Let $Z_1 = r_1 \underline{/\theta_1}$ and $Z_2 = r_2 \underline{/\theta_2}$. Then

$$Z_1 Z_1 = [r_1\underline{/\theta_1}][r_2\underline{/\theta_2}]$$
$$= [r_1(\cos\theta_1 + j\sin\theta_1)] \times [r_2(\cos\theta_2 + j\sin\theta_2)]$$
$$= r_1 r_2(\cos\theta_1\cos\theta_2 + j\sin\theta_2\cos\theta_1 + j\cos\theta_2\sin\theta_1$$
$$+ j^2\sin\theta_1\sin\theta_2)$$
$$= r_1 r_2[(\cos\theta_1\cos\theta_2 - \sin\theta_1\sin\theta_2)$$
$$+ j(\sin\theta_1\cos\theta_2 + \cos\theta_1\sin\theta_2)]$$
$$= r_1 r_2[\cos(\theta_1 + \theta_2) + j\sin(\theta_1 + \theta_2)]$$

(from compound angle formulae)

$$= r_1 r_2 \underline{/(\theta_1 + \theta_2)}$$

Hence to obtain the product of complex numbers in polar form their moduli are multiplied together and their arguments are added. This result is true for all polar complex numbers.

Thus $\quad 3\underline{/25^\circ} \times 2\underline{/32^\circ} = 6\underline{/57^\circ}, \qquad 4\underline{/11^\circ} \times 5\underline{/-18^\circ} = 20\underline{/-7^\circ},$

$2\left.\dfrac{}{}\middle|\dfrac{\pi}{3}\right. \times 7\left.\dfrac{}{}\middle|\dfrac{\pi}{6}\right. = 14\left.\dfrac{}{}\middle|\dfrac{\pi}{2}\right.$, and so on.

(b) Division

Let $Z_1 = r_1\underline{/\theta_1}$ and $Z_2 = r_2\underline{/\theta_2}$. Then

$$\frac{Z_1}{Z_2} = \frac{r_1\underline{/\theta_1}}{r_2\underline{/\theta_2}} = \frac{r_1(\cos\theta_1 + j\sin\theta_1)}{r_2(\cos\theta_2 + j\sin\theta_2)}$$

$$= \frac{r_1(\cos\theta_1 + j\sin\theta_1)}{r_2(\cos\theta_2 + j\sin\theta_2)} \times \frac{(\cos\theta_2 - j\sin\theta_2)}{(\cos\theta_2 - j\sin\theta_2)}$$

$$= \frac{r_1(\cos\theta_1\cos\theta_2 - j\sin\theta_2\cos\theta_1 + j\sin\theta_1\cos\theta_2 - j^2\sin\theta_1\sin\theta_2)}{r_2(\cos^2\theta_2 + \sin^2\theta_2)}$$

$$= \frac{r_1[(\cos\theta_1\cos\theta_2 + \sin\theta_1\sin\theta_2) + j(\sin\theta_1\cos\theta_2 - \cos\theta_1\sin\theta_2)]}{r_2(1)}$$

$$= \frac{r_1}{r_2}[\cos(\theta_1 - \theta_2) + j\sin(\theta_1 - \theta_2)]$$

(from compound angle formulae)

$$= \frac{r_1}{r_2}\underline{/(\theta_1 - \theta_2)}$$

Hence to obtain the ratio of two complex numbers in polar form their moduli are divided and their arguments subtracted. This result is true for all polar complex numbers. Thus

$$\frac{8\underline{/58^\circ}}{2\underline{/11^\circ}} = 4\underline{/47^\circ}$$

$$\frac{9\underline{/136^\circ}}{3\underline{/-60^\circ}} = 3\underline{/(136^\circ - -60^\circ)} = 3\underline{/196^\circ} = 3\underline{/-164^\circ}$$

and

$$\frac{10\left.\dfrac{}{}\middle|\dfrac{\pi}{2}\right.}{5\left.\dfrac{}{}\middle|-\dfrac{\pi}{4}\right.} = 2\left.\dfrac{}{}\middle|\dfrac{3\pi}{4}\right., \text{ and so on}$$

It may be concluded from the above that multiplication and division of complex numbers are more easily achieved by using polar form than by cartesian form. Also, addition and subtraction of complex numbers can only be achieved when the numbers are in cartesian form.

Conversion from cartesian or rectangular form to polar form, and *vice versa*, may be achieved by using the $R \to P$ and $P \to R$ conversion facility which is available on most calculators with scientific notation. This allows, of course, a great saving of time.

Figure 5

Figure 6

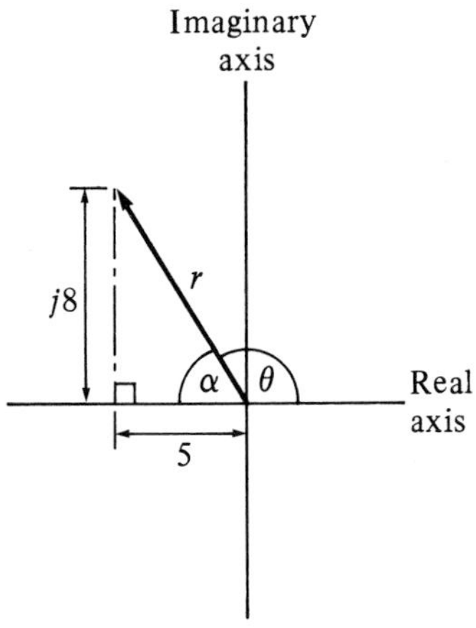

Worked problems on the polar form of complex numbers

Problem 1. Determine the modulus and argument of the complex number $Z = -5 + j8$ and express Z in polar form.

Figure 5 indicates that the complex number $-5 + j8$ lies in the second quadrant of the Argand diagram.

Modulus, $|Z| = r = \sqrt{[(-5)^2 + (8)^2]} = \mathbf{9.434}$, correct to four significant figures

$$\alpha = \arctan(8/5) = \mathbf{58^\circ}.$$

Argument, $\arg Z = \theta = 180^\circ - 58^\circ = 122^\circ$.

In polar form $-5 + j8$ is written as $\mathbf{9.434\angle 122^\circ}$.

Problem 2. Convert $5\angle -132^\circ$ into $a + jb$ form correct to four significant figures.

Figure 6 indicates that the polar complex number $5\angle -132^\circ$ lies in the third quadrant of the Argand diagram. Using trigonometrical ratios,

$$x = 5\cos 48^\circ = 3.346 \quad \text{and} \quad y = 5\sin 48^\circ = 3.716$$

Hence $\mathbf{5\angle -132^\circ = -3.346 - j3.716}$.

Alternatively,

$$\begin{aligned} 5\angle -132^\circ &= 5(\cos -132^\circ + j\sin -132^\circ) \\ &= 5\cos(-132^\circ) + j5\sin(-132^\circ) \\ &= \mathbf{-3.346 - j3.716}, \text{ as above} \end{aligned}$$

With this latter method the real and imaginary parts are obtained directly, using a calculator.

Problem 3. Evaluate

(a) $\dfrac{15\angle 30^\circ \times 8\angle 45^\circ}{4\angle 60^\circ}$

(b) $4\angle 30^\circ + 3\angle -60^\circ - 5\angle -135^\circ$

giving answers in polar and in cartesian forms, correct to three significant figures.

(a) $$\begin{aligned} \frac{15\angle 30^\circ \times 8\angle 45^\circ}{4\angle 60^\circ} &= \frac{15 \times 8}{4}\angle(30^\circ + 45^\circ - 60^\circ) \\ &= 30.0\angle 15^\circ \quad \text{in polar form} \\ &= 30.0(\cos 15^\circ + j\sin 15^\circ) \\ &= \mathbf{29.0 + j7.76} \quad \text{in cartesian form} \end{aligned}$$

(b) $4\angle 30^\circ + 3\angle -60^\circ - 5\angle -135^\circ$.

The advantages of polar form are seen when multiplying and dividing complex numbers. Addition or subtraction in polar form is not possible. Each polar complex number has to be converted into cartesian form first, which may be achieved easily using a calculator. Thus

$$\begin{aligned} 4\angle 30^\circ &= 3.464 + j2.000 \\ 3\angle -60^\circ &= 1.500 - j2.598 \\ 5\angle -135^\circ &= -3.536 - j3.536 \end{aligned}$$

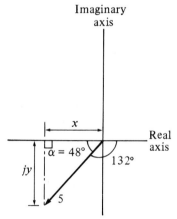

Hence

$$4\underline{/\,30°} + 3\underline{/\,-60°} - 5\underline{/\,-135°} = (3.464 + j2.000) + (1.500 - j2.598)$$
$$-(-3.536 - j3.536)$$
$$= \mathbf{8.50 + j2.94} \quad \text{in cartesian form}$$
$$= \mathbf{8.99\underline{/\,19°5'}} \quad \text{in polar form}$$

Problem 4. Two impedances in an electrical network are given by $Z_1 = 4.7\underline{/\,35°}$ and $Z_2 = 7.3\underline{/\,-48°}$. Determine in polar form the total impedance Z_T given that $Z_T = Z_1 Z_2/(Z_1 + Z_2)$.

$$Z_1 = 4.7\underline{/\,35°} = 4.7\cos 35° + j4.7\sin 35° = 3.85 + j2.70$$
$$Z_2 = 7.3\underline{/\,-48°} = 7.3\cos(-48°) + j7.3\sin(-48°) = 4.88 - j5.42$$
$$Z_1 + Z_2 = (3.85 + j2.70) + (4.88 - j5.42) = 8.73 - j2.72$$
$$= \sqrt{(8.73^2 + 2.72^2)}\left|\arctan\frac{-2.72}{8.73}\right.$$
$$= 9.14\underline{/\,-17.3°}$$

Hence

$$Z_T = \frac{Z_1 Z_2}{Z_1 + Z_2} = \frac{4.7\underline{/\,35°} \times 7.3\underline{/\,-48°}}{9.14\underline{/\,-17.3°}}$$
$$= \frac{4.7 \times 7.3}{9.14}\underline{/\,[35° - 48° - (17.3°)]} = \mathbf{3.75\underline{/\,4.3°}}$$

Further problems on the polar form of complex numbers may be found in section 9, problems 30 to 48, page 20.

8. De Moivre's theorem —powers and roots of complex numbers

From Section 7, $r\underline{/\,\theta} \times r\underline{/\,\theta} = r^2\underline{/\,2\theta}$, $r\underline{/\,\theta} \times r\underline{/\,\theta} \times r\underline{/\,\theta} = r^3\underline{/\,3\theta}$, and so on. Such results are generally expressed in de Moivre's theorem, which may be stated as

$$[r\underline{/\,\theta}]^n = r^n\underline{/\,n\theta} \quad (= r^n(\cos n\theta + j\sin n\theta))$$

This result is true for all positive, negative or fractional values of n. De Moivre's theorem is thus useful is determining powers and roots of complex numbers. For example,

$$[2\underline{/\,15°}]^6 = 2^6\underline{/\,(6 \times 15°)} = 64\underline{/\,90°} = 0 + j64$$

A square root of a complex number is determined as follows:

$$\sqrt{[r\underline{/\,\theta}]} = [r\underline{/\,\theta}]^{1/2} = r^{1/2}\underline{/\,\tfrac{1}{2}\theta}$$

However, it is important to realise that a real number has two square roots, equal in size but opposite in sign. On an Argand diagram the roots are 180° apart (see worked problem 3).

Worked problems on powers and roots of complex numbers

Problem 1. Determine the square of the complex number $(3 - j4)$, (a) in cartesian form and (b) in polar form, using de Moivre's theorem. Compare the results obtained and show the roots on an Argand diagram.

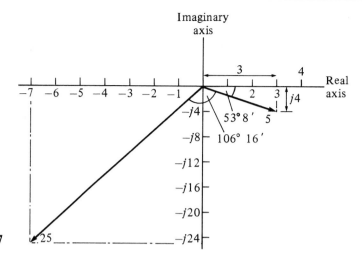

Figure 7

(a) In cartesian form,

$$(3 - j4)^2 = (3 - j4)(3 - j4) = 9 - j12 - j12 + j^2 16$$
$$= -7 - 24$$

(b) In polar form,

$$(3 - j4) = \sqrt{[(3)^2 + (4)^2]}\underline{/\arctan\tfrac{4}{3}} \quad \text{(see Fig. 7)}$$
$$= 5\underline{/-53°8'}.$$

$$[5\underline{/-53°8'}]^2 = 5^2\underline{/(2 \times -53°8')} = 25\underline{/-106°16'}$$

The complex number $(3 - j4)$, together with its square, i.e. $(3 - j4)^2$, is shown in Fig. 7.

$$25\underline{/-106° \ 16'} \text{ in cartesian form is}$$
$$25\cos(-106°16') + j25\sin(-106°16')$$

i.e. $-7 - j24$, as in part (a).

Problem 2. Determine $(-2 + j3)^5$ in polar and in cartesian form.

$Z = -2 + j3$ is situated in the second quadrant of the Argand diagram. Thus

$$r = \sqrt{[(2)^2 + (3)^2]} = \sqrt{13} \quad \text{and} \quad \alpha = \arctan\tfrac{3}{2} = 56°19'$$

Hence the argument $\theta = 180° - 56°19' = 123°41'$. Thus $-2 + j3$ in polar form is $\sqrt{13}\underline{/123°41'}$.

$$(-2 + j3)^5 = [\sqrt{13}\underline{/123°41'}]^5$$
$$= (\sqrt{13})^5 \underline{/(5 \times 123°41')} \qquad \text{from de Moivre's theorem}$$
$$= 13^{5/2}\underline{/618°25'} = 13^{5/2}\underline{/258°25'}$$
$$\text{(since } 618°25' \equiv 618°25' - 360°)$$
$$= 13^{5/2}\underline{/-101°35'} = 609.3\underline{/-101°35'}$$

In cartesian form,

$$609.3\underline{/-101°35'} = 609.3\cos(-101°35') + j609.3\sin(-101°35')$$
$$= -122.3 - j596.9$$

Problem 3. Determine the two square roots of the complex number $(12 + j5)$ in cartesian and polar form, correct to three significant figures. Show the roots on an Argand diagram.

In polar form $12 + j5 = \sqrt{(12^2 + 5^2)}\underline{/}\arctan(5/12)$, since $12 + j5$ is in the first quadrant of the Argand diagram, i.e. $12 + j5 = 13\underline{/}22°37'$.

Since we are finding the square roots of $13\underline{/}22°37'$ there will be two solutions. To obtain the second solution it is helpful to express $13\underline{/}22°37'$ also as

$$13\underline{/}(360° + 22°37'), \text{ i.e. } 13\underline{/}382°37'$$

(we have merely rotated one revolution to obtain this result). The reason for doing this is that when we divide the angles by 2 we still obtain angles less than 360°, as shown below. Hence

$$\sqrt{(12 + j5)} = \sqrt{[13\underline{/}22°37']} \quad \text{or} \quad \sqrt{[13\underline{/}382°37']}$$
$$= [13\underline{/}22°37']^{1/2} \quad \text{or} \quad [13\underline{/}382°37']^{1/2}$$
$$= 13^{1/2}\underline{/}(\tfrac{1}{2} \times 22°37') \quad \text{or} \quad 13^{1/2}\underline{/}(\tfrac{1}{2} \times 382°37'),$$
$$\text{from de Moivre's theorem,}$$
$$= \sqrt{13}\underline{/}11°19' \quad \text{or} \quad \sqrt{13}\underline{/}191°19'$$
$$= \mathbf{3.61}\underline{/}\mathbf{11°19'} \quad \text{or} \quad \mathbf{3.61}\underline{/}\mathbf{-168°\ 41'}$$

These two solutions of $\sqrt{(12 + j5)}$ are shown in the Argand diagram of Fig. 8.

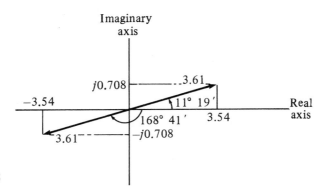

Figure 8

$3.61\underline{/}11°19'$ is in the first quadrant of the Argand diagram. Thus
$$3.61\underline{/}11°19' = 3.61\,(\cos 11°19' + j\sin 11°19') = 3.540 + j0.708$$

$3.61\underline{/}-168°41'$ is in the third quadrant of the Argand diagram. Thus
$$3.61\underline{/}-168°41' = 3.61\,[\cos(-168°41') + j\sin(-168°41')]$$
$$= -3.540 - j0.708$$

Thus in cartesian form the two roots are $\pm\,(\mathbf{3.540 + j0.708}).$

From the Argand diagram the roots are seen to be 180° apart, i.e. they lie on a straight line. This is always true when finding square roots of complex numbers.

Further problems on powers and roots of complex numbers may be found in section 9 following, problems 49 to 58, page 21.

9. Further problems

Introduction to complex numbers

In problems 1 to 5 solve the quadratic equations.

1. $x^2 + 16 = 0$

$$[x = \pm j4]$$

2. $x^2 - 2x + 2 = 0$

$$[x = 1 \pm j]$$

3. $2x^2 + 3x + 4 = 0$

$$\left[x = -\frac{3}{2} \pm j\frac{\sqrt{23}}{4} \quad \text{or} \quad -0.750 \pm j1.199\right]$$

4. $5y^2 + 2y = -3$

$$\left[y = -\frac{1}{5} \pm j\frac{\sqrt{14}}{5} \quad \text{or} \quad -0.200 \pm j0.748\right]$$

5. $4t^2 = t - 1$

$$\left[t = \frac{1}{8} \pm j\frac{\sqrt{15}}{8} \quad \text{or} \quad 0.125 \pm j0.484\right]$$

6. Evaluate (a) $-j^5$, (b) j^{13} (c) j^{56}

$$[\text{(a)} \ -j \qquad \text{(b)} \ j \qquad \text{(c)} \ 1]$$

Argand diagram

7. Show on an Argand diagram the following complex numbers:

(a) $3 + j6$ (b) $2 - j3$ (c) $-3 + j4$ (d) $-1 - j5$

8. Show the following numbers on an Argand diagram:

(a) 3 (b) $j3$ (c) $j^2 3$ (d) $j^3 3$

9. Draw an Argand diagram and plot the points representing the following complex numbers:

(a) $(1 + j2)$ (b) $j(1 + j2)$ (c) $j^2(1 + j2)$ (d) $-j(1 + j2)$

From each plotted point draw a straight line to the origin. What is the angle between each straight line?

$$[90°]$$

Operations on cartesian complex numbers

10. Write down the complex conjugates of the following complex numbers:

(a) $4 + j$ (b) $3 - j2$ (c) $-5 - j$
(a) $[4 - j]$ (b) $[3 + j2]$ (c) $[-5 + j]$

In problems 11 to 15, evaluate in $a + jb$ form assuming that $Z_1 = 2 + j3$, $Z_2 = 3 - j4$, $Z_3 = -1 + j2$ and $Z_4 = -2 - j5$.

11. (a) $Z_1 - Z_2$ (b) $Z_2 + Z_3 - Z_4$

$$[\text{(a)} \ -1 + j7 \qquad \text{(b)} \ 4 + j3]$$

12. (a) $Z_1 Z_2$ (b) $Z_3 Z_4$

$$[\text{(a)} \ 18 + j \qquad \text{(b)} \ 12 + j]$$

13. (a) $Z_1 Z_3 Z_4$ (b) $Z_2 Z_3 + Z_4$

$$[\text{(a)} \ 21 + j38 \qquad \text{(b)} \ 3 + j5]$$

14. (a) $\dfrac{Z_1}{Z_2}$ (b) $\dfrac{Z_1 + Z_2}{Z_3 + Z_4}$

$$[\text{(a)} \ -\tfrac{6}{25} + j\tfrac{17}{25} \qquad \text{(b)} \ -\tfrac{2}{3} + j]$$

15. (a) $\dfrac{Z_1 Z_2}{Z_1 + Z_2}$ (b) $Z_1 + \dfrac{Z_2}{Z_3} + Z_4$

$$\left[\text{(a) } \tfrac{89}{26} + j\tfrac{23}{26} \quad \text{(b) } -\tfrac{11}{5} - j\tfrac{12}{5}\right]$$

16. Evaluate $\left[\dfrac{(1+j)^2 - (1-j)^2}{j}\right]$

$$[4]$$

17. If $Z_1 = 4 - j3$ and $Z_2 = 2 + j$ evaluate x and y given

$$x + jy = \frac{1}{Z_1 - Z_2} + \frac{1}{Z_1 Z_2}$$

$$[x = 0.188, y = 0.216]$$

18. Evaluate (a) j^8 (b) j^{11} (c) $3/j^3$ (d) $5/j^6$

$$[\text{(a) } 1 \quad \text{(b) } -j \quad \text{(c) } j3 \quad \text{(d) } -5]$$

19. Evaluate (a) $(1+j)^4$ (b) $\dfrac{2-j}{2+j}$ (c) $\dfrac{1}{2+j3}$

$$\left[\text{(a) } -4 \quad \text{(b) } \tfrac{3}{5} - j\tfrac{4}{5} \quad \text{(c) } \tfrac{2}{13} - j\tfrac{3}{13}\right]$$

20. If $Z = \dfrac{1+j3}{1-j2}$, evaluate Z^2 in $a + jb$ form.

$$[0 - j2]$$

21. Evaluate (a) j^{33} (b) $\dfrac{1}{(2-j2)^4}$ (c) $\dfrac{1+j3}{2+j4} + \dfrac{3-j2}{5-j}$

$$\left[\text{(a) } j \quad \text{(b) } -\tfrac{1}{64} \quad \text{(c) } 1.354 - j0.169\right]$$

22. In an electrical circuit the equivalent impedance Z is given by

$$Z = Z_1 + \frac{Z_2 Z_3}{Z_2 + Z_3}$$

Determine Z is rectangular form, correct to two decimal places, when

$$Z_1 = 5.91 + j3.15, \quad Z_2 = 5 + j12 \text{ and } Z_3 = 8 - j15.$$

$$[Z = 21.62 + j8.39]$$

23. Given $Z_1 = 5 - j9$ and $Z_2 = 7 + j2$, determine in $(a + jb)$ form, correct to four decimal places

(i) $\dfrac{1}{Z_1}$ (ii) $\dfrac{1}{Z_2}$ (iii) $\dfrac{1}{Z_1} + \dfrac{1}{Z_2}$ (iv) $\dfrac{1}{(1/Z_1) + (1/Z_2)}$

$$\left[\begin{array}{l}\text{(i) } 0.0472 + j0.0849, \text{ (ii) } 0.1321 - j0.0377, \\ \text{(iii) } 0.1793 + j0.0472, \text{ (iv) } 5.2158 - j1.3730\end{array}\right]$$

Complex equations

In problems 24 to 28 solve the given complex equations:

24. $4(a + jb) = 7 - j3$

$$\left[a = \tfrac{7}{4}, b = -\tfrac{3}{4}\right]$$

25. $(3 + j4)(2 - j3) = x + jy$

$$[x = 18, y = -1]$$

26. $(1 + j)(2 - j) = j(p + jq)$

$$[p = 1, q = -3]$$

27. $(a - j3b) + (b - j2a) = 4 + j6$

$$[a = 18, b = -14]$$

28. $5+j2 = \sqrt{(e+jf)}$

$$[e=21, f=20]$$

29. An equation derived from an a.c. bridge circuit is given by

$$(R_3)\left(\frac{-j}{\omega C_1}\right) = \left(R_x - \frac{j}{\omega C_x}\right)\left(\frac{R_4(-j/(\omega C_4))}{R_4 - (j/(\omega C_4))}\right)$$

Components R_3, R_4, C_1 and C_4 have known values. Determine expressions for R_x and C_x in terms of the known components.

$$\left[R_x = \frac{R_3 C_4}{C_1}; C_x = \frac{C_1 R_4}{R_3}\right]$$

Polar form of complex numbers

In problems 30 to 32 determine the modulus and the argument of each of the complex numbers given.

30. (a) $3+j4$ (b) $2-j5$

$$[\text{(a) } 5, 53°8' \quad \text{(b) } 5.385, -68°12']$$

31. (a) $-4+j$ (b) $-5-j3$

$$[\text{(a) } 4.123, 165°58' \quad \text{(b) } 5.831, -149°2']$$

32. (a) $(2+j)^2$ (b) $j(3-j)$

$$[\text{(a) } 5, 53°8' \quad \text{(b) } 3.162, 71°34']$$

In problems 33 to 35 express the given cartesian complex numbers in polar form, leaving answers in surd form.

33. (a) $6+j5$ (b) $3-j2$ (c) -3

$$[\text{(a) } \sqrt{61}\angle 39°48' \quad \text{(b) } \sqrt{13}\angle -33°41' \quad \text{(c) } 3\angle 180° \text{ or } 3\angle \pi]$$

34. (a) $-5+j$ (b) $-4-j3$ (c) $-j2$

$$\left[\text{(a) } \sqrt{26}\angle 168°41' \quad \text{(b) } 5\angle -143°8' \quad \text{(c) } 2\angle -90° \text{ or } 2\left|-\frac{\pi}{2}\right.\right]$$

35. (a) $(-1+j)^3$ (b) $-j(1-j)$ (c) $j^3(2-j3)$

$$[\text{(a) } \sqrt{8}\angle 45° \quad \text{(b) } \sqrt{2}\angle -135° \quad \text{(c) } \sqrt{13}\angle -146°19']$$

In problems 36 to 38 convert the given polar complex numbers into $(a+jb)$ form, giving answers correct to four significant figures.

36. (a) $6\angle 30°$ (b) $4\angle 60°$ (c) $3\angle 45°$

$$[\text{(a) } 5.196+j3.000 \quad \text{(b) } 2.000+j3.464 \quad \text{(c) } 2.121+j2.121]$$

37. (a) $2\left|\frac{\pi}{2}\right.$ (b) $3\angle \pi$ (c) $5\left|\frac{5\pi}{6}\right.$

$$[\text{(a) } 0+j2.000 \quad \text{(b) } -3.000+j0 \quad \text{(c) } [-4.330+j2.500]]$$

38. (a) $8\angle 150°$ (b) $4.2\angle -120°$ (c) $3.6\angle -25°$

$$[\text{(a) } -6.928+j4.000 \quad \text{(b) } -2.100-j3.637 \quad \text{(c) } 3.263-j1.521]$$

39. Using an Argand diagram, evaluate in polar form

(a) $2\angle 30° + 3\angle 40°$ (b) $5.5\angle 120° - 2.5\angle -50°$

$$[\text{(a) } 4.982\angle 36° \quad \text{(b) } 7.974\angle 123°7']$$

In problems 40 to 42, evaluate in polar form.

40. (a) $2\angle 40° \times 5\angle 20°$ (b) $2.6\angle 72° \times 4.3\angle 45°$

$$[\text{(a) } 10\angle 60° \quad \text{(b) } 11.8\angle 117°]$$

41. (a) $5.8\angle 35° \div 2\angle -10°$ (b) $4\angle 30° \times 3\angle 70° \div 2\angle -15°$

$$[\text{(a) } 2.9\angle 45° \quad \text{(b) } 6\angle 115°]$$

42. (a) $\dfrac{4.1\angle\,20° \times 3.2\angle\,-62°}{1.2\angle\,150°}$ (b) $6\angle\,25° + 3\angle\,-36° - 4\angle\,72°$

[(a) $10.93\angle\,168°$ (b) $7.289\angle\,-24°35'$]

43. Solve the complex equations, giving answers correct to four significant figures.

(a) $\dfrac{12\angle\,(\pi/2) \times 3\angle\,(3\pi/4)}{2\angle\,-(\pi/3)} = x + jy$

(b) $15\left|\dfrac{\pi}{3}\right. + 12\left|\dfrac{\pi}{2}\right. - 6\left|-\dfrac{\pi}{3}\right. = r\angle\,\theta$

[(a) $x = 4.659, y = -17.39$ (b) $r = 30.52, \theta = 81°31'$]

44. Three vectors are represented by P, $2\angle\,30°$, Q, $3\angle\,90°$ and R, $4\angle\,-60°$. Determine in polar form the vectors represented by (a) $P + Q + R$ and (b) $P - Q - R$.

[(a) $3.770\angle\,8°10'$ (b) $1.488\angle\,100°22'$]

45. The total impedance Z_T of an electrical circuit is given by

$$Z_T = \dfrac{Z_1 \times Z_2}{Z_1 + Z_2} + Z_3$$

Determine Z_T in polar form correct to three significant figures when

$Z_1 = 3.2\angle\,-56°$, $Z_2 = 7.4\angle\,25°$ and $Z_3 = 6.3\angle\,62°$.

[$6.61\angle\,37.2°$]

46. A star-connected impedance Z_1 is given by

$$Z_1 = \dfrac{Z_A Z_B}{Z_A + Z_B + Z_C}$$

Evaluate Z_1, in both cartesian and polar form, given $Z_A = (20 + j0)\,\Omega$, $Z_B = (0 - j20)\,\Omega$ and $Z_C = (10 + j10)\,\Omega$.

[$(4 - j12)\,\Omega$ or $12.65\angle\,-71.57°\,\Omega$]

47. The current I flowing in an impedance is given by

$$I = \dfrac{(8\angle\,60°)(10\angle\,0°)}{(8\angle\,60° + 5\angle\,30°)}\ \text{A}$$

Determine the value of current in polar form correct to two decimal places.

[$6.36\angle\,11.46°\,A$]

48. A delta-connected impedance Z_A is given by

$$Z_A = \dfrac{Z_1 Z_2 + Z_2 Z_3 + Z_3 Z_1}{Z_2}$$

Determine Z_A, in both cartesian and polar form, given $Z_1 = (10 + j0)\,\Omega$, $Z_2 = (0 - j10)\,\Omega$ and $Z_3 = (10 + j10)\,\Omega$.

[$(10 + j20)\,\Omega$; $22.36\angle\,63.43°\,\Omega$]

Powers and roots of complex numbers

In problems 49 to 52, evaluate in cartesian and in polar form.

49. (a) $(2 + j3)^2$ (b) $(4 - j5)^2$

[(a) $-5 + j12$; $13\angle\,112°37'$ (b) $-9 - j40$; $41\angle\,-102°41'$]

50. (a) $(-3 + j2)^5$ (b) $(-2 - j)^3$

[(a) $597 + j122$; $609.3\angle\,11°33'$ (b) $-2 - j11$; $11.18\angle\,-100°18'$]

51. (a) $(4\angle\,32°)^4$ (b) $(2\angle\,125°)^5$

[(a) $-157.6 + j201.7$; $256\angle\,128°$ (b) $-2.789 - j31.88$; $32\angle\,-95°$]

52. (a) $\left(3\left|\!\underline{-\dfrac{\pi}{3}}\right.\right)^3$ (b) $(1.5\underline{/\!-160°})^4$

$[\text{(a)}\ -27+j0;\ 27\underline{/\!-\pi}$ (b) $0.8792+j4.986;\ 5063\underline{/\ 80°}]$

In problems 53 to 55, determine the two square roots of the given complex numbers in cartesian form and show the results on an Argand diagram.

53. (a) $2+j$ (b) $3-j2$

$[\text{(a)}\ \pm(1.455+j0.344)$ (b) $\pm(1.817-j0.550)]$

54. (a) $-3+j4$ (b) $-1-j3$

$[\text{(a)}\ \pm(1+j2)$ (b) $\pm(1.040-j1.443)]$

55. (a) $5\underline{/\ 36°}$ (b) $14\left|\!\underline{\dfrac{3\pi}{2}}\right.$

$[\text{(a)}\ \pm(2.127+j0.691)$ (b) $\pm(-2.646+j2.646)]$

56. If $Z=3\underline{/\ 30°}$ evaluate in polar form (a) Z^2 (b) $\sqrt{Z}$

$[\text{(a)}\ 9\underline{/\ 60°}$ (b) $\sqrt{3}\underline{/\ 15°}$ and $\sqrt{3}\underline{/\!-165°}]$

57. Convert $2-j$ into polar form and hence evaluate $(2-j)^7$ in polar form.

$[\sqrt{5}\underline{/\!-26°34'};\ 279.5\underline{/\ 174°3'}]$

58. Simplify, without the use of tables,

$$\frac{(\cos(\pi/9)-j\sin(\pi/9))^4}{(\cos(\pi/9)+j\sin(\pi/9))^5}$$

$[-1]$

2 Application of complex numbers to series a.c. circuits

1. Introduction

Simple a.c. circuits may be analysed by using phasor diagrams. However, when circuits become more complicated analysis is considerably simplified by using complex numbers. It is essential that the basic operations used with complex numbers, as outlined in chapter 1, are thoroughly understood before proceeding with a.c. circuit analysis.

2. Series a.c. circuits

(a) Pure resistance

In an a.c. circuit containing resistance R only (see Fig. 1(a)), the current I_R is **in phase** with the applied voltage V_R as shown in the phasor diagram of Fig. 1(b). The phasor diagram may be superimposed on the Argand diagram as shown in Fig. 1(c). The impedance Z of the circuit is given by

$$Z = \frac{V_R\angle\, 0°}{I_R\angle\, 0°} = R$$

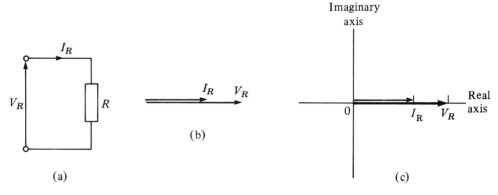

Figure 1 (a) Circuit diagram
(b) Phasor diagram
(c) Argand diagram

(b) Pure inductance

In an a.c. circuit containing pure inductance L only (see Fig. 2(a)), the current I_L **lags** the applied voltage V_L by 90° as shown in the phasor diagram of Fig. 2(b). The phasor diagram may be superimposed on the Argand diagram as shown in Fig. 2(c). The impedance Z of the circuit is

Figure 2 (a) Circuit diagram
(b) Phasor diagram
(c) Argand diagram

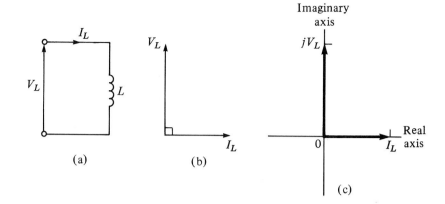

(a)   (b)   (c)

Figure 3 (a) Circuit diagram
(b) Phasor diagram
(c) Argand diagram

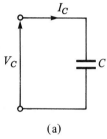

(a)

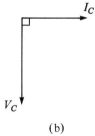

(b)

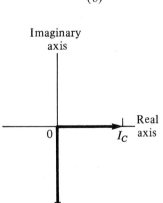

(c)

given by

$$Z = \frac{V_L\angle 90°}{I_L\angle 0°} = \frac{V_L}{I_L}\angle 90° = X_L\angle 90° \quad \text{or} \quad jX_L$$

where X_L is the **inductive reactance** given by

$$X_L = \omega L = 2\pi f L \text{ ohms,}$$

where f is the frequency in hertz and L is the inductance in henrys.

(c) Pure capacitance

In an a.c. circuit containing pure capacitance only (see Fig. 3(a)), the current I_C **leads** the applied voltage V_C by 90° as shown in the phasor diagram of Fig. 3(b). The phasor diagram may be superimposed on the Argand diagram as shown in Fig. 3(c). The impedance Z of the circuit is given by

$$Z = \frac{V_C\angle -90°}{I_C\angle 0°} = \frac{V_C}{I_C}\angle -90° = X_C\angle -90° \quad \text{or} \quad -jX_C$$

where X_C is the **capacitive reactance** given by

$$X_C = \frac{1}{\omega C} = \frac{1}{2\pi f C} \text{ ohms,}$$

where C is the capacitance in farads.

$$\left(\text{Note: } -jX_C = \frac{-j}{\omega C} = \frac{-j\,(j)}{\omega C\,(j)} = \frac{-j^2}{j\omega C} = \frac{-(-1)}{j\omega C} = \frac{1}{j\omega C} \right)$$

(d) R–L series circuit

In an a.c. circuit containing resistance R and inductance L in series (see Fig. 4(a)), the applied voltage V is the phasor sum of V_R and V_L as shown in the phasor diagram of Fig. 4(b). The current I lags the applied voltage V by an angle lying between 0° and 90°—the actual value depending on the values of V_R and V_L, which depend on the values of R and L. The circuit phase

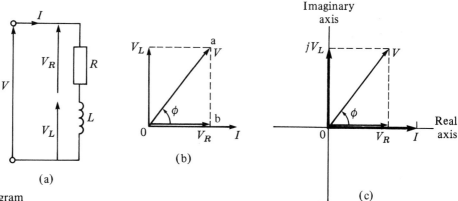

Figure 4 (a) Circuit diagram
(b) Phasor diagram
(c) Argand diagram

angle, i.e., the angle between the current and the applied voltage, is shown as angle ϕ in the phasor diagram. In any series circuit the current is common to all components and is thus taken as the reference phasor in Fig. 4(b). The phasor diagram may be superimposed on the Argand diagram as shown in Fig. 4(c), where it may be seen that in complex form the supply voltage V is given by $V = V_R + jV_L$.

Figure 5(a) shows the voltage triangle that is derived from the phasor diagram of Fig. 4(b) (i.e. triangle $0ab$). If each side of the voltage triangle is divided by current I then the impedance triangle of Fig. 5(b) is derived. The impedance triangle may be superimposed on the Argand diagram, as shown in Fig. 5(c), where it may be seen that in complex form the impedance Z is given by $Z = R + jX_L$.

Thus, for example, an impedance expressed as $(3 + j4)\,\Omega$ means that the resistance is 3 Ω and the inductive reactance is 4 Ω.

In polar form, $Z = |Z|\underline{/\phi}$ where, from the impedance triangle, the modulus of impedance $|Z| = \sqrt{(R^2 + X_L^2)}$ and the circuit phase angle $\phi = \arctan(X_L/R)$ lagging.

(e) R–C series circuit

In an a.c. circuit containing resistance R and capacitance C in series (see

Figure 5 (a) Voltage triangle
(b) Impedance triangle
(c) Argand diagram

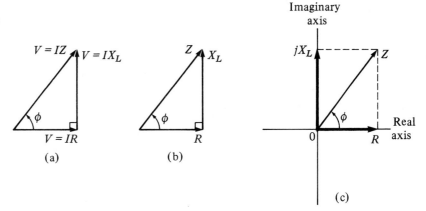

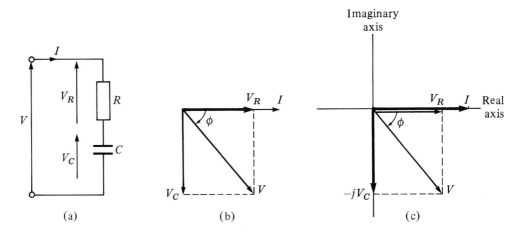

(a)   (b)   (c)

Figure 6 (a) Circuit diagram
(b) Phasor diagram
(c) Argand diagram

Fig. 6(a)), the applied voltage V is the phasor sum of V_R and V_C as shown in the phasor diagram of Fig. 6(b). The current I leads the applied voltage V by an angle lying between $0°$ and $90°$—the actual value depending on the values of V_R and V_C, which depend on the values of R and C. The circuit phase angle is shown as angle ϕ in the phasor diagram. The phasor diagram may be superimposed on the Argand diagram as shown in Fig. 6(c), where it may be seen that in complex form the supply voltage V is given by
$$V = V_R - jV_C.$$
Figure 7(a) shows the voltage triangle that is derived from the phasor diagram of Fig. 6(b). If each side of the voltage triangle is divided by current I, the impedance triangle is derived as shown in Fig. 7(b). The impedance triangle may be superimposed on the Argand diagram as shown in Fig. 7(c), where it may be seen that in complex form the impedance Z is given by
$$Z = R - jX_C.$$
Thus, for example, an impedance expressed as $(9 - j14)\,\Omega$ means that the resistance is $9\,\Omega$ and the capacitive reactance X_C is $14\,\Omega$.

In polar form, $Z = |Z|\underline{/\phi}$ where, from the impedance triangle, $|Z| = \sqrt{(R^2 + X_C^2)}$ and $\phi = \arctan(X_C/R)$ leading.

Figure 7 (a) Voltage triangle
(b) Impedance triangle
(c) Argand diagram

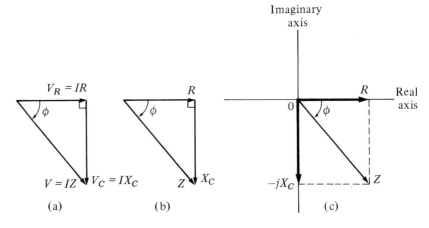

(a)   (b)   (c)

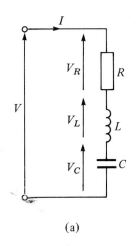

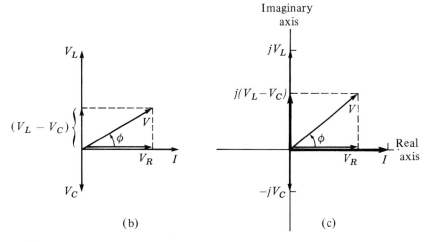

Figure 8 (a) Circuit diagram
(b) Phasor diagram
(c) Argand diagram

(f) R–L–C series circuit

In an a.c. circuit containing resistance R, inductance L and capacitance C in series (see Fig. 8(a)), the applied voltage V is the phasor sum of V_R, V_L and V_C as shown in the phasor diagram of Fig. 8(b) (where the condition $V_L > V_C$ is shown). The phasor diagram may be superimposed on the Argand diagram as shown in Fig. 8(c), where it may be seen that in complex form the supply voltage V is given by $V = V_R + j(V_L - V_C)$.

From the voltage triangle the impedance triangle is derived and superimposing this on the Argand diagram gives, in complex form,

$$\text{impedance } Z = R + j(X_L - X_C)$$

or

$$Z = |Z| \underline{/\phi}$$

where $|Z| = \sqrt{[R^2 + (X_L - X_C)^2]}$ and $\phi = \arctan((X_L - X_C)/R)$.

When $V_L = V_C$, $X_L = X_C$, and the applied voltage V and the current I are in phase. This effect is called **series resonance** (see chapter 6).

(g) General series circuit

In an a.c. circuit containing several impedances connected in series, say, Z_1, Z_2, $Z_3, \ldots, Z_n$, then the total equivalent impedance Z_T is given by

$$Z_T = Z_1 + Z_2 + Z_3 + \cdots + Z_n$$

Worked problems on the application of complex numbers to series a.c. circuits

Problem 1. Determine the values of the resistance and the series-connected inductance or capacitance for each of the following impedances:

(a) $(12 + j5)\,\Omega$ (b) $-j40\,\Omega$ (c) $30 \underline{/60°}\,\Omega$
(d) $2.20 \times 10^6 \underline{/-30°}\,\Omega$.

Assume for each a frequency of 50 Hz.

(a) From section 2(d), for an R–L series circuit, impedance $Z = R + jX_L$. Thus $Z = (12 + j5)\,\Omega$ represents a resistance of $12\,\Omega$ and an inductive reactance of $5\,\Omega$ in series.

Since inductive reactance $X_L = 2\pi fL$,

$$\text{inductance } L = \frac{X_L}{2\pi f} = \frac{5}{2\pi(50)} = 0.0159\,\text{H}$$

i.e., the inductance is $15.9\,\text{mH}$.

Thus an impedance $(12 + j5)\,\Omega$ represents a resistance of $12\,\Omega$ in series with an inductance of $15.9\,\text{mH}$.

(b) From section 2(c), for a purely capacitive circuit, impedance $Z = -jX_C$. Thus $Z = -j40\,\Omega$ represents zero resistance and a capacitive reactance of $40\,\Omega$.

Since capacitive reactance $X_C = 1/(2\pi fC)$,

$$\text{capacitance } C = \frac{1}{2\pi fX_C} = \frac{1}{2\pi(50)(40)}\,\text{F}$$

$$= \frac{10^6}{2\pi(50)(40)}\,\mu\text{F} = 79.6\,\mu\text{F}$$

Thus an impedance $-j40\,\Omega$ represents a pure capacitor of capacitance $79.6\,\mu\text{F}$.

(c) $30\underline{/\,60^\circ} = 30(\cos 60^\circ + j\sin 60^\circ) = 15 + j25.98$.   Thus $Z = 30\underline{/\,60^\circ}\,\Omega = (15 + j25.98)\,\Omega$ represents a resistance of $15\,\Omega$ and an inductive reactance of $25.98\,\Omega$ in series (from section 2(d)).

Since $X_L = 2\pi fL$,

$$\text{inductance } L = \frac{X_L}{2\pi f} = \frac{25.98}{2\pi(50)} = 0.0827\,\text{H}$$

Thus an impedance $30\underline{/\,60^\circ}\,\Omega$ represents a resistance of $15\,\Omega$ in series with an inductance of $82.7\,\text{mH}$.

(d) $2.20 \times 10^6\underline{/-30^\circ} = 2.20 \times 10^6\,[\cos(-30^\circ) + j\sin(-30^\circ)]$
$= 1.905 \times 10^6 - j1.10 \times 10^6$.

Thus $Z = 2.20 \times 10^6\underline{/-30^\circ}\,\Omega = (1.905 \times 10^6 - j1.10 \times 10^6)\,\Omega$ represents a resistance of $1.905 \times 10^6\,\Omega$ (i.e. $1.905\,\text{M}\Omega$) and a capacitive reactance of $1.10 \times 10^6\,\Omega$ in series (from section 2(e)).

Since capacitive reactance $X_C = 1/(2\pi fC)$,

$$\text{capacitance } C = \frac{1}{2\pi fX_C} = \frac{1}{2\pi(50)(1.10 \times 10^6)}\,\text{F}$$

$$= 2.894 \times 10^{-9}\,\text{F}\quad\text{or}\quad 2.894\,\text{nF}$$

Thus an impedance $2.2 \times 10^6\underline{/-30^\circ}\,\Omega$ represents a resistance of $1.905\,\text{M}\Omega$ in series with a $2.894\,\text{nF}$ capacitor.

Problem 2. Determine, in polar and rectangular forms, the current flowing in an inductor of negligible resistance and inductance $159.2\,\text{mH}$ when it is connected to a $250\,\text{V}$, $50\,\text{Hz}$ supply.

Inductive reactance $X_L = 2\pi fL = 2\pi(50)(159.2 \times 10^{-3}) = 50\,\Omega$. Thus circuit impedance $Z = (0 + j50)\,\Omega = 50\underline{/\,90^\circ}\,\Omega$.

Supply voltage, $V = 250\underline{/\,0^\circ}\text{V}$ (or $(250 + j0)$ V). Hence

$$\text{current } I = \frac{V}{Z} = \frac{250\underline{/\,0^\circ}}{50\underline{/\,90^\circ}} = \frac{250}{50}\underline{/\,0-90^\circ} = \mathbf{5\underline{/-90^\circ}\text{A}}$$

(Note: since the voltage is given as 250 V, this is assumed to mean $250\underline{/\,0°}$ V or $(250 + j0)$ V).

Alternatively,

$$I = \frac{V}{Z} = \frac{(250 + j0)}{(0 + j50)} = \frac{250(-j50)}{j50(-j50)} = \frac{-j(50)(250)}{50^2} = \mathbf{-j5\,A}$$

which is the same as $5\underline{/-90°}$ A

Problem 3. A 3 μF capacitor is connected to a supply of frequency 1 kHz and a current of $2.83\underline{/\,90°}$ A flows. Determine the value of the supply p.d.

$$\text{Capacitive reactance } X_C = \frac{1}{2\pi fC} = \frac{1}{2\pi(1000)(3 \times 10^{-6})} = 53.05\,\Omega$$

Hence circuit impedance $Z = (0 - j53.05)\,\Omega = 53.05\underline{/-90°}$

$$\text{Current } I = 2.83\underline{/\,90°}\,A \text{ (or } (0 + j2.83)\,A)$$

$$\text{Supply p.d., } V = IZ = (2.83\underline{/\,90°})(53.05\underline{/-90°})$$

i.e.

$$\textbf{p.d.} = \mathbf{150\underline{/\,0°}\,V}$$

Alternatively,

$$V = IZ = (0 + j2.83)(0 - j53.05)$$
$$= -j^2(2.83)(53.05) = \mathbf{150\,V}$$

Problem 4. The impedance of an electrical circuit is $(30 - j50)$ ohms. Determine (a) the resistance, (b) the capacitance, (c) the modulus of the impedance, and (d) the current flowing and its phase angle, when the circuit is connected to a 240 V, 50 Hz supply.

(a) Since impedance $Z = (30 - j50)\,\Omega$, **the resistance is 30 ohms** and the capacitance reactance is $50\,\Omega$.
(b) Since $X_C = 1/(2\pi fC)$,

$$\textbf{capacitance, } \mathbf{C} = \frac{1}{2\pi fX_C} = \frac{1}{2\pi(50)(50)} = \mathbf{63.66\,\mu F}$$

(c) The modulus of impedance, $|Z| = \sqrt{(R^2 + X_C^2)} = \sqrt{(30^2 + 50^2)}$
$$= \mathbf{58.31\,\Omega}$$

(d) Impedance $Z = (30 - j50)\,\Omega = 58.31\underline{/\,\arctan\dfrac{X_C}{R}} = 58.31\underline{/-59.04°}\,\Omega$.

Hence

$$\text{current } I = \frac{V}{Z} = \frac{240\underline{/\,0°}}{58.31\underline{/-59.04°}} = \mathbf{4.12\underline{/\,59.04°}\,A}$$

Problem 5. A 200 V, 50 Hz supply is connected across a coil of negligible resistance and inductance 0.15 H connected in series with a $32\,\Omega$ resistor. Determine (a) the impedance of the coil, (b) the current and circuit phase angle, (c) the p.d. across the $32\,\Omega$ resistor, and (d) the p.d. across the coil.

(a) Inductive reactance $X_L = 2\pi fL = 2\pi(50)(0.15) = 47.1\,\Omega$
Impedance $Z = R + jX_L = (32 + j47.1)\,\Omega$ or $57.0\underline{/\,55.81°}\,\Omega$
The circuit diagram is shown in Fig. 9.

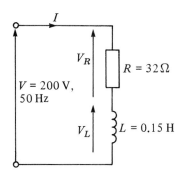

$V = 200$ V,
50 Hz

V_R

$R = 32\,\Omega$

V_L

$L = 0.15$ H

Figure 9

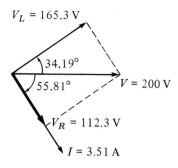

Figure 10

(b) Current $I = \dfrac{V}{Z} = \dfrac{200\angle 0°}{57.0\angle 55.81°} = \mathbf{3.51\angle -55.81° A}$

i.e., **the current is 3.51 A lagging the voltage by 55.81°**.

(c) P.d. across the $32\,\Omega$ resistor,

$$V_R = IR = (3.51\angle -55.81°)(32\angle 0°)$$

i.e.

$$V_R = \mathbf{112.3\angle -55.81° V}$$

(d) P.d. across the coil,

$$V_L = IX_L = (3.51\angle -55.81°)(47.1\angle 90°)$$

i.e.

$$V_L = \mathbf{165.3\angle 34.19° V}$$

The phasor sum of V_R and V_L is the supply voltage V as shown in the phasor diagram of Fig. 10.

$V_R = 112.3\angle -55.81° = (63.11 - j92.89)\,V$
$V_L = 165.3\angle 34.19° V = (136.73 + j92.89)\,V$

Hence

$$V = V_R + V_L = (63.11 - j92.89) + (136.73 + j92.89)$$
$$= (200 + j0)\,V \text{ or } 200\angle 0° V, \text{ correct to three significant figures.}$$

Problem 6. Determine the value of impedance if a current of $(7 + j16)\,A$ flows in a circuit when the supply voltage is $(120 + j200)\,V$. If the frequency of the supply is 5 MHz, determine the value of the components forming the series circuit.

$$\text{Impedance } Z = \frac{V}{I} = \frac{(120 + j200)}{(7 + j16)} = \frac{233.24\angle 59.04°}{17.464\angle 66.37°}$$
$$= 13.36\angle -7.33°\,\Omega \quad \text{or} \quad (13.25 - j1.705)\,\Omega$$

The series circuit thus consists of a **$13.25\,\Omega$ resistor** and a capacitor of capacitive reactance $1.705\,\Omega$.

Since $X_C = 1/(2\pi f C)$,

$$\text{capacitance } C = \frac{1}{2\pi f X_C} = \frac{1}{2\pi(5 \times 10^6)(1.705)}$$
$$= 1.867 \times 10^{-8}\,F = \mathbf{18.67\,nF}$$

Problem 7. For the circuit shown in Fig. 11, determine the value of impedance Z_2.

Note the symbol for a voltage source (see BS3939:1985).

Total circuit impedance

$$Z = \frac{V}{I} = \frac{70\angle 30°}{3.5\angle -20°}$$
$$= 20\angle 50°\,\Omega \quad \text{or} \quad (12.86 + j15.32)\,\Omega$$

Total impedance $Z = Z_1 + Z_2$ (see section 2(g)). Hence

$$(12.86 + j15.32) = (4.36 - j2.10) + Z_2$$

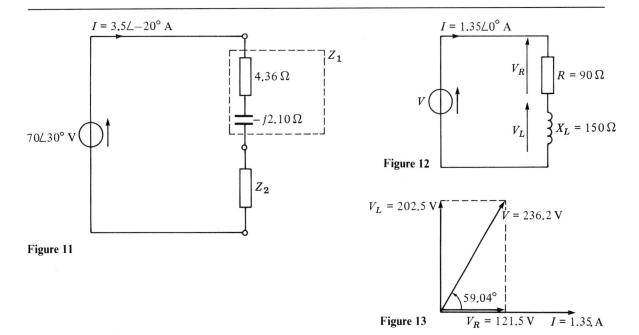

Figure 11

Figure 12

Figure 13

from which

$$\text{impedance } Z_2 = (12.86 + j15.32) - (4.36 - j2.10)$$
$$= \mathbf{(8.50 + j17.42)\,\Omega} \quad \text{or} \quad \mathbf{19.38\underline{/\,63.99°}\,\Omega}$$

Problem 8. A circuit comprises a resistance of $90\,\Omega$ in series with an inductor of inductive reactance $150\,\Omega$. If the supply current is $1.35\underline{/\,0°}\,\text{A}$, determine (a) the supply voltage, (b) the voltage across the $90\,\Omega$ resistance, (c) the voltage across the inductance, and (d) the circuit phase angle. Draw the phasor diagram.

The circuit diagram is shown in Fig. 12.
(a) Circuit impedance $Z = R + jX_L$
$$= (90 + j150)\,\Omega \quad \text{or} \quad 174.93\underline{/\,59.04°}\,\Omega.$$
Supply voltage, $V = IZ = (1.35\underline{/\,0°})(174.93\underline{/\,59.04°})$
$$= \mathbf{236.2\underline{/\,59.04°}\,V} \quad \text{or} \quad \mathbf{(121.5 + j202.5)\,V}$$
(b) Voltage across $90\,\Omega$ resistor, $V_R = \mathbf{121.5\,V}$ (since $V = V_R + jV_L$).
(c) Voltage across inductance, $V_L = \mathbf{202.5\,V}$ leading V_R by $90°$.
(d) Circuit phase angle is the angle between the supply current and voltage, i.e., **59.04° lagging.** The phasor diagram is shown in Fig. 13.

Problem 9. A coil of resistance $25\,\Omega$ and inductance $20\,\text{mH}$ has an alternating voltage given by $v = 282.8\sin(628.4t + (\pi/3))$ volts applied across it. Determine (a) the rms value of voltage (in polar form), (b) the circuit impedance, (c) the rms current flowing, and (d) the circuit phase angle.

(a) Voltage $v = 282.8\sin(628.4t + (\pi/3))$ volts means $V_m = 282.8\,\text{V}$, hence

$$\text{rms voltage } V = 0.707 \times 282.8 \left(\text{or} \, \frac{1}{\sqrt{2}} \times 282.8 \right)$$

$$\text{i.e., } V = 200\,\text{V}$$

In complex form the rms voltage may be expressed as $\mathbf{200\left/\dfrac{\pi}{3}\right.}$ **V** or $\mathbf{200\left/\underline{60°}\right.}$ **V**.

(b) $\omega = 2\pi f = 628.4 \text{ rad/s}$ hence frequency $f = 628.4/(2\pi) = 100 \text{ Hz}$.

Inductive reactance $X_L = 2\pi f L = 2\pi(100)(20 \times 10^{-3}) = 12.57\,\Omega$

Hence

circuit impedance $Z = R + jX_L = (25 + j12.57)\,\Omega$ or $27.98\left/\underline{26.69°}\right.\,\Omega$

(c) rms current, $I = \dfrac{V}{Z} = \dfrac{200\left/\underline{60°}\right.}{27.98\left/\underline{26.69°}\right.} = \mathbf{7.148\left/\underline{33.31°}\right. A.}$

(d) Circuit phase angle is the angle between current I and voltage V, i.e., $60° - 33.31° = \mathbf{26.69°\ lagging.}$

Problem 10. A 240 V, 50 Hz voltage is applied across a series circuit comprising a coil of resistance $12\,\Omega$ and inductance 0.10 H, and a $120\,\mu\text{F}$ capacitor. Determine the current flowing in the circuit.

The circuit diagram is shown in Fig. 14.

Inductive reactance, $X_L = 2\pi f L = 2\pi(50)(0.10) = 31.4\,\Omega$

Capacitive reactance, $X_C = \dfrac{1}{2\pi f C} = \dfrac{1}{2\pi(50)(120 \times 10^{-6})} = 26.5\,\Omega$

Impedance $Z = R + j(X_L - X_C)$ (see section 2(f))

i.e.

$$Z = 12 + j(31.4 - 26.5) = (12 + j4.9)\,\Omega \quad \text{or} \quad 13.0\left/\underline{22.2°}\right.\,\Omega$$

Current flowing, $I = \dfrac{V}{Z} = \dfrac{240\left/\underline{0°}\right.}{13.0\left/\underline{22.2°}\right.} = \mathbf{18.5\left/\underline{-22.2°}\right. A,}$

i.e., the current flowing is 18.5 A, lagging the voltage by 22.2°.
The phasor diagram is shown on the Argand diagram in Fig. 15.

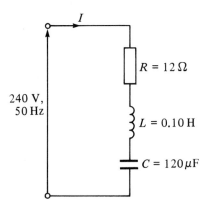

240 V, 50 Hz

I

$R = 12\,\Omega$

$L = 0.10\ \text{H}$

$C = 120\,\mu\text{F}$

Figure 14

Problem 11. A coil of resistance R ohms and inductance L henrys is connected in series with a $50\,\mu\text{F}$ capacitor. If the supply voltage is 225 V at 50 Hz and the current flowing in the circuit is $1.5\left/\underline{-30°}\right.$ A, determine the values of R and L. Determine also the voltage across the coil and the voltage across the capacitor.

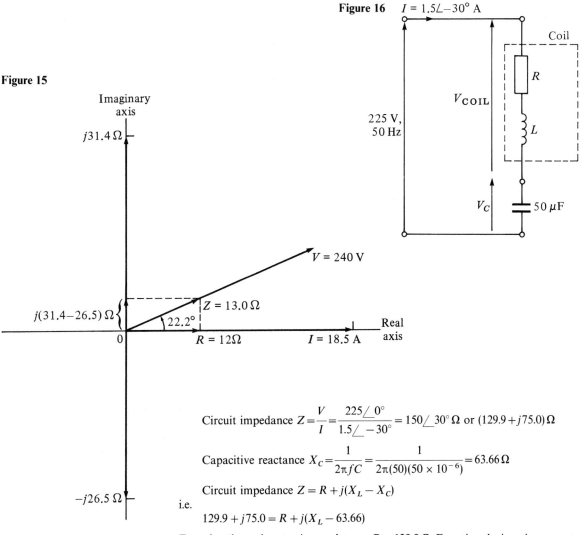

Figure 16 $I = 1.5\angle -30°$ A

Figure 15

Circuit impedance $Z = \dfrac{V}{I} = \dfrac{225\angle\,0°}{1.5\angle\,-30°} = 150\angle\,30°\,\Omega$ or $(129.9 + j75.0)\,\Omega$

Capacitive reactance $X_C = \dfrac{1}{2\pi fC} = \dfrac{1}{2\pi(50)(50\times 10^{-6})} = 63.66\,\Omega$

Circuit impedance $Z = R + j(X_L - X_C)$

i.e.
$$129.9 + j75.0 = R + j(X_L - 63.66)$$

Equating the real parts gives **resistance $R = 129.9\,\Omega$.** Equating the imaginary parts gives $75.0 = X_L - 63.66$, from which

$$X_L = 75.0 + 63.66 = 138.66\,\Omega$$

Since $X_L = 2\pi fL$,

$$\textbf{inductance } L = \dfrac{X_L}{2\pi f} = \dfrac{138.66}{2\pi(50)} = \textbf{0.441 H}$$

The circuit diagram is shown in Fig. 16.

Voltage across coil, $V_{\text{COIL}} = IZ_{\text{COIL}}$

$Z_{\text{COIL}} = R + jX_L = (129.9 + j138.66)\,\Omega$ or $190\angle\,46.87°\,\Omega$. Hence
$$V_{\text{COIL}} = (1.5\angle\,-30°)(190\angle\,46.87°)$$
$$= \textbf{285}\angle\,\textbf{16.87° V}\quad\text{or}\quad\textbf{(272.74} + \textbf{j82.71) V}$$

Voltage across capacitor, $V_C = IX_C = (1.5 \underline{/-30°})(63.66 \underline{/-90°})$
$$= \mathbf{95.49 \underline{/-120°} \, V} \quad \text{or} \quad (\mathbf{-47.75 - j82.70}) \, V$$

(Check: Supply voltage $V = V_{\text{COIL}} + V_C$
$$= (272.74 + j82.71) + (-47.75 - j82.70)$$
$$= (225 + j0) \, V \text{ or } 225 \underline{/0°} \, V)$$

Problem 12. When two impedances $Z_1 = 5 \underline{/30°} \, \Omega$ and $Z_2 = 10 \underline{/-45°} \, \Omega$ are connected in series across a certain a.c. supply, a current of $6.11 \underline{/60°}$ A flows. Determine (a) the total equivalent circuit impedance, (b) the supply voltage, (c) the circuit phase angle, (d) the p.d. across impedance Z_1, and (e) the p.d. across impedance Z_2.

(a) Total circuit impedance $Z = Z_1 + Z_2 = 5 \underline{/30°} + 10 \underline{/-45°}$
$$= (4.33 + j2.5) + (7.07 - j7.07)$$
$$= (\mathbf{11.40 - j4.57}) \, \Omega \text{ or } \mathbf{12.28 \underline{/-21.84°} \, \Omega}$$

(b) Supply voltage $V = IZ = (6.11 \underline{/60°})(12.28 \underline{/-21.84°})$
$$= \mathbf{75 \underline{/38.16°} \, V}$$

(c) Circuit phase angle is the angle between current and voltage, i.e. $60° - 38.16°$, i.e. **21.84° leading.**

(d) P.d. across impedance Z_1,
$$V_1 = IZ_1 = (6.11 \underline{/60°})(5 \underline{/30°})$$
$$= \mathbf{30.55 \underline{/90°} \, V} \quad \text{or} \quad (\mathbf{0 + j30.55}) \, V$$

(e) P.d. across impedance Z_2,
$$V_2 = IZ_2 = (6.11 \underline{/60°})(10 \underline{/-45°})$$
$$= \mathbf{61.10 \underline{/15°} \, V} \quad \text{or} \quad (\mathbf{59.02 + j15.81}) \, V$$

(Check: Supply voltage
$$V = V_1 + V_2$$
$$= (0 + j30.55) + (59.02 + j15.81) = (59.02 + j46.36) \, V$$
$$= 75 \underline{/38.15} \, \dot{V})$$

Problem 13. For the circuit shown in Fig. 17, determine the values of voltages V_1 and V_2 if the supply frequency is 4 kHz. Determine also the value of the supply voltage V and the circuit phase angle. Draw the phasor diagram.

For impedance Z_1,
$$X_C = \frac{1}{2\pi fC} = \frac{1}{2\pi(4000)(2.653 \times 10^{-6})} = 15 \, \Omega$$

Hence
$$Z_1 = (8 - j15) \, \Omega \quad \text{or} \quad 17 \underline{/-61.93°} \, \Omega$$
$$\mathbf{Voltage} \, V_1 = IZ_1 = (6 \underline{/0°})(17 \underline{/-61.93°})$$
$$= \mathbf{102 \underline{/-61.93°} \, V} \quad \text{or} \quad (\mathbf{48 - j90}) \, V.$$

For impedance Z_2,
$$X_L = 2\pi fL = 2\pi(4000)(0.477 \times 10^{-3}) = 12 \, \Omega$$

Hence
$$Z_2 = (5 + j12) \, \Omega \quad \text{or} \quad 13 \underline{/67.38°} \, \Omega$$

Figure 17 $I = 6\angle 0°$ A

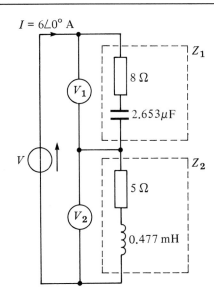

Figure 18 $V_2 = 78$ V

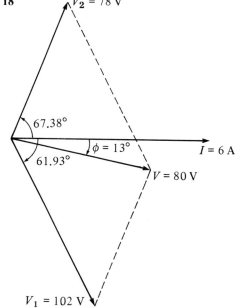

Voltage, $V_2 = IZ_2 = (6\angle 0°)(13\angle 67.38°)$
$$= 78\angle 67.38° \text{ V} \quad \text{or} \quad (30+j72) \text{ V}.$$

Supply voltage, $V = V_1 + V_2 = (48 - j90) + (30 + j72)$
$$= (78 - j18) \text{ V} \quad \text{or} \quad 80\angle -13° \text{ V}$$

Circuit phase angle, $\phi = 13°$ **leading.** The phasor diagram is shown in Fig. 18.

Further problems on the application of complex numbers to series a.c. circuits may be found in the following section (3), problems 1 to 24.

3. Further problems

1. Determine the resistance R and series inductance L (or capacitance C) for each of the following impedances, assuming the frequency to be 50 Hz.
   (a) $(4 + j7)\,\Omega$ (b) $(3 - j20)\,\Omega$ (c) $j10\,\Omega$

   (d) $-j3\,\text{k}\Omega$ (e) $15\angle(\pi/3)\,\Omega$ (f) $6\angle -45°\,\text{M}\Omega$

   $$\begin{bmatrix} \text{(a)} \ R = 4\,\Omega, \ L = 22.3\,\text{mH} & \text{(b)} \ R = 3\,\Omega, \ C = 159.2\,\mu\text{F} \\ \text{(c)} \ R = 0, \ L = 31.8\,\text{mH} & \text{(d)} \ R = 0, \ C = 1.061\,\mu\text{F} \\ \text{(e)} \ R = 7.5\,\Omega, \ L = 41.3\,\text{mH} & \text{(f)} \ R = 4.243\,\text{M}\Omega, \ C = 0.750\,\text{nF} \end{bmatrix}$$

2. A $0.4\,\mu$F capacitor is connected to a 250 V, 2 kHz supply. Determine the current flowing.

   $$[1.257\angle 90° \text{ A} \quad \text{or} \quad j1.257 \text{ A}]$$

3. Two voltages in a circuit are represented by $(15 + j10)$ V and $(12 - j4)$ V. Determine the magnitude of the resultant voltage when these voltages are added.

   $$[27.66 \text{ V}]$$

4. A current of $2.5\angle -90°$ A flows in a coil of inductance 314.2 mH and negligible resistance when connected across a 50 Hz supply. Determine the value of the supply p.d.

   $$[246.8\angle 0° \text{ V}]$$

5. The impedance of a coil is given by $(50 + j120)\,\text{m}\Omega$.
Explain the terms in the expression. What information is given by this expression but is not given by the statement that the impedance of the coil is $130\,\text{m}\Omega$?

$$[R = 50\,\text{m}\Omega, \; X_L = 120\,\text{m}\Omega]$$

6. A voltage $(75 + j90)\,\text{V}$ is applied across an impedance and a current of $(5 + j12)\,A$ flows. Determine (a) the value of the circuit impedance, and (b) the values of the components comprising the circuit if the frequency is $1\,\text{kHz}$.

$$\begin{bmatrix} \text{(a)} \; Z = (8.61 - j2.66)\,\Omega \text{ or } 9.01\underline{/-17.19°}\,\Omega \\ \text{(b)} \; R = 8.61\,\Omega, \; C = 59.83\,\mu F \end{bmatrix}$$

7. Determine, in polar form, the complex impedances for the circuits shown in Fig. 19 if the frequency in each case is $50\,\text{Hz}$.

$$[\text{(a)} \; 44.53\underline{/-63.31°} \quad \text{(b)} \; 19.77\underline{/52.62°}\,\Omega \quad \text{(c)} \; 113.5\underline{/-58.08°}]$$

8. For the circuit shown in Fig. 20 determine the impedance Z in polar and rectangular forms.

$$[Z = (1.85 + j6.2)\,\Omega \text{ or } 6.47\underline{/73.39°}\,\Omega]$$

9. A $30\,\mu F$ capacitor is connected in series with a resistance R at a frequency of $200\,\text{Hz}$. The resulting current leads the voltage by $30°$. Determine the magnitude of R.

$$[45.95\,\Omega]$$

10. The current in a circuit is $(4 + j3)\,\text{mA}$ when the applied voltage is $(7 - j3)\,\text{V}$. Determine the impedance of the circuit and the phase angle between current and voltage.

$$[1.52\,\text{k}\Omega; \; 60.07° \text{ leading}]$$

11. A coil has a resistance of $40\,\Omega$ and an inductive reactance of $75\,\Omega$. The current in the coil is $1.70\underline{/0°}\,A$. Determine the value of (a) the supply voltage, (b) the p.d. across the $40\,\Omega$ resistance, (c) the p.d. across the inductive part of the coil, and (d) the circuit phase angle. Draw the phasor diagram.

$$\begin{bmatrix} \text{(a)} \; (68 + j127.5)\,\text{V} \text{ or } 144.5\underline{/61.93°}\,\text{V} \\ \text{(b)} \; 68\underline{/0°}\,\text{V} \quad \text{(c)} \; 127.5\underline{/90°} \\ \text{(d)} \; 61.93° \text{ lagging} \end{bmatrix}$$

12. An alternating voltage of $100\,\text{V}$, $50\,\text{Hz}$ is applied across an impedance of $(20 - j30)\,\Omega$. Calculate (a) the resistance, (b) the capacitance, (c) the current, and (d) the phase angle between current and voltage.

$$[\text{(a)} \; 20\,\Omega \quad \text{(b)} \; 106.1\,\mu F \quad \text{(c)} \; 2.774\,A \quad \text{(d)} \; 56.31° \text{ leading}]$$

13. A resistance of $45\,\Omega$ is connected in series with a $42\,\mu F$ capacitor. If the applied voltage is $250\,\text{V}$ at $50\,\text{Hz}$, determine (a) the capacitive reactance, (b) the impedance, (c) the current and its phase relative to the applied voltage, (d) the voltage across the resistance, and (e) the voltage across the capacitor. Draw the phasor diagram.

$$\begin{bmatrix} \text{(a)} \; 75.79\,\Omega \quad \text{(b)} \; 88.14\underline{/-59.3°}\,\Omega \quad \text{(c)} \; 2.836\,A, \text{ leading V by } 59.3° \\ \text{(d)} \; 127.6\underline{/59.3°}\,\text{V} \quad \text{(e)} \; 214.9\underline{/-30.7°}\,\text{V} \end{bmatrix}$$

14. A capacitor C is connected in series with a coil of resistance R and inductance $30\,\text{mH}$. The current flowing in the circuit is $2.5\underline{/-40°}\,A$ when the supply p.d. is $200\,\text{V}$ at $400\,\text{Hz}$. Determine the value of (a) resistance R, (b) capacitance C, (c) the p.d. across C, and (d) the p.d. across the coil.

$$\begin{bmatrix} \text{(a)} \; 61.28\,\Omega \quad \text{(b)} \; 16.59\,\mu F \\ \text{(c)} \; 59.95\underline{/-130°}\,\text{V} \quad \text{(d)} \; 242.9\underline{/10.9°}\,\text{V} \end{bmatrix}$$

15. A series circuit consists of a $10\,\Omega$ resistor, a coil of inductance $0.09\,\text{H}$ and negligible resistance, and a $150\,\mu F$ capacitor, and is connected to a $100\,\text{V}$, $50\,\text{Hz}$ supply. Calculate the current flowing and its phase relative to the supply voltage.

$$[8.17\,A \text{ lagging V by } 35.2°]$$

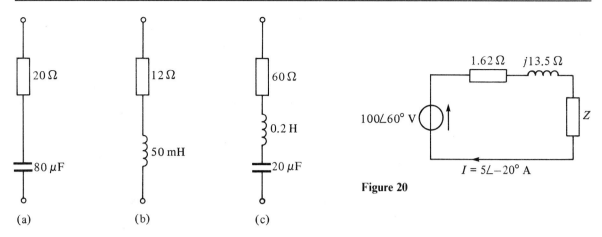

Figure 19

(a)   (b)   (c)

Figure 20

16. A 150 mV, 5 kHz source supplies an a.c. circuit consisting of a coil of resistance 25 Ω and inductance 5 mH connected in series with a capacitance of 177 nF.
    Determine the current flowing and its phase angle relative to the source voltage.

$$[4.44 \underline{/\ 42.3^\circ}\text{ mA}]$$

17. Two impedances $Z_1 = (2 + j6)\,\Omega$ and $Z_2 = (5 - j2)\,\Omega$ are connected in series to a 100 V supply voltage. Determine the current flowing in the circuit.

$$[12.40\underline{/\ -29.74^\circ}\text{ A}]$$

18. Two impedances, $Z_1 = 5\underline{/\ 30^\circ}\,\Omega$ and $Z_2 = 10\underline{/\ 45^\circ}\,\Omega$, draw a current of 3.36 A when connected in series to a certain a.c. supply. Determine (a) the supply voltage, (b) the phase angle between the voltage and current, (c) the p.d. across Z_1 and (d) the p.d. across Z_2.

$$[\text{(a) } 50\text{ V} \quad \text{(b) } 40.01^\circ \quad \text{(c) } 16.8\underline{/\ 30^\circ}\text{ V} \quad \text{(d) } 33.6\underline{/\ 45^\circ}\text{ V}]$$

19. A 4500 pF capacitor is connected in series with a 50 Ω resistor across an alternating voltage $v = 212.1 \sin(\pi 10^6 t + \pi/4)$ volts. Calculate (a) the rms value of the voltage, (b) the circuit impedance, (c) the rms current flowing, (d) the circuit phase angle, (e) the voltage across the resistor, and (f) the voltage across the capacitor.

$$\begin{bmatrix} \text{(a) } 150\underline{/\ 45^\circ}\text{ V} & \text{(b) } 86.63\underline{/\ -54.75^\circ}\,\Omega \\ \text{(c) } 1.73\underline{/\ 99.75^\circ}\text{ A} & \text{(d) } 54.75^\circ \text{ leading} \\ \text{(e) } 86.5\underline{/\ 99.75^\circ}\text{ V} & \text{(f) } 122.4\underline{/\ 9.75^\circ}\text{ V} \end{bmatrix}$$

20. If the p.d. across a coil is $(30 + j20)$ V at 60 Hz and the coil consists of 50 mH inductance and 10 Ω resistance, determine the value of current flowing (in polar and cartesian forms).

$$[1.69\underline{/\ -28.36^\circ}\text{ A; } (1.49 - j0.80)\text{ A}]$$

21. Three impedances are connected in series across a 120 V, 10 kHz supply. The impedances are:
    (i) Z_1, a coil of inductance 200 μH and resistance 8 Ω
    (ii) Z_2, a resistance of 12 Ω
    (iii) Z_3, a 0.50 μF capacitor in series with a 15 Ω resistor.
    Determine (a) the circuit impedance, (b) the circuit current, (c) the circuit phase angle, and (d) the p.d. across each impedance.

$$\begin{bmatrix} \text{(a) } 39.95\underline{/\ -28.82^\circ}\,\Omega & \text{(b) } 3.00\underline{/\ 28.82^\circ}\text{ A} \\ \text{(c) } 28.82^\circ \text{ leading} & \text{(d) } V_1 = 44.70\underline{/\ 86.35^\circ}\text{ V}, V_2 = 36\underline{/\ 28.82^\circ}\text{ V} \\ & V_3 = 105.6\underline{/\ -35.95^\circ}\text{ V} \end{bmatrix}$$

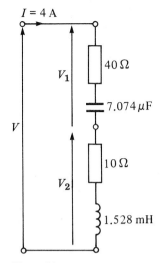

Figure 21

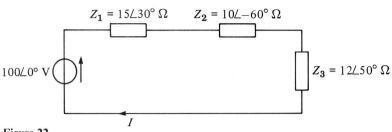

Figure 22

22. Determine the value of voltages V_1 and V_2 in the circuit shown in Fig. 21, if the frequency of the supply is 2.5 kHz. Find also the value of the supply voltage V and the circuit phase angle. Draw the phasor diagram.

$$\left[\begin{array}{l} V_1 = 164\angle -12.68° \text{ V or } (160 - j36)\text{ V} \\ V_2 = 104\angle\ 67.38° \text{ V or } (40 + j96)\text{ V} \\ V\ = 208.8\angle\ 16.7° \text{ V or } (200 + j60)\text{ V} \\ \text{Phase angle} = 16.7° \text{ lagging.} \end{array}\right]$$

23. A circuit comprises a coil of inductance 40 mH and resistance 20 Ω in series with a variable capacitor. The supply voltage is 120 V at 50 Hz. Determine the value of capacitance needed to cause a current of 2.0 A to flow in the circuit.

$$[46.04\,\mu\text{F}]$$

24. (a) Explain how alternating quantities can be represented by complex operators.

   (b) For the circuit shown in Fig. 22, determine (i) the circuit current I flowing, and (ii) the p.d. across each impedance.

$$\left[\begin{array}{ll} \text{(b)} & \text{(i)}\ \ 3.71\angle -17.35° \text{ A} \\ & \text{(ii)}\ \ V_1 = 55.65\angle\ 12.65° \text{ V} \\ & \qquad V_2 = 37.1\angle -77.35° \text{ V} \\ & \qquad V_3 = 44.52\angle\ 32.65° \text{ V} \end{array}\right]$$

3 Application of complex numbers to parallel a.c. networks

1. Introduction

As with series circuits, parallel networks may be analysed by using phasor diagrams. However, with parallel networks containing more than two branches this can become very complicated. It is with parallel a.c. network analysis in particular that the full benefit of using complex numbers may be appreciated. Before analysing such networks it is necessary to define admittance, conductance and susceptance.

2. Admittance, conductance and susceptance

Admittance is defined as the current I flowing in an a.c. circuit divided by the supply voltage V (i.e. it is the reciprocal of impedance Z). The symbol for admittance is Y. Thus

$$Y = \frac{I}{V} = \frac{1}{Z}$$

The unit of admittance is the **siemen, S.**

An impedance may be resolved into a real part R and an imaginary part X, giving $Z = R \pm jX$. Similarly, an admittance may be resolved into two parts—the real part being called the **conductance G**, and the imaginary part being called the **susceptance B**—and expressed in complex form. Thus

$$\text{admittance } Y = G + jB.$$

When an a.c. circuit contains:
(a) **pure resistance**, then

$$Z = R \quad \text{and} \quad Y = \frac{1}{Z} = \frac{1}{R} = G$$

(b) **pure inductance**, then

$$Z = jX_L \quad \text{and} \quad Y = \frac{1}{Z} = \frac{1}{jX_L} = \frac{-j}{(jX_L)(-j)} = \frac{-j}{X_L} = -jB_L$$

Thus a negative sign is associated with inductive susceptance, B_L.
(c) **pure capacitance** then

$$Z = -jX_C \quad \text{and} \quad Y = \frac{1}{Z} = \frac{1}{-jX_C} = \frac{j}{(jX_C)(j)} = \frac{j}{X_C} = +jB_C.$$

Thus a positive sign is associated with capacitive susceptance, B_C.

(d) **resistance and inductance in series**, then

$$Z = R + jX_L \quad \text{and} \quad Y = \frac{1}{Z} = \frac{1}{R + jX_L} = \frac{(R - jX_L)}{R^2 + X_L^2}$$

i.e.

$$Y = \frac{R}{R^2 + X_L^2} - j\frac{X_L}{R^2 + X_L^2} \quad \text{or} \quad Y = \frac{R}{|Z|^2} - j\frac{X_L}{|Z|^2}$$

Thus conductance, $G = R/|Z|^2$ and inductive susceptance, $B_L = -X_L/|Z|^2$. (Note that in an inductive circuit, the imaginary term of the impedance, X_L, is positive, whereas the imaginary term of the admittance, B_L, is negative.)

(e) **resistance and capacitance in series**, then

$$Z = R - jX_C \quad \text{and} \quad Y = \frac{1}{Z} = \frac{1}{R - jX_C} = \frac{R + jX_C}{R^2 + X_C^2}$$

i.e.

$$Y = \frac{R}{R^2 + X_C^2} + j\frac{X_C}{R^2 + X_C^2} \quad \text{or} \quad Y = \frac{R}{|Z|^2} + j\frac{X_C}{|Z|^2}$$

(Thus conductance, $G = R/|Z|^2$ and capacitive susceptance, $B_C = X_C/|Z|^2$. (Note that in a capacitive circuit, the imaginary term of the impedance, X_C, is negative, whereas the imaginary term of the admittance, B_C, is positive.)

(f) **resistance and inductance in parallel**, then

$$\frac{1}{Z} = \frac{1}{R} + \frac{1}{jX_L} = \frac{jX_L + R}{(R)(jX_L)},$$

from which

$$Z = \frac{(R)(jX_L)}{R + jX_L} \quad \left(\text{i.e. } \frac{\text{product}}{\text{sum}} \right)$$

and

$$Y = \frac{1}{Z} = \frac{R + jX_L}{jRX_L} = \frac{R}{jRX_L} + \frac{jX_L}{jRX_L}$$

i.e.,

$$Y = \frac{1}{jX_L} + \frac{1}{R} = \frac{(-j)}{(jX_L)(-j)} + \frac{1}{R}$$

or

$$Y = \frac{1}{R} - \frac{j}{X_L}$$

Thus conductance, $G = 1/R$ and inductance susceptance, $B_L = -1/X_L$.

(g) **resistance and capacitance in parallel**, then

$$Z = \frac{(R)(-jX_C)}{R - jX_C} \quad \left(\text{i.e., } \frac{\text{product}}{\text{sum}} \right)$$

and

$$Y = \frac{1}{Z} = \frac{R - jX_C}{-jRX_C} = \frac{R}{-jRX_C} - \frac{jX_C}{-jRX_C}$$

i.e.

$$Y = \frac{1}{-jX_C} + \frac{1}{R} = \frac{(j)}{(-jX_C)(j)} + \frac{1}{R}$$

or

$$Y = \frac{1}{R} + \frac{j}{X_C} \qquad (1)$$

Thus conductance, $G = 1/R$ and capacitive susceptance, $B_C = 1/X_C$.

The conclusions that may be drawn from sections (d) to (g) above are:

(i) that a **series** circuit is more easily represented by an **impedance,**

(ii) that a **parallel** circuit is often more easily represented by an **admittance,** especially when more than two parallel impedances are involved.

Worked problems on admittance, conductance and susceptance

Problem 1. Determine the admittance, conductance and susceptance of the following impedances: (a) $-j5\,\Omega$ (b) $(25 + j40)\,\Omega$ (c) $(3 - j2)\,\Omega$ (d) $50 \underline{/\,40^\circ}\,\Omega$.

(a) If impedance $Z = -j5\,\Omega$, then

$$\text{admittance } Y = \frac{1}{Z} = \frac{1}{-j5} = \frac{j}{(-j5)(j)} = \frac{j}{5} = j0.2\text{S or } 0.2\underline{/\,90^\circ}\,\text{S}$$

Since there is no real part, **conductance, $G = 0$,** and **capacitive susceptance, $B_C = 0.2\,\text{S}$.**

(b) If impedance $Z = (25 + j40)\,\Omega$ then

$$\text{admittance } Y = \frac{1}{Z} = \frac{1}{(25 + j40)} = \frac{25 - j40}{25^2 + 40^2}$$

$$= \frac{25}{2225} - \frac{j40}{2225} = (0.0112 - j0.0180)\,\text{S}$$

Thus **conductance, $G = 0.0112\,\text{S}$** and **inductive susceptance, $B_L = 0.0180\,\text{S}$.**

(c) If impedance $Z = (3 - j2)\,\Omega$, then

$$\text{admittance } Y = \frac{1}{Z} = \frac{1}{(3 - j2)} = \frac{3 + j2}{3^2 + 2^2}$$

$$= \left(\frac{3}{13} + j\frac{2}{13}\right)\text{S} \quad \text{or} \quad (0.231 + j0.154)\,\text{S}$$

Thus **conductance, $G = 0.231\,\text{S}$** and **capacitive susceptance, $B_C = 0.154\,\text{S}$.**

(d) If impedance $Z = 50\underline{/\,40^\circ}\,\Omega$, then

$$\text{admittance } Y = \frac{1}{Z} = \frac{1}{50\underline{/\,40^\circ}} = \frac{1}{50}\underline{/\,-40^\circ}$$

$$= 0.02\underline{/\,-40^\circ}\,\text{S} \quad \text{or} \quad (0.0153 - j0.0129)\,\text{S}$$

Thus **conductance, $G = 0.0153\,\text{S}$** and **inductive susceptance, $B_L = 0.0129\,\text{S}$.**

Figure 1 (a) Circuit diagram
(b) Phasor diagram

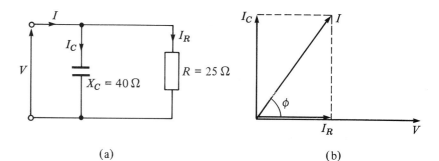

(a)   (b)

Problem 2. Determine expressions for the impedance of the following admittances:
(a) $0.004 \underline{/ \ 30°}$ S   (b) $(0.001 - j0.002)$ S   (c) $(0.05 + j0.08)$ S.

(a) Since admittance $Y = 1/Z$, impedance $Z = 1/Y$. Hence

$$\text{impedance } Z = \frac{1}{0.004 \underline{/ \ 30°}} = \mathbf{250 \underline{/ -30°} \ \Omega} \quad \text{or} \quad \mathbf{(216.5 - j125) \ \Omega}$$

(b) Impedance $Z = \dfrac{1}{(0.001 - j0.002)} = \dfrac{0.001 + j0.002}{(0.001)^2 + (0.002)^2}$

$$= \mathbf{(200 + j400) \ \Omega} \quad \text{or} \quad \mathbf{447.2 \underline{/ \ 63.43°} \ \Omega}$$

(c) Admittance $Y = (0.05 + j0.08)$ S $= 0.094 \underline{/ \ 57.99°}$. Hence

$$\text{impedance } Z = \frac{1}{0.094 \underline{/ \ 57.99°}} = \mathbf{10.64 \underline{/ -57.99°} \ \Omega} \quad \text{or} \quad \mathbf{(5.64 - j9.02) \ \Omega}$$

Problem 3. The admittance of a circuit is $(0.040 + j0.025)$ S. Determine the values of
the resistance and the capacitive reactance of the circuit if they are connected (a) in
parallel, (b) in series. Draw the phasor diagram for each of the circuits.

(a) Parallel connection

Admittance $Y = (0.040 + j0.025)$ S, therefore conductance, $G = 0.040$ S and
capacitive susceptance, $B_C = 0.025$ S. From equation (1) of section 2(g), when a
circuit consists of resistance R and capacitive reactance in parallel, then $Y = (1/R) + (j/X_C)$. Hence

$$\text{resistance } R = \frac{1}{G} = \frac{1}{0.040} = \mathbf{25 \ \Omega}$$

and

$$\text{capacitive reactance } X_C = \frac{1}{B_C} = \frac{1}{0.025} = \mathbf{40 \ \Omega}$$

The circuit and phasor diagrams are shown in Fig. 1.

(b) Series connection

Admittance $Y = (0.040 + j0.025)$ S, therefore

$$\text{impedance } Z = \frac{1}{Y} = \frac{1}{0.040 + j0.025} = \frac{0.040 - j0.025}{(0.040)^2 + (0.025)^2} = \mathbf{(17.98 - j11.24) \ \Omega}$$

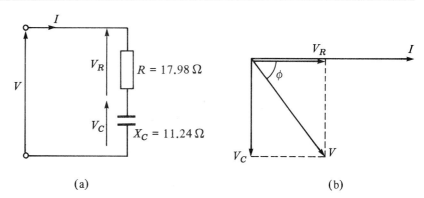

Figure 2 (a) Circuit diagram
(b) Phasor diagram

(a)   (b)

Thus the **resistance, $R = 17.98\,\Omega$** and **capacitive reactance, $X_C = 11.24\,\Omega$.** The circuit and phasor diagrams are shown in Fig. 2.

The circuits shown in Figs. 1(a) and 2(a) are equivalent in that they take the same supply current I for a given supply voltage V; the phase angle ϕ between the current and voltage is the same in each of the phasor diagrams shown in Figs. 1(b) and 2(b).

Further problems on admittance, conductance and susceptance may be found in section 4, problems 1 to 8, page 49.

3. Parallel a.c. networks

Figure 3 shows a circuit diagram containing three impedances, Z_1, Z_2 and Z_3 connected in parallel. The potential difference across each impedance is the same, i.e. the supply voltage V. Current $I_1 = V/Z_1$, $I_2 = V/Z_2$ and $I_3 = V/Z_3$. If Z_T is the total equivalent impedance of the circuit then $I = V/Z_T$.

The supply current, $I = I_1 + I_2 + I_3$ (phasorially). Thus

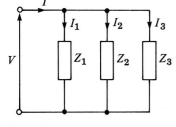

Figure 3

$$\frac{V}{Z_T} = \frac{V}{Z_1} + \frac{V}{Z_2} + \frac{V}{Z_3} \quad \text{and} \quad \frac{1}{Z_T} = \frac{1}{Z_1} + \frac{1}{Z_2} + \frac{1}{Z_3}$$

or total admittance, $Y_T = Y_1 + Y_2 + Y_3$.

In general, for n impedances connected in parallel,

$$Y_T = Y_1 + Y_2 + Y_3 + \cdots + Y_n \quad \text{(phasorially)}$$

It is in parallel circuit analysis that the use of admittance has its greatest advantage.

Current division in a.c. circuits

For the special case of two impedances, Z_1 and Z_2, connected in parallel (see Fig. 4),

$$\frac{1}{Z_T} = \frac{1}{Z_1} + \frac{1}{Z_2} = \frac{Z_2 + Z_1}{Z_1 Z_2}$$

The total impedance, $Z_T = Z_1 Z_2/(Z_1 + Z_2)$ (i.e. product/sum). From Fig. 4,

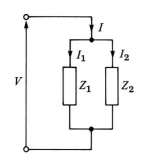

Figure 4

$$\text{supply voltage, } V = IZ_T = I\left(\frac{Z_1 Z_2}{Z_1 + Z_2}\right)$$

Also,

$$V = I_1 Z_1 \quad (\text{and } V = I_2 Z_2)$$

Thus

$$I_1 Z_1 = I\left(\frac{Z_1 Z_2}{Z_1 + Z_2}\right)$$

i.e.,

$$\textbf{current } I_1 = I\left(\frac{Z_2}{Z_1 + Z_2}\right)$$

Similarly,

$$\textbf{current } I_2 = I\left(\frac{Z_1}{Z_1 + Z_2}\right)$$

Note that all of the above circuit symbols infer complex quantities either in cartesian or polar form.

The following worked problems show how complex numbers are used to analyse parallel a.c. networks.

Worked problems on parallel a.c. networks

Problem 1. Determine the values of currents I_1, I_2 and I shown in the network of Fig. 5.

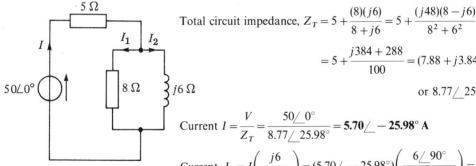

Figure 5

Total circuit impedance, $Z_T = 5 + \dfrac{(8)(j6)}{8 + j6} = 5 + \dfrac{(j48)(8 - j6)}{8^2 + 6^2}$

$$= 5 + \frac{j384 + 288}{100} = (7.88 + j3.84)\,\Omega$$

$$\text{or } 8.77\underline{/25.98°}\,\Omega$$

Current $I = \dfrac{V}{Z_T} = \dfrac{50\underline{/0°}}{8.77\underline{/25.98°}} = \textbf{5.70}\underline{/-25.98°}\,\textbf{A}$

Current, $I_1 = I\left(\dfrac{j6}{8 + j6}\right) = (5.70\underline{/-25.98°})\left(\dfrac{6\underline{/90°}}{10\underline{/36.87°}}\right) = \textbf{3.42}\underline{/27.15°}\,\textbf{A}$

Current $I_2 = I\left(\dfrac{8}{8 + j6}\right) = (5.70\underline{/-25.98°})\left(\dfrac{8\underline{/0°}}{10\underline{/36.87°}}\right) = \textbf{4.56}\underline{/-62.85°}\,\textbf{A}$

(*Note*: $I = I_1 + I_2 = 3.42\underline{/27.15°} + 4.56\underline{/-62.85°}$
$= (3.043 + j1.561) + (2.081 - j4.058)$
$= (5.124 - j2.497)\,\text{A} = 5.70\underline{/-25.98°}\,\text{A})$

Problem 2. For the parallel network shown in Fig. 6, determine the value of supply current I and its phase relative to the 40 V supply.

Impedance $Z_1 = (5 + j12)\,\Omega$, $Z_2 = (3 - j4)\,\Omega$ and $Z_3 = 8\,\Omega$

Supply current $I = \dfrac{V}{Z_T} = V Y_T$,

where Z_T = total circuit impedance, and Y_T = total circuit admittance.

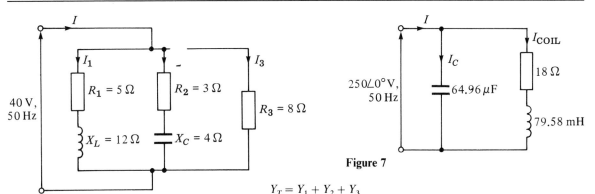

Figure 6

Figure 7

$$Y_T = Y_1 + Y_2 + Y_3$$

$$= \frac{1}{Z_1} + \frac{1}{Z_2} + \frac{1}{Z_3} = \frac{1}{(5+j12)} + \frac{1}{(3-j4)} + \frac{1}{8}$$

$$= \frac{5-j12}{5^2 + 12^2} + \frac{3+j4}{3^2 + 4^2} + \frac{1}{8}$$

$$= (0.0296 - j0.0710) + (0.1200 + j0.1600) + (0.1250)$$

i.e.,

$$Y_T = (0.2746 + j0.0890)\,\text{S or } 0.289\underline{/\ 17.96°}\,\text{S}$$

$$\text{Current } I = V Y_T = (40\underline{/\ 0°})(0.289\underline{/\ 17.96°}) = 11.6\underline{/\ 17.96°}\,\text{A}$$

Hence the current I is 11.6 A and is leading the 40 V supply by 17.96°.
Alternatively, current $I = I_1 + I_2 + I_3$.

$$\text{Current } I_1 = \frac{40\underline{/\ 0°}}{5+j12} = \frac{40\underline{/\ 0°}}{13\underline{/\ 67.38°}} = 3.077\underline{/\ -67.38°}\,\text{A or } (1.183 - j2.840)\,\text{A}$$

$$\text{Current } I_2 = \frac{40\underline{/\ 0°}}{3-j4} = \frac{40\underline{/\ 0°}}{5\underline{/\ -53.13°}} = 8\underline{/\ 53.13°}\,\text{A or } (4.80 + j6.40)\,\text{A}$$

$$\text{Current } I_3 = \frac{40\underline{/\ 0°}}{8\underline{/\ 0°}} = 5\underline{/\ 0°}\,\text{A or } (5 + j0)\,\text{A}.$$

Thus current $I = I_1 + I_2 + I_3$

$$= (1.183 - j2.840) + (4.80 + j6.40) + (5 + j0)$$

$$= 10.983 + j3.560 = \mathbf{11.6\underline{/\ 17.96°}\,A}, \text{ as previously obtained.}$$

Problem 3. An a.c. network consists of a coil, of inductance 79.58 mH and resistance 18 Ω, in parallel with a capacitor of capacitance 64.96 μF. If the supply voltage is $250\underline{/\ 0°}$ V at 50 Hz, determine (a) the total equivalent circuit impedance, (b) the supply current, (c) the circuit phase angle, (d) the current in the coil, and (e) the current in the capacitor.

The circuit diagram is shown in Fig. 7.
    Inductive reactance, $X_L = 2\pi f L = 2\pi(50)(79.58 \times 10^{-3}) = 25\,\Omega$. Hence the impedance of the coil,

$$Z_{\text{COIL}} = (R + jX_L) = (18 + j25)\,\Omega \text{ or } 30.81\underline{/\ 54.25°}\,\Omega$$

$$\text{Capacitive reactance, } X_C = \frac{1}{2\pi f C} = \frac{1}{2\pi(50)(64.96 \times 10^{-6})} = 49\,\Omega$$

In complex form, the impedance presented by the capacitor, Z_C is $-jX_C$, i.e., $-j49\,\Omega$ or $49\underline{/-90°}\,\Omega$.

(a) Total equivalent circuit impedance,

$$Z_T = \frac{Z_{\text{COIL}}Z_C}{Z_{\text{COIL}}+Z_C}\left(\text{i.e., } \frac{\text{product}}{\text{sum}}\right) = \frac{(30.81\underline{/\,54.25°})(49\underline{/-90°})}{(18+j25)+(-j49)}$$

$$= \frac{(30.81\underline{/\,54.25°})(49\underline{/-90°})}{18-j24} = \frac{(30.81\underline{/\,54.25°})(49\underline{/-90°})}{30\underline{/-53.13°}}$$

$$= 50.32\underline{/\,(54.25°-90°-(-53.13°))} = \mathbf{50.32\underline{/\,17.38°}\,\Omega \text{ or } (48.02+j15.03)\,\Omega}$$

(b) Supply current $I = \dfrac{V}{Z_T} = \dfrac{250\underline{/\,0°}}{50.32\underline{/\,17.38°}} = \mathbf{4.97\underline{/-17.38°}\,A}$

(c) Circuit phase angle = **17.38° lagging**, i.e., the current I lags the voltage V by 17.38°.

(d) Current in the coil, $I_{\text{COIL}} = \dfrac{V}{Z_{\text{COIL}}} = \dfrac{250\underline{/\,0°}}{30.81\underline{/\,54.25°}} = \mathbf{8.11\underline{/-54.25°}\,A}$

(e) Current in the capacitor, $I_C = \dfrac{V}{Z_C} = \dfrac{250\underline{/\,0°}}{49\underline{/-90°}} = \mathbf{5.10\underline{/\,90°}\,A}$.

Problem 4. A parallel a.c. network is shown in Fig. 8. Determine (a) the admittance of each branch, (b) the total network admittance, (c) the total network conductance, (d) the total circuit susceptance, (e) the equivalent circuit impedance, (f) the supply current I, and (g) the current flowing in branch EF.

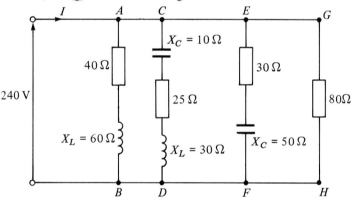

Figure 8

(a)

Admittance of branch AB, $Y_{AB} = \dfrac{1}{40+j60} = \dfrac{40-j60}{40^2+60^2}$

$$= \mathbf{(0.007\,69-j0.011\,54)\,S}$$

Admittance of branch CD, $Y_{CD} = \dfrac{1}{25-j10+j30} = \dfrac{25-j20}{25^2+20^2}$

$$= \mathbf{(0.024\,39-j0.019\,51)\,S}$$

Admittance of branch EF, $Y_{EF} = \dfrac{1}{30-j50} = \dfrac{30+j50}{30^2+50^2}$

$$= \mathbf{(0.008\,82+j0.014\,71)\,S}$$

Admittance of branch GH, $Y_{GH} = \tfrac{1}{80} = \mathbf{(0.012\,50+j0)\,S}$

(b) Total circuit admittance, $Y_T = Y_{AB} + Y_{CD} + Y_{EF} + Y_{GH}$

$$= (0.0534 - j0.0163)\,\text{S} \quad \text{or} \quad 0.0558\underline{/-16.97°}\,\text{S}.$$

(c) Total circuit **conductance**, $G = 0.0534\,\text{S}.$

(d) Total circuit **susceptance**, $B = 0.0163\,\text{S}.$

Since the sign of the imaginary part of the total admittance is negative, the circuit is inductive (see section 2).

(e) Equivalent series circuit impedance,

$$Z_T = \frac{1}{Y_T} = \frac{1}{0.0558\underline{/-16.97°}}$$

$$= 17.92\underline{/16.97°}\,\Omega \quad \text{or} \quad (17.14 + j5.23)\,\Omega$$

Thus the four-branch parallel circuit shown in Fig. 8 is equivalent to a $17.14\,\Omega$ resistor connected in series with an inductor of inductive reactance $5.23\,\Omega$.

(f) Supply current, $I = \dfrac{V}{Z_T} = \dfrac{240\underline{/0°}}{17.92\underline{/16.97°}} = 13.39\underline{/-16.97°}\,\text{A}.$

(Alternatively, $I = VY_T = (240\underline{/0°})(0.0558\underline{/-16.97°}) = 13.39\underline{/-16.97°}\,\text{A}$)

(g) Current flowing in branch EF,

$$I_{EF} = \frac{V}{Z_{EF}} = \frac{240\underline{/0°}}{30 - j50} = \frac{240\underline{/0°}}{58.31\underline{/59.04°}}$$

$$= 4.12\underline{/-59.04°}\,\text{A}.$$

Problem 5. (a) For the network diagram of Fig. 9, determine the value of impedance Z_1. (b) If the supply frequency is $5\,\text{kHz}$, determine the value of the components comprising impedance Z_1.

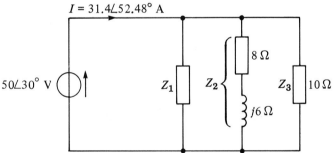

Figure 9

(a) Total circuit admittance, $Y_T = \dfrac{I}{V} = \dfrac{31.4\underline{/52.48°}}{50\underline{/30°}}$

$$= 0.628\underline{/22.48°}\,\text{S} \quad \text{or} \quad (0.58 + j0.24)\,\text{S}$$

$$Y_T = Y_1 + Y_2 + Y_3$$

Thus

$$(0.58 + j0.24) = Y_1 + \frac{1}{(8 + j6)} + \frac{1}{10}$$

$$= Y_1 + \frac{8 - j6}{8^2 + 6^2} + 0.1$$

i.e., $0.58 + j0.24 = Y_1 + 0.08 - j0.06 + 0.1$. Hence

$$Y_1 = (0.58 - 0.08 - 0.1) + j(0.24 + 0.06)$$
$$= (0.4 + j0.3)\,\text{S} \quad \text{or} \quad 0.5\underline{/\,36.87°}\,\text{S}.$$

Thus

$$\textbf{Impedance, } Z_1 = \frac{1}{Y_1} = \frac{1}{0.5\underline{/\,36.87°}} = 2\underline{/-36.87°}\,\Omega \quad \text{or} \quad \textbf{(1.6} - j\textbf{1.2)}\,\Omega$$

(b) Since $Z_1 = (1.6 - j1.2)\,\Omega$, **resistance** $= \textbf{1.6}\,\Omega$ and capacitive reactance, $X_C = 1.2\,\Omega$. Since

$$X_C = \frac{1}{2\pi f C},$$

$$\text{capacitance, } C = \frac{1}{2\pi f X_C} = \frac{1}{2\pi(5000)(1.2)}\,\text{F}$$

i.e., **capacitance** $= \textbf{26.53}\,\mu\textbf{F}$.

Problem 6. For the series-parallel arrangement shown in Fig. 10, determine (a) the equivalent series circuit impedance, (b) the supply current, I, (c) the circuit phase angle, (d) the values of voltages V_1 and V_2, and (e) the values of currents I_A and I_B.

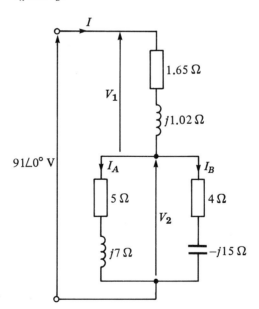

Figure 10

(a) The impedance, Z, of the two branches connected in parallel is given by

$$Z = \frac{(5 + j7)(4 - j15)}{(5 + j7) + (4 - j15)} = \frac{20 - j75 + j28 - j^2 105}{9 - j8} = \frac{125 - j47}{9 - j8}$$

$$= \frac{133.54\underline{/-20.61°}}{12.04\underline{/-41.63°}} = 11.09\underline{/(-20.61° - (-41.63°))}$$

$$= 11.09\underline{/\,21.02°}\,\Omega \quad \text{or} \quad (10.35 + j3.98)\,\Omega$$

Equivalent series circuit impedance, $Z_T = (1.65 + j1.02) + (10.35 + j3.98)$
$$= (12 + j5)\,\Omega \quad \text{or} \quad 13\underline{/\ 22.62°}\,\Omega$$

(b) Supply current,
$$I = \frac{V}{Z} = \frac{91\underline{/\ 0°}}{13\underline{/\ 22.62°}} = 7\underline{/\ -22.62°}\,\text{A}$$

(c) Circuit phase angle = **22.62° lagging**

(d) Voltage $V_1 = IZ_1$, where $Z_1 = (1.65 + j1.02)\,\Omega$ or $1.94\underline{/\ 31.72°}\,\Omega$. Hence
$$V_1 = (7\underline{/\ -22.62°})(1.94\underline{/\ 31.72°}) = \mathbf{13.58\underline{/\ 9.10°}\,V}$$

Voltage $V_2 = IZ$, where Z is the equivalent impedance of the two branches connected in parallel. Hence
$$V_2 = (7\underline{/\ -22.62°})(11.09\underline{/\ 21.02°}) = \mathbf{77.63\underline{/\ -1.60°}\,V}$$

(e) Current $I_A = V_2/Z_A$, where $Z_A = (5 + j7)\,\Omega$ or $8.60\underline{/\ 54.46°}\,\Omega$. Thus
$$I_A = \frac{77.63\underline{/\ -1.6°}}{8.60\underline{/\ 54.46°}} = \mathbf{9.03\underline{/\ -56.06°}\,A}$$

Current $I_B = V_2/Z_B$, where $Z_B = (4 - j15)\,\Omega$ or $15.524\underline{/\ -75.07°}\,\Omega$. Thus
$$I_B = \frac{77.63\underline{/\ -1.6°}}{15.524\underline{/\ -75.07°}} = \mathbf{5.00\underline{/\ 73.47°}\,A}$$

$$\left[\text{Alternatively, by current division,}\right.$$

$$I_A = I\left(\frac{Z_B}{Z_A + Z_B}\right) = 7\underline{/\ -22.62°}\left(\frac{15.524\underline{/\ -75.07°}}{(5 + j7) + (4 - j15)}\right)$$

$$= 7\underline{/\ -22.62°}\left(\frac{15.524\underline{/\ -75.07°}}{(9 - j8)}\right)$$

$$= 7\underline{/\ -22.62°}\left(\frac{15.524\underline{/\ -75.07°}}{12.04\underline{/\ -41.63°}}\right) = \mathbf{9.03\underline{/\ -56.06°}\,A}$$

$$\left. I_B = I\left(\frac{Z_A}{Z_A + Z_B}\right) = 7\underline{/\ -22.62°}\left(\frac{8.60\underline{/\ 54.46°}}{12.04\underline{/\ -41.63°}}\right) = \mathbf{5.00\underline{/\ 73.47°}\,A}\right]$$

Further problems on parallel a.c. networks may be found in section 4 following, problems 9 to 28, page 50.

4. Further problems

Admittance, conductance and susceptance

1. Determine the admittance (in polar form), conductance and susceptance of the following impedances:
   (a) $j10\,\Omega$ (b) $-j40\,\Omega$ (c) $32\underline{/\ -30°}\,\Omega$ (d) $(5 + j9)\,\Omega$
   (e) $(16 - j10)\,\Omega$

$$\left[\begin{array}{l}\text{(a)}\ 0.1\underline{/\ -90°}\,\text{S},\ 0,\ 0.1\,\text{S}\\\text{(b)}\ 0.025\underline{/\ 90°}\,\text{S},\ 0,\ 0.025\,\text{S}\\\text{(c)}\ 0.03125\underline{/\ 30°}\,\text{S},\ 0.0271\,\text{S},\ 0.0156\,\text{S}\\\text{(d)}\ 0.0971\underline{/\ -60.95°}\,\text{S},\ 0.0471\,\text{S},\ 0.0849\,\text{S}\\\text{(e)}\ 0.0530\underline{/\ 32.01°}\,\text{S},\ 0.0449\,\text{S},\ 0.0281\,\text{S}\end{array}\right]$$

Figure 11

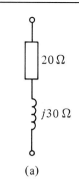

(a)

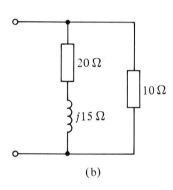

(b)

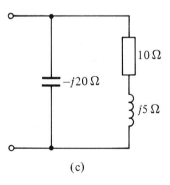

(c)

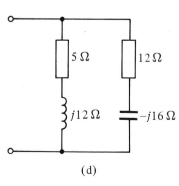

(d)

2. Derive expressions, in polar form, for the impedances of the following admittances: (a) $0.05 \angle 40°$ S   (b) $0.0016 \angle -25°$ S   (c) $(0.1 + j0.4)$ S   (d) $(0.025 - j0.040)$ S.

$$\begin{bmatrix} \text{(a) } 20 \angle -40° \, \Omega & \text{(b) } 625 \angle 25° \, \Omega \\ \text{(c) } 2.425 \angle -75.96° \, \Omega & \text{(d) } 21.20 \angle 57.99° \, \Omega \end{bmatrix}$$

3. Calculate the admittance of a circuit containing an inductance of 0.25 mH connected in series with a resistance of $50 \, \Omega$, assuming a frequency of 100 kHz.

$$[0.00607 \angle -72.34° \, \text{S} \quad \text{or} \quad (0.00184 - j0.00578) \, \text{S}]$$

4. The admittance of a series circuit is $(0.010 - j0.004)$ S. Determine the values of the circuit components if the frequency is 50 Hz.

$$[R = 86.21 \, \Omega; \, L = 109.8 \, \text{mH}]$$

5. The admittance of a network is $(0.05 - j0.08)$ S. Determine the values of resistance and reactance in the circuit if they are connected (a) in series, (b) in parallel.

$$[\text{(a) } R = 5.62 \, \Omega, \, X_L = 8.99 \, \Omega \quad \text{(b) } R = 20 \, \Omega, \, X_L = 12.5 \, \Omega]$$

6. The admittance of a two-branch parallel network is $(0.02 + j0.05)$ S. Determine the circuit components if the frequency is 1 kHz.

$$[R = 50 \, \Omega, \, C = 7.958 \, \mu\text{F}]$$

7. Determine the total admittance, in rectangular and polar forms, of each of the networks shown in Fig. 11.

$$\begin{bmatrix} \text{(a) } (0.0154 - j0.0231) \, \text{S} & \text{or} & 0.0278 \angle -56.31° \, \text{S} \\ \text{(b) } (0.132 - j0.024) \, \text{S} & \text{or} & 0.134 \angle -10.30° \, \text{S} \\ \text{(c) } (0.08 + j0.01) \, \text{S} & \text{or} & 0.0806 \angle 7.125° \, \text{S} \\ \text{(d) } (0.0596 - j0.031) \, \text{S} & \text{or} & 0.0672 \angle -27.48° \, \text{S} \end{bmatrix}$$

8. If the impedance of a circuit is expressed in the form $R - jX$, determine an expression for the corresponding admittance in terms of R and X in $(a + jb)$ form.

$$\left[Y = \frac{R}{Z^2} + j\frac{X_C}{Z^2} \right]$$

Parallel a.c. networks

9. Determine the equivalent circuit impedances of the parallel networks shown in Fig. 12.

$$\begin{bmatrix} \text{(a) } (4 - j8) \, \Omega & \text{or} & 8.94 \angle -63.43° \, \Omega \\ \text{(b) } (7.56 + j1.95) \, \Omega & \text{or} & 7.81 \angle 14.47° \, \Omega \\ \text{(c) } (14.04 - j0.74) \, \Omega & \text{or} & 14.06 \angle -3.02° \, \Omega \end{bmatrix}$$

10. Determine the value and phase of currents I_1 and I_2 in the network shown in Fig. 13.

$$[I_1 = 8.94 \angle -10.3° \, \text{A}, \, I_2 = 17.89 \angle 79.7° \, \text{A}]$$

11. For the series-parallel network shown in Fig. 14, determine (a) the total network impedance across AB, and (b) the supply current flowing if a supply of alternating voltage $30 \angle 20°$ V is connected across AB.

$$[\text{(a) } 10 \angle 36.87° \, \Omega \quad \text{(b) } 3 \angle -16.87° \, \text{A}]$$

12. For the parallel network shown in Fig. 15, determine (a) the equivalent circuit impedance, (b) the supply current I, (c) the circuit phase angle, and (d) currents I_1 and I_2.

$$\begin{bmatrix} \text{(a) } 10.32 \angle -6.29° \, \Omega & \text{(b) } 4.84 \angle 6.29° \, \text{A} \\ \text{(c) } 6.29° \text{ leading}, & \text{(d) } I_1 = 0.953 \angle -73.39° \, \text{A} \\ & I_2 = 4.766 \angle 17.64° \, \text{A} \end{bmatrix}$$

Figure 12

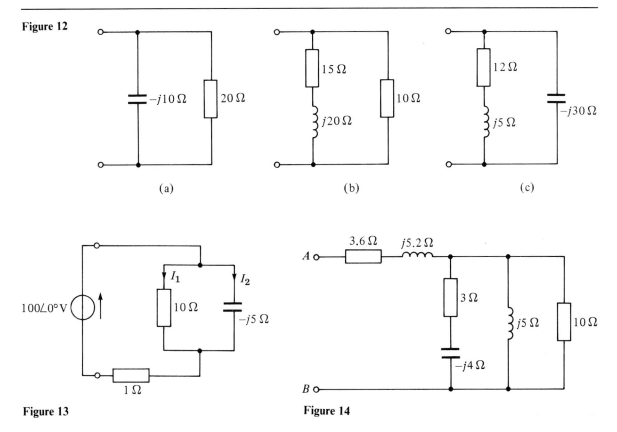

(a)   (b)   (c)

Figure 13   **Figure 14**

13. Two impedances, $Z_1 = (2 + j6)\,\Omega$ and $Z_2 = (5 - j2)\,\Omega$, are connected in parallel across a 100 V a.c. supply. Determine the supply current and its phase relative to the 100 V supply.

[23.67 A; 20.02° lagging]

14. For the network shown in Fig. 16, determine (a) current I_1, (b) current I_2, (c) current I, (d) the equivalent input impedance, and (e) the supply phase angle.

$$
\begin{bmatrix}
\text{(a)} & 15.08\underline{/\,90°}\,\text{A} & \text{(b)} & 3.39\underline{/\,-45.15°}\,\text{A} \\
\text{(c)} & 12.90\underline{/\,79.33°}\,\text{A} & \text{(d)} & 9.30\underline{/\,-79.33°}\,\Omega \\
\text{(e)} & 79.33°\ \text{leading}
\end{bmatrix}
$$

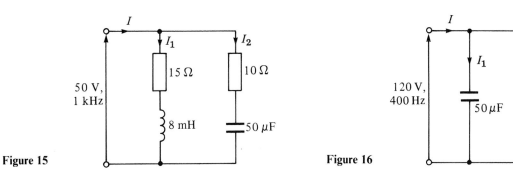

Figure 15   **Figure 16**

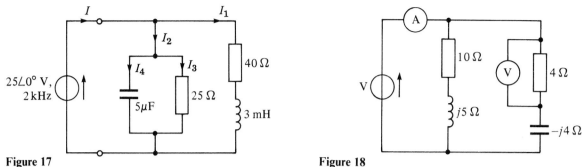

Figure 17   **Figure 18**

15. Determine, for the network shown in Fig. 17, (a) the total network admittance, (b) the total network impedance, (c) the supply current I, (d) the network phase angle, and (e) currents I_1, I_2, I_3 and I_4.

$$\begin{bmatrix} \text{(a) } 0.0733\underline{/\ 43.40°}\,\text{S} \quad\quad \text{(b) } 13.65\underline{/\ -43.40°}\,\Omega \\ \text{(c) } 1.83\underline{/\ 43.40°}\,\text{A} \quad\quad \text{(d) } 43.40°\ \text{leading} \\ \text{(e) } I_1 = 0.455\underline{/\ -43.30°}\,\text{A},\ I_2 = 1.862\underline{/\ 57.51°}\,\text{A}, \\ \quad I_3 = 1\underline{/\ 0°}\,\text{A}. \quad\quad I_4 = 1.570\underline{/\ 90°}\,\text{A}. \end{bmatrix}$$

16. Three impedances of $(90 + j120)\,\Omega, (120 - j50)\,\Omega$ and $(106.1 + j106.1)\,\Omega$ are connected in parallel to a 200 V a.c. supply. Calculate (a) the input admittance and (b) the supply current.

$$\begin{bmatrix} \text{(a) } (0.01581 - j0.00709)\,\text{S} \quad \text{or} \quad 0.0173\underline{/\ -24.14°}\,\text{S} \\ \text{(b) } 3.46\underline{/\ -24.14°}\,\text{A} \end{bmatrix}$$

17. Four impedances of $(10 - j20)\,\Omega, (30 + j0)\,\Omega, (2 - j15)\,\Omega$ and $(25 + j12)\,\Omega$ are connected in parallel across a 250 V a.c. supply. Find the supply current and its phase angle.

$$[32.62\underline{/\ 43.54°}\,\text{A}]$$

18. In the network shown in Fig. 18, the voltmeter indicates 24 V. Determine the reading on the ammeter.

$$[7.53\,\text{A}]$$

19. Three impedances are connected in parallel to a 100 V, 50 Hz supply. The first impedance is $(10 + j12.5)\,\Omega$ and the second impedance is $(20 + j8)\,\Omega$. Determine the third impedance if the total current is $20\underline{/\ -25°}\,\text{A}$.

$$[(9.75 + j1.82)\,\Omega \quad \text{or} \quad 9.92\underline{/\ 10.57°}\,\Omega]$$

20. For the four-branch parallel network shown in Fig. 19, determine (a) the total network impedance, and (b) the supply current I.

$$[\text{(a) } 3.195\underline{/\ 18.05°}\,\Omega \quad\quad \text{(b) } 12.52\underline{/\ 11.95°}\,\text{A}]$$

21. For each of the network diagrams shown in Fig. 20, determine the supply current I and their phase relative to the applied voltages.

$$\begin{bmatrix} \text{(a) } 1.632\underline{/\ -17.1°}\,\text{A} \\ \text{(b) } 5.411\underline{/\ -8.45°}\,\text{A} \end{bmatrix}$$

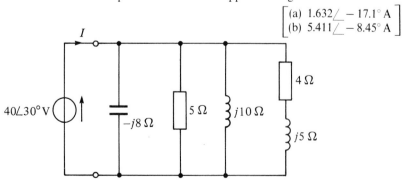

Figure 19

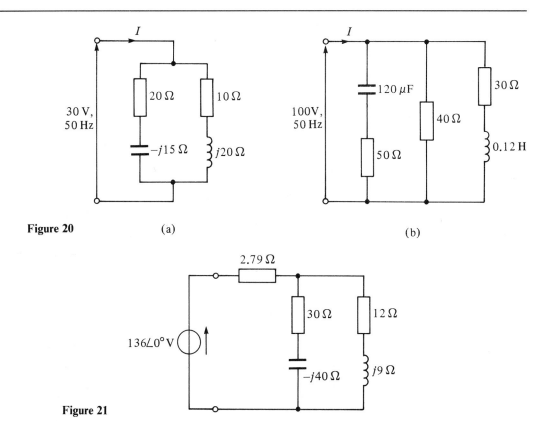

Figure 20   (a)   (b)

Figure 21

22. Determine the value of current flowing in the $(12 + j9)\,\Omega$ impedance in the network shown in Fig. 21.

$$[7.66 \underline{/-33.41°}\,\text{A}]$$

23. In the series-parallel network shown in Fig. 22 the p.d. between points A and B is $50 \underline{/-68.13°}$ V. Determine (a) the supply current I, (b) the equivalent input impedance, (c) the supply voltage V, (d) the supply phase angle, (e) the p.d. across points B and C, and (f) the value of currents I_1 and I_2.

$$\begin{bmatrix} \text{(a)} \ 11.99 \underline{/-31.81°}\,\text{A} & \text{(b)} \ 8.54 \underline{/\ 20.56°}\,\Omega \\ \text{(c)} \ 102.4 \underline{/-11.25°}\,\text{V} & \text{(d)} \ 20.56°\,\text{lagging} \\ \text{(e)} \ 86.0 \underline{/\ 17.90°}\,\text{V} & \text{(f)} \ I_1 = 7.37 \underline{/-13.06°}\,\text{A} \\ & \quad\quad I_2 = 5.54 \underline{/-57.17°}\,\text{A} \end{bmatrix}$$

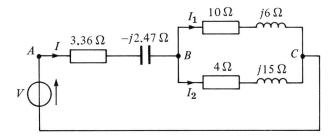

Figure 22

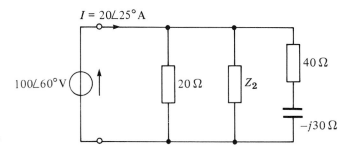

Figure 23

24. For the network shown in Fig. 23, determine (a) the value of impedance Z_2, (b) the current flowing in Z_2, and (c) the components comprising Z_2 if the supply frequency is 2 kHz.

$$\left[\begin{array}{ll} \text{(a)} \ \ 6.25 \underline{/\ 52.33°}\ \Omega & \text{(b)} \ \ 16.0 \underline{/\ 7.67°}\ A \\ \text{(c)} \ \ R = 3.82\,\Omega, \ L = 0.394\,\text{mH} \end{array} \right]$$

25. A four-branch parallel network is shown in Fig. 24. Determine (a) currents I_1, I_2, I_3 and I_4, (b) the supply current I, (c) the total equivalent impedance, (d) the total network admittance, (e) the total conductance, and (f) the total susceptance.

$$\left[\begin{array}{l} \text{(a)} \ \ I_1 = 4.47 \underline{/\ -26.57}\ A, \ I_2 = 2.50 \underline{/\ 0°}\ A, \\ \qquad I_3 = 4.096 \underline{/\ 55.01°}\ A, \ I_4 = 1.88 \underline{/\ -55.71°}\ A \\ \text{(b)} \ \ 9.90 \underline{/\ -1.13°}\ A \qquad \text{(c)} \ \ 5.05 \underline{/\ -1.13°}\ \Omega \\ \text{(d)} \ \ (0.1981 - j0.0039)\,S \quad \text{or} \quad 0.198 \underline{/\ -1.13°}\ S \\ \text{(e)} \ \ 0.1981\,S \qquad \text{(f)} \ \ 0.0039\,S \end{array} \right]$$

26. Coils of impedance $(5 + j8)\,\Omega$ and $(12 + j16)\,\Omega$ are connected in parallel. In series with this combination is an impedance of $(15 - j40)\,\Omega$. If the alternating supply p.d. is $150 \underline{/\ 0°}$ V, determine (a) the equivalent network impedance, (b) the supply current, (c) the supply phase angle, (d) the current in the $(5 + j8)\,\Omega$ impedance, and (e) the current in the $(12 + j16)\,\Omega$ impedance.

$$\left[\begin{array}{ll} \text{(a)} \ \ 39.31 \underline{/\ -61.84°}\ \Omega & \text{(b)} \ \ 3.816 \underline{/\ 61.84°}\ A \\ \text{(c)} \ \ 61.84° \ \text{leading} & \text{(d)} \ \ 2.595 \underline{/\ 60.28°}\ A \\ \text{(e)} \ \ 1.224 \underline{/\ 65.18°}\ A \end{array} \right]$$

27. For the circuit shown in Fig. 25, determine (a) the input impedance, (b) the source voltage V, (c) the p.d. between points A and B, and (d) the current in the $10\,\Omega$ resistor.

$$\left[\begin{array}{ll} \text{(a)} \ \ 10.0 \underline{/\ 36.87°}\ \Omega & \text{(b)} \ \ 150 \underline{/\ 66.87°}\ V \\ \text{(c)} \ \ 90 \underline{/\ 51.91°}\ V & \text{(d)} \ \ 2.50 \underline{/\ 18.22°}\ A \end{array} \right]$$

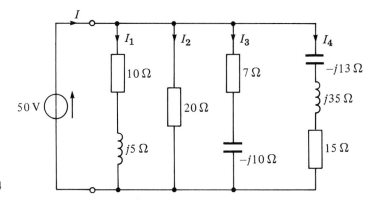

Figure 24

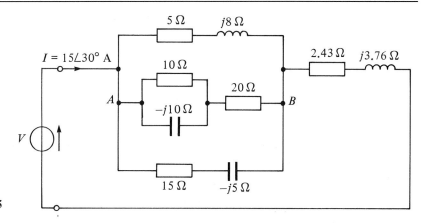

Figure 25

28. A circuit consists of a 25 Ω resistor in series with a coil having an inductance of 100 mH and a resistance of 12 Ω. A 50 μF capacitor is connected in parallel with the coil. The complete network is connected across a 250 V, 50 Hz supply. Determine (a) the supply current, (b) the current in the coil, and (c) the current in the capacitor.

$$\begin{bmatrix} \text{(a)} \ 3.09 \diagdown -35.27° \text{ A} & \text{(b)} \ 5.72 \diagdown -55.69° \text{ A} \\ \text{(c)} \ 3.02 \diagdown 103.41° \text{ A} & \end{bmatrix}$$

4 Power in a.c. circuits

1. Introduction

Alternating currents and voltages change their polarity during each cycle. It is not surprising therefore to find that power also pulsates with time.

The product of voltage v and current i at any instant of time is called instantaneous power p, and is given by

$$p = vi$$

2. Determination of power in a.c. circuits

(a) Purely resistive a.c. circuits

Let a voltage $v = V_m \sin \omega t$ be applied to a circuit comprising resistance only. The resulting current is $i = I_m \sin \omega t$, and the corresponding instantaneous power, p, is given by

$$p = vi = (V_m \sin \omega t)(I_m \sin \omega t)$$

i.e.,

$$p = V_m I_m \sin^2 \omega t$$

From trigonometrical double angle formulae, $\cos 2A = 1 - 2\sin^2 A$, from which

$$\sin^2 A = \tfrac{1}{2}(1 - \cos 2A)$$

Thus

$$\sin^2 \omega t = \tfrac{1}{2}(1 - \cos 2\omega t)$$

Figure 1 The waveforms of v, i and p

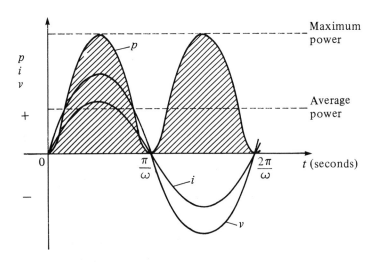

Then power $p = V_m I_m [\frac{1}{2}(1 - \cos 2\omega t)]$, i.e., $\boldsymbol{p = \frac{1}{2} V_m I_m (1 - \cos 2\omega t)}$.

The waveforms of v, i and p are shown in Fig. 1. The waveform of power repeats itself after π/ω seconds and hence the power has a frequency twice that of voltage and current. The power is always positive, having a maximum value of $V_m I_m$. The average or mean value of the power is $\frac{1}{2} V_m I_m$.

The rms value of voltage $V = 0.707 V_m$, i.e. $V = V_m/\sqrt{2}$, from which $V_m = \sqrt{2} V$. Similarly, the rms value of current, $I = I_m/\sqrt{2}$, from which, $I_m = \sqrt{2} I$. Hence the average power, P, developed in a purely resistive a.c. circuit is given by

$$P = \frac{1}{2} V_m I_m = \frac{1}{2}(\sqrt{2} V)(\sqrt{2} I) = VI \text{ watts}$$

Also, power $P = I^2 R$ or V^2/R as for a d.c. circuit, since $V = IR$.

Summarising, the average power P in a purely resistive a.c. circuit is given by

$$\boldsymbol{P = VI = I^2 R = \frac{V^2}{R} \text{ watts}}$$

where V and I are rms values.

(b) Purely inductive a.c. circuits

Let a voltage $v = V_m \sin \omega t$ be applied to a circuit containing pure inductance (theoretical case). The resulting current is $i = I_m \sin(\omega t - (\pi/2))$ since current lags voltage by $90°$ in a purely inductive circuit, and the corresponding instantaneous power, p, is given by

$$p = vi = (V_m \sin \omega t) I_m \sin\left(\omega t - \frac{\pi}{2}\right)$$

i.e.,

$$p = V_m I_m \sin \omega t \sin\left(\omega t - \frac{\pi}{2}\right)$$

However, $\sin(\omega t - \pi/2) = -\cos \omega t$. Thus

$$p = -V_m I_m \sin \omega t \cos \omega t$$

Rearranging gives $p = -\frac{1}{2} V_m I_m (2 \sin \omega t \cos \omega t)$. However, from the double-angle formulae, $2 \sin \omega t \cos \omega t = \sin 2\omega t$. Thus

$$\boldsymbol{\text{power, } p = -\frac{1}{2} V_m I_m \sin 2\omega t}$$

The waveforms of v, i and p are shown in Fig. 2. The frequency of power is twice that of voltage and current. For the power curve shown in Fig. 2, the area above the horizontal axis is equal to the area below, thus over a complete cycle the average power P is zero. It is noted that when v and i are both positive, power p is positive and energy is delivered from the source to the inductance; when v and i have opposite signs, power p is negative and energy is returned from the inductance to the source.

In general, when the current through an inductance is increasing, energy is transferred from the circuit to the magnetic field, but this energy is returned when the current is decreasing.

Summarising, the average power P in a purely inductive a.c. circuit is zero.

Figure 2 Power in a purely inductive a.c. circuit

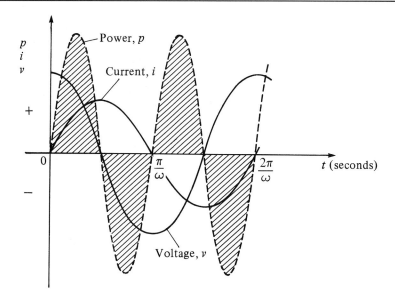

(c) Purely capacitive a.c. circuits

Let a voltage $v = V_m \sin \omega t$ be applied to a circuit containing pure capacitance. The resulting current is $i = I_m(\omega t + (\pi/2))$, since current leads voltage by $90°$ in a purely capacitive circuit, and the corresponding instantaneous power, p, is given by

$$p = vi = (V_m \sin \omega t) I_m \sin\left(\omega t + \frac{\pi}{2} \right)$$

i.e.,

$$p = V_m I_m \sin \omega t \sin\left(\omega t + \frac{\pi}{2} \right)$$

However, $\sin(\omega t + (\pi/2)) = \cos \omega t$. Thus

$$p = V_m I_m \sin \omega t \cos \omega t$$

Rearranging gives $p = \frac{1}{2} V_m I_m (2 \sin \omega t \cos \omega t)$. Thus

$$\text{power, } \boldsymbol{p = \tfrac{1}{2} V_m I_m \sin 2\omega t}$$

The waveforms of v, i and p are shown in Fig. 3. Over a complete cycle **the average power P is zero.** When the voltage across a capacitor is increasing, energy is transferred from the circuit to the electric field, but this energy is returned when the voltage is decreasing.

Summarising, the average power P in a purely capacitive a.c. circuit is zero.

(d) R–L or R–C a.c. circuits

Let a voltage $v = V_m \sin \omega t$ be applied to a circuit containing resistance and inductance or resistance and capacitance. Let the resulting current be $i = I_m \sin(\omega t + \phi)$, where phase angle ϕ will be positive for an R–C

Figure 3 Power in a purely capacitive a.c. circuit

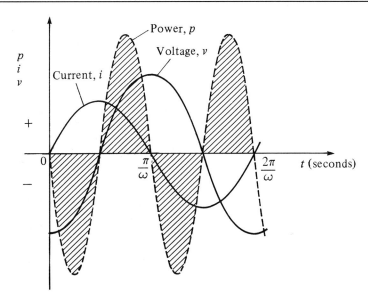

circuit and negative for an R–L circuit. The corresponding instantaneous power, p, is given by

$$p = vi = (V_m \sin \omega t)(I_m \sin(\omega t + \phi))$$

i.e.

$$p = V_m I_m \sin \omega t \sin(\omega t + \phi)$$

Products of sine functions may be changed into differences of cosine functions by using

$$\sin A \sin B = -\tfrac{1}{2}[\cos(A + B) - \cos(A - B)]$$

Substituting $\omega t = A$ and $(\omega t + \phi) = B$ gives

$$\text{power } p = V_m I_m \{-\tfrac{1}{2}[\cos(\omega t + \omega t + \phi) - \cos(\omega t - (\omega t + \phi))]\}$$

i.e.,

$$p = \tfrac{1}{2} V_m I_m [\cos(-\phi) - \cos(2\omega t + \phi)]$$

However, $\cos(-\phi) = \cos \phi$. Thus

$$\boldsymbol{p = \tfrac{1}{2} V_m I_m [\cos \phi - \cos(2\omega t + \phi)]}$$

The instantaneous power p thus consists of

(i) a sinusoidal term, $-\tfrac{1}{2} V_m I_m \cos(2\omega t + \phi)$, which has a mean value over a cycle of zero, and

(ii) a constant term, $\tfrac{1}{2} V_m I_m \cos \phi$ (since ϕ is constant for a particular circuit).

Thus the average value of power, $P = \tfrac{1}{2} V_m I_m \cos \phi$. Since $V_m = \sqrt{2}\, V$ and $I_m = \sqrt{2}\, I$, average power,

$$P = \tfrac{1}{2}(\sqrt{2}\, V)(\sqrt{2}\, I) \cos \phi$$

i.e.,

$$\boldsymbol{P = VI \cos \phi \text{ watts}}$$

Figure 4 Power in a.c. circuit containing resistance and inductive reactance

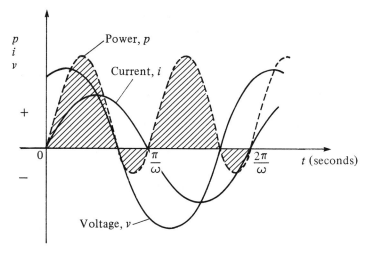

The waveforms of v, i and p, are shown in Fig. 4. for an $R-L$ circuit. The waveform of power is seen to pulsate at twice the supply frequency. The areas of the power curve (shown shaded) above the horizontal time axis represent power supplied to the load; the small areas below the axis represent power being returned to the supply from the inductance as the magnetic field collapses.

A similar shape of power curve is obtained for an $R-C$ circuit, the small areas below the horizontal axis representing power being returned to the supply from the charged capacitor. The difference between the areas above and below the horizontal axis represents the heat loss due to the circuit resistance. Since power is dissipated only in a pure resistance, the alternative equations for power, $P = I_R^2 R$, may be used, where I_R is the rms current flowing through the resistance.

Summarising, the average power P in a circuit containing resistance and inductance and/or capacitance, whether in series or in parallel, is given by $P = V I \cos \phi$ or $P = I_R^2 R$ (V, I and I_R being rms values).

3. Power triangle and power factor

A phasor diagram in which the current I lags the applied voltage V by angle ϕ (i.e., an inductive circuit) is shown in Fig. 5(a). The horizontal component of V is $V \cos \phi$, and the vertical component of V is $V \sin \phi$. If each of the voltage phasors of triangle $0ab$ is multiplied by I, Fig. 5(b) is produced and is known as the **"power triangle"**. Each side of the triangle represents a particular type of power:

True or active power $P = V I \cos \phi$ watts (W)

Apparent power $S = V I$ voltamperes (VA)

Reactive power $Q = V I \sin \phi$ vars (var)

The power triangle is **not** a phasor diagram since quantities P, Q and S are mean values and not rms values of sinusoidally varying quantities.

Superimposing the power triangle on an Argand diagram produces a relationship between P, S and Q in complex form, i.e., $S = P + jQ$.

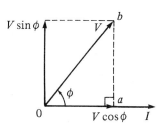

Figure 5 (a) Phasor diagram

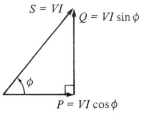

$S = VI$

$Q = VI \sin \phi$

ϕ

$P = VI \cos \phi$

(b) Power triangle for inductive circuit

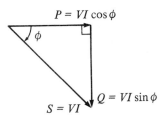

$P = VI \cos \phi$

ϕ

$Q = VI \sin \phi$

$S = VI$

Figure 6 Power triangle for capacitive circuit

4. Use of complex numbers for determination of power

Apparent power, S, is an important quantity since a.c. apparatus, such as generators, transformers and cables, is usually rated in voltamperes rather than in watts. The allowable output of such apparatus is usually limited not by mechanical stress but by temperature rise, and hence by the losses in the device. The losses are determined by the voltage and current and are almost independent of the power factor. Thus the amount of electrical equipment installed to supply a certain load is essentially determined by the voltamperes of the load rather than by the power alone. The **rating** of a machine is defined as the maximum apparent power that it is designed to carry continuously without overheating.

The reactive power, Q, contributes nothing to the net energy transfer and yet it causes just as much loading of the equipment as if it did so. Reactive power is a term much used in power generation, distribution and utilisation of electrical energy.

Inductive reactive power, by convention, is defined as positive reactive power; capacitive reactive power, by convention, is defined as negative reactive power.

The above relationships derived from the phasor diagram of an inductive circuit may be shown to be true for a capacitive circuit, the power triangle being as shown in Fig. 6.

Power factor is defined as

$$\text{power factor} = \frac{\text{active power } P}{\text{apparent power } S}$$

For sinusoidal voltages and currents,

$$\text{power factor} = \frac{P}{S} = \frac{VI \cos \phi}{VI} = \cos \phi = \frac{R}{Z} \text{ (from the impedance triangle)}$$

A circuit in which current lags voltage (i.e., an inductive circuit) is said to have a lagging power factor, and indicates a lagging reactive power Q. A circuit in which current leads voltage (i.e.,a capacitive circuit) is said to have a leading power factor, and indicates a leading reactive power Q.

Let a circuit be supplied by an alternating voltage $V \angle \alpha$, where

$$V \angle \alpha = V(\cos \alpha + j \sin \alpha) = V \cos \alpha + jV \sin \alpha = a + jb \qquad (1)$$

Let the current flowing in the circuit be $I \angle \beta$, where

$$I \angle \beta = I(\cos \beta + j \sin \beta) = I \cos \beta + jI \sin \beta = c + jd \qquad (2)$$

From sections 2 and 3, power $P = VI \cos \phi$, where ϕ is the angle between the voltage V and current I. If the voltage is $V \angle \alpha°$ and the current is $I \angle \beta°$, then the angle between voltage and current is $(\alpha - \beta)°$. Thus

$$\text{power } P = VI \cos(\alpha - \beta)$$

From compound angle formulae, $\cos(\alpha - \beta) = \cos \alpha \cos \beta + \sin \alpha \sin \beta$. Hence power $P = VI[\cos \alpha \cos \beta + \sin \alpha \sin \beta]$.

Rearranging gives $P = (V \cos \alpha)(I \cos \beta) + (V \sin \alpha)(I \sin \beta)$, i.e., $P = (a)(c) + (b)(d)$ from equations (1) and (2).

Summarising, if $V = (a + jb)$ **and** $I = (c + jd)$, then

$$\text{power, } P = ac + bd \tag{3}$$

Thus power may be calculated from the sum of the products of the real components and imaginary components of voltage and current.

$$\text{Reactive power, } Q = VI\sin(\alpha - \beta)$$

From compound angle formulae, $\sin(\alpha - \beta) = \sin\alpha\cos\beta - \cos\alpha\sin\beta$. Thus

$$Q = VI[\sin\alpha\cos\beta - \cos\alpha\sin\beta]$$

Rearranging gives $Q = (V\sin\alpha)(I\cos\beta) - (V\cos\alpha)(I\sin\beta)$ i.e., $Q = (b)(c) - (a)(d)$ from equations (1) and (2).

Summarising, if $V = (a + jb)$ **and** $I = (c + jd)$. then

$$\text{reactive power } Q = bc - ad \tag{4}$$

Expressions (3) and (4) provide an alternative method of determining true power P and reactive power Q when the voltage and current are complex quantities.

From section 3, apparent power $S = P + jQ$. However, merely multiplying V by I in complex form will not give this result, i.e. (from above)

$$S = VI = (a + jb)(c + jd) = (ac - bd) + j(bc + ad).$$

Here the real part is not the expression for power as given in equation (3) and the imaginary part is not the expression for reactive power given in equation (4).

The correct expression may be derived by multiplying the voltage V by the conjugate of the current, i.e. $(c - jd)$, denoted by I^*. Thus

$$\text{apparent power } S = VI^* = (a + jb)(c - jd)$$

$$= (ac + bd) + j(bc - ad)$$

i.e., $S = P + jQ$, from equations (3) and (4).

Thus the active and reactive powers may be determined if, and only if, the voltage V is multiplied by the conjugate of current I. As stated in section 3, a positive value of Q indicates an inductive circuit, i.e., a circuit having a lagging power factor, whereas a negative value of Q indicates a capacitive circuit, i.e., a circuit having a leading power factor.

Worked problems on power in a.c. circuits

Problem 1. A coil of resistance $5\,\Omega$ and inductive reactance $12\,\Omega$ is connected across a supply voltage of $52\underline{/\,30°}$ volts. Determine the active power in the circuit.

The circuit diagram is shown in Fig. 7.

$$\text{Impedance } Z = (5 + j12)\,\Omega \quad \text{or} \quad 13\underline{/\,67.38°}\,\Omega$$

$$\text{Voltage } V = 52\underline{/\,30°}\,V \quad \text{or} \quad (45.03 + j26.0)\,V$$

$$\text{Current } I = \frac{V}{Z} = \frac{52\underline{/\,30°}}{13\underline{/\,67.38°}} = 4\underline{/\,-37.38°}\,A \quad \text{or} \quad (3.18 - j2.43)\,A$$

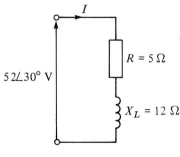

I

$R = 5\,\Omega$

$52\underline{/30°}\,V$

$X_L = 12\,\Omega$

Figure 7

There are three methods of calculating power.

Method 1. Active power, $P = VI \cos\phi$, where ϕ is the angle between voltage V and current I. Hence

$$P = (52)(4) \cos [30 - (-37.38°)] = (52)(4) \cos 67.38° = \textbf{80 W.}$$

Method 2. Active power, $P = I_R^2 R = (4)^2(5) = \textbf{80 W.}$

Method 3. Since $V = (45.03 + j26.0)$ V and $I = (3.18 - j2.43)$ A, then active power, $P = (45.03)(3.18) + (26.0)(-2.43)$ from equation (3), i.e., $P = 143.2 - 63.2 = \textbf{80 W.}$

Problem 2. A current of $(15 + j8)$ A flows in a circuit whose supply voltage is $(120 + j200)$ V. Determine (a) the active power, and (b) the reactive power.

(a) *Method 1.* Active power $P = (120)(15) + (200)(8)$, from equation (3), i.e., $P = 1800 + 1600 = \textbf{3400 W}$ or $\textbf{3.4 kW.}$

  Method 2. Current $I = (15 + j8)$ A $= 17 \underline{/28.07°}$ A and
  Voltage $V = (120 + j200)$ V $= 233.24 \underline{/59.04°}$ V.
  Angle between voltage and current $= 59.04 - 28.07 = 30.97°$.
  Hence power, $P = VI \cos\phi = (233.24)(17) \cos 30.97° = \textbf{3.4 kW.}$

(b) *Method 1.* Reactive power, $Q = (200)(15) - (120)(8)$ from equation (4), i.e., $Q = 3000 - 960 = \textbf{2040 var}$ or $\textbf{2.04 kvar.}$

  Method 2. Reactive power, $Q = VI \sin\phi = (233.24)(17) \sin 30.97° = \textbf{2.04 kvar.}$

Alternatively, parts (a) and (b) could have been obtained directly, using

$$\text{Apparent power, } S = VI^* = (120 + j200)(15 - j8)$$

$$= (1800 + 1600) + j(3000 - 960)$$

$$= 3400 + j2040 = P + jQ$$

from which **power** $P = \textbf{3400 W}$ and **reactive power,** $Q = \textbf{2040 var.}$

Problem 3. A series circuit possesses resistance R and capacitance C. The circuit dissipates a power of 1.732 kW and has a power factor of 0.866 leading. If the applied voltage is given by $v = 141.4 \sin (10^4 t + (\pi/9))$ volts, determine (a) the current flowing and its phase, (b) the value of resistance R, and (c) the value of capacitance C.

(a) Since $v = 141.4 \sin (10^4 t + (\pi/9))$ volts, then 141.4 V represents the maximum value, from which the rms voltage, $V = 141.4/\sqrt{2} = 100$ V, and phase angle of voltage $= +\pi/9$ rad $= 20°$ leading. Hence as a phasor the voltage V is written as $\textbf{100} \underline{/\textbf{20° V.}}$

  Power factor $= 0.866 = \cos\phi$, from which $\phi = \text{arc} \cos 0.866 = 30°$. Hence the angle between voltage and current is $30°$.

  Power $P = VI \cos\phi$. Hence $1732 = (100)I \cos 30°$ from which

$$\text{current, } I = \frac{1732}{(100)(0.866)} = \textbf{20 A}$$

Since the power factor is leading, the current phasor leads the voltage—in this case by $30°$. Since the voltage has a phase angle of $20°$,

$$\text{current, } I = 20 \underline{/(20° + 30°)} \text{ A} = \textbf{20} \underline{/\textbf{50° A}}$$

(b) Impedance $Z = \dfrac{V}{I} = \dfrac{100 \underline{/20°}}{20 \underline{/50°}} = 5 \underline{/-30°}$ Ω or $(4.33 - j2.5)\Omega$

Hence the **resistance,** $R = \textbf{4.33}\,\Omega$ and the capacitive reactance, $X_C = 2.5\,\Omega$.

Alternatively, the resistance may be determined from active power, $P = I^2 R$. Hence $1732 = (20)^2 R$, from which,

$$\text{resistance } R = \frac{1732}{(20)^2} = \mathbf{4.33\,\Omega}$$

(c) Since $v = 141.4 \sin(10^4 t + (\pi/9))$ volts, angular velocity $\omega = 10^4$ rad/s. Capacitive reactance, $X_C = 2.5\,\Omega$, thus

$$2.5 = \frac{1}{2\pi f C} = \frac{1}{\omega C}$$

from which

$$\text{capacitance, } C = \frac{1}{2.5\omega} = \frac{1}{(2.5)(10^4)} = \mathbf{40\,\mu F}$$

Problem 4. For the circuit shown in Fig. 8, determine the active power developed between points (a) A and B, (b) C and D, (c) E and F.

$$\text{Circuit impedance, } Z = 5 + \frac{(3 + j4)(-j10)}{(3 + j4 - j10)} = 5 + \frac{(40 - j30)}{(3 - j6)}$$

$$= 5 + \frac{50\angle -36.87°}{6.71\angle -63.43°} = 5 + 7.45\angle 26.56°$$

$$= 5 + 6.66 + j3.33$$
$$= (11.66 + j3.33)\,\Omega \quad \text{or}$$
$$12.13\angle 15.94°\,\Omega$$

$$\text{Current } I = \frac{V}{Z} = \frac{100\angle 0°}{12.13\angle 15.94°} = 8.24\angle -15.94°\,\text{A}.$$

(a) Active power developed between points A and $B = I^2 R = (8.24)^2(5) = \mathbf{339.5\,W}.$
(b) Active power developed between points C and D **is zero**, since no power is developed in a pure capacitor.

(c) Current, $I_1 = I\left(\dfrac{Z_{CD}}{Z_{CD} + Z_{EF}}\right) = 8.24\angle -15.94°\left(\dfrac{-j10}{3 - j6}\right)$

$$= 8.24\angle -15.94°\left(\frac{10\angle -90°}{6.71\angle -63.43°}\right) = 12.28\angle -42.51°\,\text{A}$$

Hence the active power developed between points E and $F = I_1^2 R = (12.28)^2(3)$
$$= \mathbf{452.4\,W}.$$

(Check: Total active power developed $= 339.5 + 452.4 = 791.9$ W or 792 W, correct to three significant figures.

Total active power, $P = I^2 R_T = (8.24)^2(11.66) = 792$ W (since $11.66\,\Omega$ is the total circuit equivalent resistance)

or $P = VI\cos\phi = (100)(8.24)\cos 15.94° = 792$ W.)

Problem 5. The circuit shown in Fig. 9 dissipates an active power of 400 W and has a power factor of 0.766 lagging. Determine (a) the apparent power, (b) the reactive power, (c) the value and phase of current I, and (d) the value of impedance Z.

Since power factor $= 0.766$ lagging, the circuit phase angle $\phi = \arccos 0.766$, i.e., $\phi = 40°$ lagging which means that the current I lags voltage V by $40°$.

Figure 8

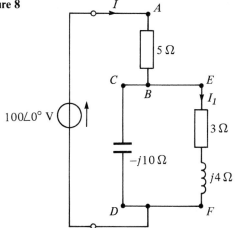

Figure 9

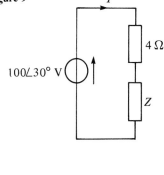

(a) Since power, $P = VI \cos \phi$, the magnitude of apparent power,

$$S = VI = \frac{P}{\cos \phi} = \frac{400}{0.766} = \textbf{522.2 VA}$$

(b) Reactive power $Q = VI \sin \phi = (522.2)(\sin 40°) = \textbf{335.7 var lagging}$. (The reactive power is lagging since the circuit is inductive, which is indicated by the lagging power factor.) The power triangle is shown in Fig. 10.

(c) Since $VI = 522.2$ VA,

$$\text{magnitude of current } I = \frac{522.2}{V} = \frac{522.2}{100} = \textbf{5.222 A}$$

Since the voltage is at a phase angle of 30° (see Fig. 9) and current lags voltage by 40°, the phase angle of current is $30° - 40° = -10°$. Hence **current** $I = \textbf{5.222} \underline{/-10°}$ **A.**

(d) Total circuit impedance $Z_T = \dfrac{V}{I} = \dfrac{100 \underline{/\,30°}}{5.222 \underline{/\,-10°}}$

$$= 19.15 \underline{/\,40°}\ \Omega \quad \text{or} \quad (14.67 + j12.31)\,\Omega$$

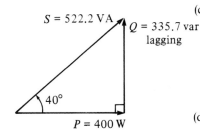

Figure 10

Hence impedance $Z = Z_T - 4 = (14.67 + j12.31) - 4$

$$= (\textbf{10.67} + \textbf{j12.31})\,\Omega \quad \text{or} \quad \textbf{16.29} \underline{/\,\textbf{49.08°}}\ \Omega.$$

Problem 6. For the series-parallel network shown in Fig. 11, determine (a) the equivalent circuit impedance, (b) the supply current I, (c) the circuit power factor,

Figure 11

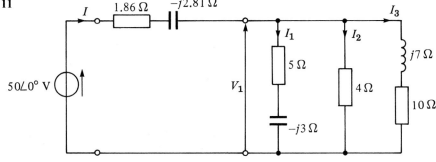

(d) the value of currents I_1, I_2 and I_3, (e) the total active power, apparent power and reactive power.

(a) For the three branches connected in parallel, the total admittance Y is given by

$$Y = \frac{1}{(5-j3)} + \frac{1}{4} + \frac{1}{(10+j7)}$$

$$= \frac{(5+j3)}{5^2+3^2} + 0.25 + \frac{10-j7}{10^2+7^2}$$

$$= (0.1471 + j0.0882) + (0.25) + (0.0671 - j0.0470)$$

i.e., $Y = (0.4642 + j0.0412)\,\text{S}$ or $0.466\angle\,5.07°\,\text{S}$. Hence total impedance of the three branches in parallel, Z, is given by

$$Z = \frac{1}{Y} = \frac{1}{0.466\angle\,5.07°} = 2.15\angle -5.07°\,\Omega \quad \text{or} \quad (2.14 - j0.19)\,\Omega$$

Thus the total circuit impedance,

$$Z_T = (2.14 - j0.19) + (1.86 - j2.81)$$

$$= (4 - j3)\,\Omega \quad \text{or} \quad 5\angle -36.87°\,\Omega$$

(b) Supply current, $I = \dfrac{V}{Z_T} = \dfrac{50\angle\,0°}{5\angle -36.87°} = 10\angle\,36.87°\,\text{A} \quad \text{or} \quad (8 + j6)\,\text{A}.$

(c) Circuit phase angle $\phi = 36.87°$ leading. Hence

circuit power factor $= \cos\phi = \cos 36.87° = \textbf{0.8 leading}$

(d) Current $I_1 = V_1/Z_1$. Voltage, $V_1 = IZ$, where Z is the equivalent impedance of the three parallel branches, i.e.,

$$V_1 = (10\angle\,36.87°)(2.15\angle -5.07°) = 21.5\angle\,31.80°\,\text{V}$$

Impedance, $Z_1 = (5 - j3)\,\Omega = 5.831\angle -30.96°$. Hence

$$\textbf{Current } I_1 = \frac{V_1}{Z_1} = \frac{21.5\angle\,31.80°}{5.831\angle -30.96°} = \textbf{3.687}\angle\,\textbf{62.76°\,A}$$

$$\textbf{Current } I_2 = \frac{V_1}{Z_2} = \frac{21.5\angle\,31.80°}{4\angle\,0°} = \textbf{5.375}\angle\,\textbf{31.80°\,A}$$

$$\textbf{Current } I_3 = \frac{V_1}{Z_3} = \frac{21.5\angle\,31.80°}{(10+j7)} = \frac{21.5\angle\,31.80°}{12.207\angle\,34.99°} = \textbf{1.761}\angle -\textbf{3.19°\,A}$$

(e) In complex form, apparent power $S = VI^*$, where I^* is the conjugate of current I. Since $I = 10\angle\,36.87°\,\text{A}$, $I^* = 10\angle -36.87°\,\text{A}$. Thus

$$S = VI^* = (50\angle\,0°)(10\angle -36.87°)$$

$$= 500\angle -36.87°\,\text{VA} \quad \text{or} \quad (400 - j300)\,\text{VA}$$

However, $S = P + jQ$.

Thus **total active power, $P = 400$ W, total apparent power, $S = 500$ VA** and **total reactive power, $Q = 300$ var**, i.e. $Q = 300$ var leading. The power triangle is shown in Fig. 12.

There are several alternative methods of determining P, Q and S. For example, power $P = VI\cos\phi$ or $P = I^2 R_T$ (where $R_T = 4\,\Omega$ from the total circuit impedance) or $P = ac + bd$ from equation (3). Similarly, reactive power,

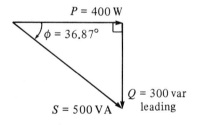

$P = 400$ W

$\phi = 36.87°$

$Q = 300$ var leading

$S = 500$ VA

Figure 12

Figure 13 (a) Circuit diagram
(b) Phasor diagram

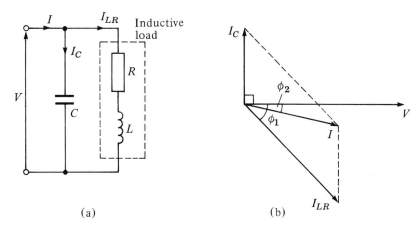

(a)   (b)

$Q = VI \sin \phi$ or $Q = bc - ad$ from equation (4). Also, apparent power, $S = VI$ or $S = \sqrt{(P^2 + Q^2)}$, and so on.

Further problems on power in a.c. circuits may be found in section 6, problems 1 to 15, page 69.

5. Power factor improvement

Figure 14 Effect of connecting capacitance in parallel with the inductive load

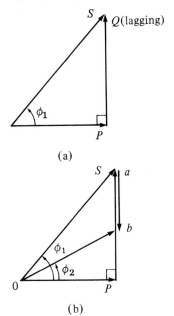

For a particular active power supplied, a high power factor reduces the current flowing in a supply system and therefore reduces the cost of cables, transformers, switchgear and generators. Supply authorities use tariffs which encourage consumers to operate at a reasonably high power factor. One method of improving the power factor of an inductive load is to connect a bank of capacitors in parallel with the load. Capacitors are rated in reactive voltamperes and the effect of the capacitors is to reduce the reactive power of the system without changing the active power. Most residential and industrial loads on a power system are inductive, i.e. they operate at a lagging power factor.

A simplified circuit diagram is shown in Fig. 13(a) where a capacitor C is connected across an inductive load. Before the capacitor is connected the circuit current is I_{LR} and is shown lagging voltage V by angle ϕ_1 in the phasor diagram of Fig. 13(b). When the capacitor C is connected it takes a current I_C which is shown in the phasor diagram leading voltage V by 90°. The supply current I in Fig. 13(a) is now the phasor sum of currents I_{LR} and I_C as shown in Fig. 13(b). The circuit phase angle, i.e., the angle between V and I, has been reduced from ϕ_1 to ϕ_2 and the power factor has been improved from $\cos \phi_1$ to $\cos \phi_2$.

Figure 14(a) shows the power triangle for an inductive circuit with a lagging power factor of $\cos \phi_1$. In Fig. 14(b), the angle ϕ_1 has been reduced to ϕ_2, i.e., the power factor has been improved from $\cos \phi_1$ to $\cos \phi_2$, by introducing leading reactive voltamperes (shown as length ab) which is achieved by connecting capacitance in parallel with the inductive load. The power factor has been improved by reducing the reactive voltamperes; the active power P has remained unaffected.

Power factor correction results in the apparent power S decreasing (from $0a$ to $0b$ in Fig. 14(b)) and thus the current decreasing, so that the power distribution system is used more efficiently.

Another method of power factor improvement, besides the use of static capacitors, is by using synchronous motors; such machines can be made to operate at leading power factors.

Worked problems on power factor improvement

Problem 1. A 300 kVA transformer is at full load with an overall power factor of 0.70 lagging. The power factor is improved by adding capacitors in parallel with the transformer until the overall power factor becomes 0.90 lagging. Determine the rating (in kilovars) of the capacitors required.

At full load, active power, $P = VI\cos\phi = (300)(0.70) = 210\,\text{kW}$.
Circuit phase angle $\phi = \arccos 0.70 = 45.57°$.
Reactive power, $Q = VI\sin\phi = (300)(\sin 45.57°) = 214.2\,\text{kvar lagging}$.
   The power triangle is shown as triangle $0ab$ in Fig. 15. When the power factor is 0.90, the circuit phase angle $\phi = \arccos 0.90 = 25.84°$. The capacitor rating needed to improve the power factor to 0.90 is given by length bd in Fig. 15.
   Tan $25.94 = ad/210$, from which $ad = 210\tan 25.84° = 101.7\,\text{kvar}$. Hence the **capacitor rating**, i.e., $bd = ab - ad = 214.2 - 101.7 = \textbf{112.5 kvar leading.}$

Problem 2. A circuit has an impedance $Z = (3 + j4)\,\Omega$ and a source p.d. of $50\angle 30°$ V at a frequency of 1.5 kHz. Determine (a) the supply current, (b) the active, apparent and reactive power, (c) the rating of a capacitor to be connected in parallel with impedance Z to improve the power factor of the circuit to 0.966 lagging, and (d) the value of capacitance needed to improve the power factor to 0.966 lagging.

(a) Supply current, $I = \dfrac{V}{Z} = \dfrac{50\angle 30°}{(3 + j4)} = \dfrac{50\angle 30°}{5\angle 53.13°} = \textbf{10}\angle \textbf{- 23.13° A.}$

(b) Apparent power, $S = VI^* = (50\angle 30°)(10\angle 23.13°)$
$$= 500\angle 53.13°\,\text{VA} = (300 + j400)\,\text{VA} = P + jQ.$$
   Hence **active power, $P = 300$ W**
   apparent power, $S = 500$ VA and
   reactive power, $Q = 400$ var lagging.
   The power triangle is shown in Fig. 16.

(c) A power factor of 0.966 means that $\cos\phi = 0.966$. Hence angle $\phi = \arccos 0.966 = 15°$.
   To improve the power factor from $\cos 53.13°$, i.e. 0.60, to 0.966, the power triangle will need to change from $0cb$ (see Fig. 17) to $0ab$, the length ca representing the rating of a capacitor connected in parallel with the circuit.
   From Fig. 17, $\tan 15° = ab/300$, from which $ab = 300\tan 15° = 80.38\,\text{var}$.

Figure 15

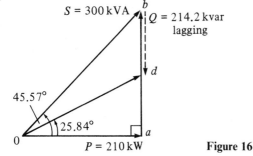

$S = 300\,\text{kVA}$
$Q = 214.2\,\text{kvar lagging}$
$45.57°$
$25.84°$
0
$P = 210\,\text{kW}$

Figure 16

$S = 500\,\text{VA}$
$Q = 400\,\text{var lagging}$
$53.13°$
$P = 300\,\text{W}$

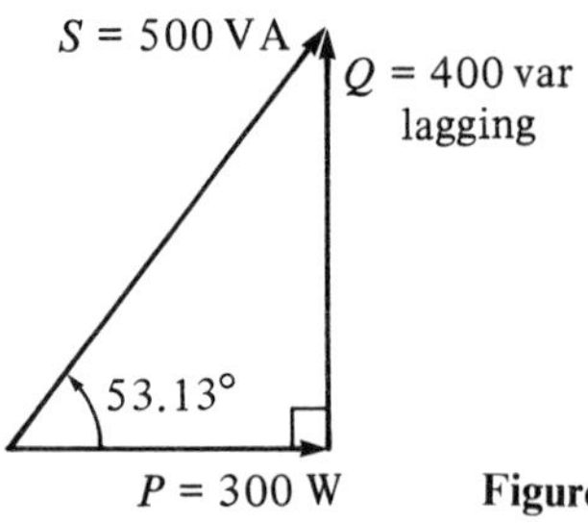

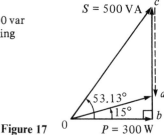

Figure 17

$S = 500\,\text{VA}$
$53.13°$
$15°$
0
$P = 300\,\text{W}$

Hence the **rating of the capacitor**, $ca = cb - ab$
$$= 400 - 80.38 = \textbf{319.6 var leading.}$$

(d) Current in capacitor, $I_C = \dfrac{Q}{V} = \dfrac{319.6}{50} = 6.39 \text{ A.}$

Capacitive reactance, $X_C = \dfrac{V}{I_C} = \dfrac{50}{6.39} = 7.82\,\Omega.$

Thus $7.82 = 1/(2\pi f C)$, from which

$$\textbf{required capacitance } C = \frac{1}{2\pi(1500)(7.82)}\text{F} \equiv \textbf{13.57}\,\boldsymbol{\mu}\textbf{F}$$

Further problems on power factor improvement may be found in the following section (6), problems 16 to 20, page 70.

6. Further problems

Power in a.c. circuits

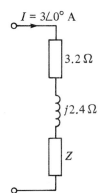

$I = 3\angle 0°$ A

$3.2\,\Omega$

$j2.4\,\Omega$

Z

Figure 18

1. When the voltage applied to a circuit is given by $(2 + j5)$ V, the current flowing is given by $(8 + j4)$ A. Determine the power dissipated in the circuit.

    [36 W]

2. A current of $(12 + j5)$ A flows in a circuit when the supply voltage is $(150 + j220)$ V. Determine (a) the active power, (b) the reactive power, and (c) the apparent power. Draw the power triangle.

    [(a) 2.90 kW   (b) 1.89 kvar lagging   (c) 3.46 kVA]

3. A capacitor of capacitive reactance $40\,\Omega$ and a resistance of $30\,\Omega$ are connected in series to a supply voltage of $200\angle 60°$ V. Determine the active power in the circuit.

    [480 W]

4. The circuit shown in Fig. 18 takes 81 VA at a power factor of 0.8 lagging. Determine the value of impedance Z.

    [$(4 + j3)\,\Omega$ or $5\angle 36.87°\,\Omega$]

5. The current in an a.c. circuit is given by $(6 + j15)$ A when the applied voltage is $(120 + j180)$ V. Determine (a) the circuit impedance, (b) the active power, and (c) the power factor.

    [(a) $13.39\angle -11.89°\,\Omega$ (b) 3420 W   (c) 0.979 leading]

6. A series circuit possesses inductance L and resistance R. The circuit dissipates a power of 2.898 kW and a power factor of 0.966 lagging. If the applied voltage is given by $v = 169.7 \sin(100t - (\pi/4))$ volts, determine (a) the current flowing and its phase, (b) the value of resistance R, and (c) the value of inductance L.

    [(a) $25\angle -60°$ A   (b) $4.64\,\Omega$ (c) 12.4 mH]

7. The p.d. across and the current in a certain circuit are represented by $(190 + j40)$ V and $(9 - j4)$ A respectively. Determine the active power and the reactive power, stating whether the latter is leading or lagging.

    [1550 W; 1120 var lagging]

8. Two impedances, $Z_1 = 6\angle 40°\,\Omega$ and $Z_2 = 10\angle 30°\,\Omega$ are connected in series and have a total reactive power of 1650 var lagging. Determine (a) the average power, (b) the apparent power, and (c) the power factor.

    [(a) 2469 W   (b) 2970 VA   (c) 0.83 lagging]

9. A current $i = 7.5 \sin(\omega t - (\pi/4))$ A flows in a circuit which has an applied voltage $v = 180 \sin(\omega t + (\pi/12))$ V. Determine (a) the circuit impedance, (b) the active power, (c) the reactive power, and (d) the apparent power. Draw the power triangle.

    [(a) $24\angle 60°\,\Omega$ (b) 337.5 W   (c) 584.6 var lagging   (d) 675 VA]

Figure 19

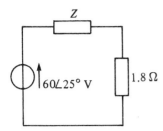

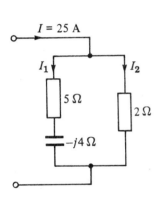

Figure 20

10. The circuit shown in Fig. 19 has a power of 480 W and a power factor of 0.8 leading. Determine (a) the apparent power, (b) the reactive power, and (c) the value of impedance Z.

$$\left[\begin{array}{ll}\text{(a) 600 VA} & \text{(b) 360 var leading}\\ \text{(c) } (3 - j3.6)\,\Omega & \text{or} \quad 4.69\underline{/-50.19°}\,\Omega\end{array}\right]$$

11. For the network shown in Fig. 20, determine (a) the values of currents I_1 and I_2, (b) the total active power, (c) the reactive power, and (d) the apparent power.

$$\left[\begin{array}{ll}\text{(a) } I_1 = 6.20\underline{/29.74°}\text{ A, } I_2 = 19.86\underline{/-8.92°}\text{ A}\\ \text{(b) 980.8 W} \quad \text{(c) 153.9 var leading} \quad \text{(d) 992.8 VA}\end{array}\right]$$

12. A circuit consists of an impedance $5\underline{/-45°}\,\Omega$ in parallel with a resistance of $10\,\Omega$. The supply current is 4 A. Determine for the circuit (a) the active power, (b) the reactive power, and (c) the power factor.

[(a) 49.34 W   (b) 28.90 var leading   (c) 0.863 leading]

13. For the series-parallel network shown in Fig. 21, determine (a) the total impedance, (b) the supply current I, (c) the circuit power factor, (d) currents I_1, I_2 and I_3, (e) the total active power, (f) the total apparent power, and (g) the total reactive power.

$$\left[\begin{array}{llll}\text{(a) } 5\underline{/36.87°}\,\Omega & \text{(b) } 16\underline{/-36.87°}\text{ A} & \text{(c) 0.8 lagging}\\ \text{(d) } I_1 = 7.20\underline{/-84.87°}\text{ A, } I_2 = 3.72\underline{/25.58°}\text{ A, } I_3 = 9.68\underline{/-24.61°}\text{ A}\\ \text{(e) 1024 W} & \text{(f) 1280 VA} & \text{(g) 768 var lagging}\end{array}\right]$$

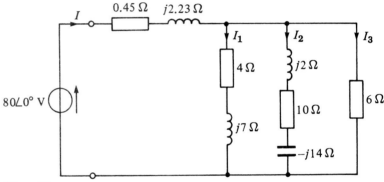

Figure 21

14. For the network shown in Fig. 22, determine the active power developed between points (a) A and B, (b) C and D, (c) E and F.

[(a) 254.1 W   (b) 0   (c) 65.92 W]

15. A voltage of $150\underline{/30°}$ V is applied to a circuit consisting of two branches in parallel. The currents in the two branches are $25\underline{/60°}$ A and $30\underline{/30°}$ A. Find (a) the apparent power in each branch, (b) the active power in each branch, (c) the total circuit apparent power, (d) the total circuit active power, and (e) the total supply current.

$$\left[\begin{array}{lll}\text{(a) 3.75 kVA; 4.50 kVA} & \text{(b) 3.25 kW; 4.50 kW}\\ \text{(c) 7.97 kVA} & \text{(d) 7.75 kW} & \text{(e) } 53.14\underline{/43.60°}\text{ A}\end{array}\right]$$

Power factor improvement

16. A 600 kVA transformer is at full load with an overall power factor of 0.64 lagging. The power factor is improved by adding capacitors in parallel with the transformer until the overall power factor becomes 0.95 lagging. Determine the rating (in vars) of the capacitors needed.

[334.8 var leading]

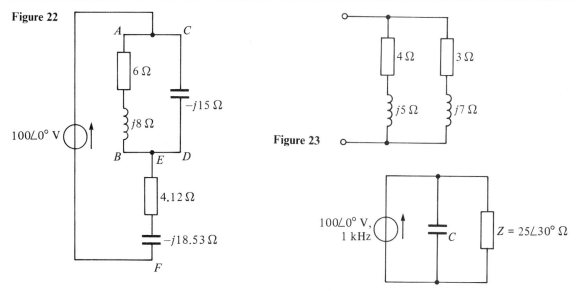

Figure 22

Figure 23

Figure 24

17. A source p.d. of $130\angle 40°$ V at 2 kHz is applied to a circuit having an impedance of $(5 + j12)\,\Omega$. Determine (a) the supply current, (b) the active, apparent and reactive powers, (c) the rating of the capacitor to be connected in parallel with the impedance to improve the power factor of the circuit to 0.940 lagging, and (d) the value of the capacitance of the capacitor required.

$$\left[\begin{array}{ll} \text{(a)} \ 10\angle -27.38°\,\text{A} & \text{(b)} \ 500\,\text{W, } 1300\,\text{VA, } 1200\,\text{var lagging} \\ \text{(c)} \ 1019\,\text{var leading} & \text{(d)} \ 0.78\,\mu\text{F} \end{array}\right]$$

18. The network shown in Fig. 23 has a total active power of 2253 W. Determine (a) the total impedance, (b) the supply current, (c) the apparent power, (d) the reactive power, (e) the circuit power factor, (f) the capacitance of the capacitor to be connected in parallel with the network to improve the power factor to 0.90 lagging, if the supply frequency is 50 Hz.

$$\left[\begin{array}{ll} \text{(a)} \ 3.51\angle 58.4°\,\Omega & \text{(b)} \ 35.0\,\text{A} \quad \text{(c)} \ 4300\,\text{VA} \quad \text{(d)} \ 3662\,\text{var lagging} \\ \text{(e)} \ 0.524\,\text{lagging} & \text{(f)} \ 43.33\,\mu\text{F} \end{array}\right]$$

19. For the network shown in Fig. 24, determine the value of capacitance C necessary to achieve a power factor of 0.96 lagging.

$$[25.73\,\mu\text{F}]$$

20. The power factor of a certain load is improved to 0.92 lagging with the addition of a 30 kvar bank of capacitors. If the resulting supply apparent power is 200 kVA, determine (a) the active power, (b) the reactive power before power factor correction, and (c) the power factor before correction.

$$[\text{(a)} \ 184\,\text{kW} \quad \text{(b)} \ 108.4\,\text{kvar lagging} \quad \text{(c)} \ 0.862\,\text{lagging}]$$

5 A.C. bridges

1. Introduction

A.C. bridges are electrical networks, based upon an extension of the Wheatstone bridge principle, used for the determination of an unknown impedance by comparison with known impedances and for the determination of frequency. In general, they contain four impedance arms, an a.c. power supply and a balance detector which is sensitive to alternating currents. It is more difficult to achieve balance in an a.c. bridge than in a d.c. bridge because both the magnitude and the phase angle of impedances are related to the balance condition. Balance equations are derived by using complex numbers. A.C. bridges provide precise methods of measurement of inductance and capacitance, as well as resistance.

2. Balance conditions for an a.c. bridge

The majority of well known a.c. bridges are classified as four-arm bridges and consist of an arrangement of four impedances (in complex form, $Z = R \pm jX$) as shown in Fig. 1. As with the d.c. Wheatstone bridge circuit, an a.c. bridge is said to be "balanced" when the current through the detector is zero (i.e., when no current flows between B and D of Fig. 1). If the current through the detector is zero, then the current I_1 flowing in impedance Z_1 must also flow in impedance Z_2. Also, at balance, the current I_4 flowing in impedance Z_4, must also flow through Z_3.

At balance: (i) the volt drop between A and B is equal to the volt drop between A and D, i.e., $V_{AB} = V_{AD}$, i.e.,

$$I_1 Z_1 = I_4 Z_4 \text{ (both in magnitude and in phase)} \qquad (1)$$

(ii) the volt drop between B and C is equal to the volt drop between D and C, i.e., $V_{BC} = V_{DC}$ i.e.,

$$I_1 Z_2 = I_4 Z_3 \text{ (both in magnitude and in phase)} \qquad (2)$$

Dividing equation (1) by equation (2) gives

$$\frac{I_1 Z_1}{I_1 Z_2} = \frac{I_4 Z_4}{I_4 Z_3}$$

from which

$$\frac{Z_1}{Z_2} = \frac{Z_4}{Z_3}$$

or

$$Z_1 Z_3 = Z_2 Z_4 \qquad (3)$$

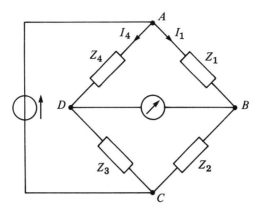

Figure 1 Four-arm bridge

Equation (3) shows that at balance the products of the impedances of opposite arms of the bridge are equal.

If in polar form, $Z_1 = |Z_1|\underline{/\alpha_1}$, $Z_2 = |Z_2|\underline{/\alpha_2}$, $Z_3 = |Z_3|\underline{/\alpha_3}$, and $Z_4 = |Z_4|\underline{/\alpha_4}$, then, from equation (3)

$$(|Z_1|\underline{/\alpha_1})(|Z_3|\underline{/\alpha_3}) = (|Z_2|\underline{/\alpha_2})(|Z_4|\underline{/\alpha_4}),$$

which shows that there are **two** conditions to be satisfied simultaneously for balance in an a.c. bridge, i.e.,

$$|Z_1||Z_3| = |Z_2||Z_4| \quad \text{and} \quad \alpha_1 + \alpha_3 = \alpha_2 + \alpha_4$$

When deriving balance equations of a.c. bridges, where at least two of the impedances are in complex form, it is important to appreciate that for a complex equation

$$a + jb = c + jd$$

the real parts are equal, i.e., $a = c$, and the imaginary parts are equal, i.e., $b = d$.

Usually one arm of an a.c. bridge circuit contains the unknown impedance while the other arms contain known fixed or variable components. Normally only two components of the bridge are variable. When balancing a bridge circuit, the current in the detector is gradually reduced to zero by successive adjustments of the two variable components. At balance, the unknown impedance can be expressed in terms of the fixed and variable components.

Procedure for determining the balance equations of any a.c. bridge circuit

(i) Determine for the bridge circuit the impedance in each arm in complex form and write down the balance equation as in equation (3). Equations are usually easier to manipulate if L and C are initially expressed as X_L and X_C, rather than ωL or $1/(\omega C)$.

(ii) Isolate the unknown terms on the left-hand side of the equation in the form $a + jb$.

(iii) Manipulate the terms on the right-hand side of the equation into the form $c + jd$.

(iv) Equate the real parts of the equation, i.e., $a = c$, and equate the imaginary parts of the equation, i.e., $b = d$.

(v) Substitute ωL for X_L and $1/(\omega C)$ for X_C where appropriate and express the final equations in their simplest form.

Types of detector used with a.c. bridges vary with the type of bridge and with the frequency at which it is operated. Common detectors used include:

(i) a C.R.O., which is suitable for use with a very wide range of frequencies;

(ii) earphones (or telephone headsets), which are suitable for frequencies up to about 10 kHz and are used often at about 1 kHz, in which region the human ear is very sensitive;

(iii) various electronic detectors, which use tuned circuits to detect current at the correct frequency; and

(iv) vibration galvanometers, which are usually used for mains-operated bridges. This type of detector consists basically of a narrow moving coil which is suspended on a fine phosphor bronze wire between the poles of a magnet. When a current of the correct frequency flows through the coil, it is set into vibration. This is because the mechanical resonant frequency of the suspension is purposely made equal to the electrical frequency of the coil current. A mirror attached to the coil reflects a spot of light on to a scale, and when the coil is vibrating the spot appears as an extended beam of light. When the band reduces to a spot the bridge is balanced. Vibration galvanometers are available in the frequency range 10 Hz to 300 Hz.

3. Types of a.c. bridge circuit

A large number of bridge circuits have been developed, each of which has some particular advantage under certain conditions. Some of the most important a.c. bridges include the Maxwell, Hay, Owen and Maxwell–Wien bridges for measuring inductance, the De Sauty and Schering bridges for measuring capacitance, and the Wien bridge for measuring frequency. Obviously a large number of combinations of components in bridges is possible.

In many bridges it is found that two of the balancing impedances will be of the same nature, and often consist of standard non-inductive resistors.

For a bridge to balance quickly the requirement is either:

(i) the adjacent arms are both pure components (i.e. either both resistors, or both pure capacitors, or one of each)—this type of bridge being called a **ratio-arm bridge** (see, for example, paras (a), (c), (e) and (g) below); or

(ii) a pair of opposite arms are pure components—this type of bridge being called a **product-arm bridge** (see, for example, paras (b), (d) and (f) below).

A ratio-arm bridge can only be used to measure reactive quantities of the same type. When using a product-arm bridge, the reactive component of the balancing impedance must be of opposite sign to the unknown reactive component.

A commercial or universal bridge is available and can be used to measure resistance, inductance or capacitance.

(a) The simple Maxwell bridge

This bridge is used to measure the resistance and inductance of a coil having a high Q-factor (where Q-factor $= \omega L/R$, see chapter 6).

A coil having unknown resistance R_x and inductance L_x is shown in the circuit diagram of a simple Maxwell bridge in Fig. 2. R_4 and L_4 represent a standard coil having known variable values. At balance, expressions for R_x and L_x may be derived in terms of known components R_2, R_3, R_4 and L_4.

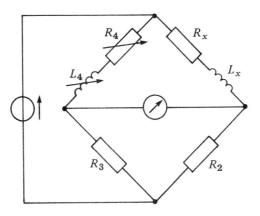

Figure 2 Simple Maxwell bridge

The procedure for determining the balance equations given in section 2 may be followed.

(i) From Fig. 2, $Z_x = R_x + jX_{L_x}$, $Z_2 = R_2$, $Z_3 = R_3$ and $Z_4 = R_4 + jX_{L_4}$. At balance, $(Z_x)(Z_3) = (Z_2)(Z_4)$, from equation (3), i.e.,
$(R_x + jX_{L_x})(R_3) = (R_2)(R_4 + jX_{L_4})$.

(ii) Isolating the unknown impedance on the left-hand side of the equation gives

$$(R_x + jX_{L_x}) = \frac{R_2}{R_3}(R_4 + jX_{L_4})$$

(iii) Manipulating the right-hand side of the equation into $(a + jb)$ form gives

$$(R_x + jX_{L_x}) = \frac{R_2 R_4}{R_3} + j\frac{R_2 X_{L_4}}{R_3}$$

(iv) Equating the real parts gives $R_x = \dfrac{R_2 R_4}{R_3}$

Equating the imaginary parts gives $X_{L_x} = \dfrac{R_2 X_{L_4}}{R_3}$

(v) Since $X_L = \omega L$, then

$$\omega L_x = \frac{R_2(\omega L_4)}{R_3} \quad \text{from which} \quad L_x = \frac{R_2 L_4}{R_3}$$

Thus at balance the unknown components in the simple Maxwell bridge are given by

$$R_x = \frac{R_2 R_4}{R_3} \quad \text{and} \quad L_x = \frac{R_2 L_4}{R_3}$$

These are known as the **"balance equations"** for the bridge.

(b) The Hay bridge

This bridge is used to measure the resistance and inductance of a coil having a very high Q-factor. A coil having unknown resistance R_x and inductance L_x is shown in the circuit diagram of a Hay bridge in Fig. 3. Following the procedure of section 2 gives:

(i) From Fig. 3, $Z_x = R_x + jX_{L_x}$, $Z_2 = R_2$, $Z_3 = R_3 - jX_{C_3}$ and $Z_4 = R_4$.
   At balance $(Z_x)(Z_3) = (Z_2)(Z_4)$, from equation (3), i.e.,
   $(R_x + jX_{L_x})(R_3 - jX_{C_3}) = (R_2)(R_4)$.

(ii) $$(R_x + jX_{L_x}) = \frac{R_2 R_4}{R_3 - jX_{C_3}}$$

(iii) Rationalising the right-hand side gives

$$(R_x + jX_{L_x}) = \frac{R_2 R_4 (R_3 + jX_{C_3})}{(R_3 - jX_{C_3})(R_3 + jX_{C_3})} = \frac{R_2 R_4 (R_3 + jX_{C_3})}{R_3^2 + X_{C_3}^2}$$

i.e.,

$$(R_x + jX_{L_x}) = \frac{R_2 R_3 R_4}{R_3^2 + X_{C_3}^2} + j\frac{R_2 R_4 X_{C_3}}{R_3^2 + X_{C_3}^2}$$

(iv) Equating the real parts gives $R_x = \dfrac{R_2 R_3 R_4}{R_3^2 + X_{C_3}^2}$

Equating the imaginary parts gives $X_{L_x} = \dfrac{R_2 R_4 X_{C_3}}{R_3^2 + X_{C_3}^2}$

(v) Since $X_{C_3} = \dfrac{1}{\omega C_3}$,

$$R_x = \frac{R_2 R_3 R_4}{R_3^2 + (1/(\omega^2 C_3^2))} = \frac{R_2 R_3 R_4}{(\omega^2 C_3^2 R_3^2 + 1)/(\omega^2 C_3^2)}$$

i.e.,

$$R_x = \frac{\omega^2 C_3^2 R_2 R_3 R_4}{1 + \omega^2 C_3^2 R_3^2}$$

Since $X_{L_x} = \omega L_x$,

$$\omega L_x = \frac{R_2 R_4 (1/(\omega C_3))}{(\omega^2 C_3^2 R_3^2 + 1)/(\omega^2 C_3^2)} = \frac{\omega^2 C_3^2 R_2 R_4}{\omega C_3 (1 + \omega^2 C_3^2 R_3^2)}$$

i.e.,

$$L_x = \frac{C_3 R_2 R_4}{(1 + \omega^2 C_3^2 R_3^2)} \quad \text{by cancelling.}$$

Thus at balance the unknown components in the Hay bridge are given by

$$R_x = \frac{\omega^2 C_3^2 R_2 R_3 R_4}{(1 + \omega^2 C_3^2 R_3^2)} \quad \text{and} \quad L_x = \frac{C_3 R_2 R_4}{(1 + \omega^2 C_3^2 R_3^2)}$$

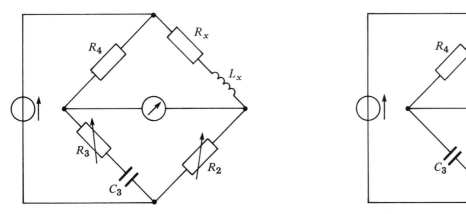

Figure 3 Hay bridge   **Figure 4** Owen bridge

Since $\omega(=2\pi f)$ appears in the balance equations, the bridge is **frequency-dependent.**

(c) The Owen bridge

This bridge is used to measure the resistance and inductance of coils possessing a large value of inductance. A coil having unknown resistance R_x and inductance L_x is shown in the circuit diagram of an Owen bridge in Fig. 4, from which $Z_x = R_x + jX_{L_x}, Z_2 = R_2 - jX_{C_2}, Z_3 = -jX_{C_3}$ and $Z_4 = R_4$.

At balance $(Z_x)(Z_3) = (Z_2)(Z_4)$, from equation (3), i.e.,
$(R_x + jX_{L_x})(-jX_{C_3}) = (R_2 - jX_{C_2})(R_4)$. Rearranging gives

$$R_x + jX_{L_x} = \frac{(R_2 - jX_{C_2})R_4}{-jX_{C_3}}$$

By rationalising and equating real and imaginary parts it may be shown that at balance the unknown components in the Owen bridge are given by

$$R_x = \frac{R_4 C_3}{C_2} \quad \text{and} \quad L_x = R_2 R_4 C_3$$

(d) The Maxwell–Wien bridge

This bridge is used to measure the resistance and inductance of a coil having a low Q-factor. A coil having unknown resistance R_x and inductance L_x is shown in the circuit diagram of a Maxwell–Wien bridge in Fig. 5, from which $Z_x = R_x + jX_{L_x}$, $Z_2 = R_2$ and $Z_4 = R_4$.

Arm 3 consists of two parallel-connected components. The equivalent impedance Z_3, is given either

(i) by $\dfrac{\text{product}}{\text{sum}}$, i.e., $Z_3 = \dfrac{(R_3)(-jX_{C_3})}{(R_3 - jX_{C_3})}$, or

(ii) by using the reciprocal impedance expression,

$$\frac{1}{Z_3} = \frac{1}{R_3} + \frac{1}{-jX_{C_3}}$$

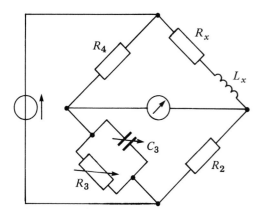

Figure 5 Maxwell–Wien bridge

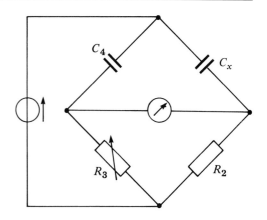

Figure 6 De Sauty bridge

from which

$$Z_3 = \frac{1}{(1/R_3) + (1/(-jX_{C_3}))} = \frac{1}{(1/R_3) + (j/X_{C_3})}$$

or

$$Z_3 = \frac{1}{\dfrac{1}{R_3} + j\omega C_3}, \quad \text{since } X_{C_3} = \frac{1}{\omega C_3}$$

Whenever an arm of an a.c. bridge consists of two branches in parallel, either method of obtaining the equivalent impedance may be used.
    For the Maxwell–Wien bridge of Fig. 5, at balance

$$(Z_x)(Z_3) = (Z_2)(Z_4), \text{ from equation 3,}$$

i.e.,

$$(R_x + jX_{L_x})\frac{(R_3)(-jX_{C_3})}{(R_3 - jX_{C_3})} = R_2 R_4$$

using method (i) for Z_3. Hence

$$(R_x + jX_{L_x}) = R_2 R_4 \frac{(R_3 - jX_{C_3})}{(R_3)(-jX_{C_3})}$$

By rationalising and equating real and imaginary parts it may be shown that at balance the unknown components in the Maxwell–Wien bridge are given by

$$R_x = \frac{R_2 R_4}{R_3} \quad \text{and} \quad L_x = C_3 R_2 R_4$$

(e) The de Sauty bridge

This bridge provides a very simple method of measuring a capacitance by comparison with another known capacitance. In the de Sauty bridge shown in Fig. 6, C_x is an unknown capacitance and C_4 is a standard capacitor.

At balance

$$(Z_x)(Z_3) = (Z_2)(Z_4)$$

i.e.,

$$(-jX_{C_x})(R_3) = (R_2)(-jX_{C_4})$$

Hence

$$(X_{C_x})(R_3) = (R_2)(X_{C_4})$$

$$\left(\frac{1}{\omega C_x}\right)(R_3) = (R_2)\left(\frac{1}{\omega C_4}\right)$$

from which

$$\frac{R_3}{C_x} = \frac{R_2}{C_4} \quad \text{or} \quad C_x = \frac{R_3 C_4}{R_2}$$

This simple bridge is usually inadequate in most practical cases. The power factor of the capacitor under test is significant because of internal dielectric losses—these losses being the dissipation within a dielectric material when an alternating voltage is applied to a capacitor.

(f) The Schering bridge

This bridge is used to measure the capacitance and equivalent series resistance of a capacitor. From the measured values the power factor of insulating materials and dielectric losses may be determined. In the circuit diagram of a Schering bridge shown in Fig. 7, C_x is the unknown capacitance and R_x its equivalent series resistance.

From Fig. 7,

$$Z_x = R_x - jX_{C_x}, \qquad Z_2 = -jX_{C_2},$$

$$Z_3 = \frac{(R_3)(-jX_{C_3})}{(R_3 - jX_{C_3})} \quad \text{and} \quad Z_4 = R_4.$$

At balance, $(Z_x)(Z_3) = (Z_2)(Z_4)$ from equation (3), i.e.,

$$(R_x - jX_{C_x})\frac{(R_3)(-jX_{C_3})}{R_3 - jX_{C_3}} = (-jX_{C_2})(R_4)$$

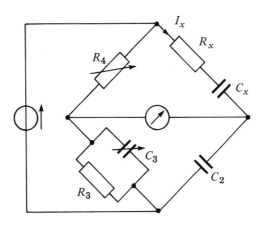

Figure 7 Schering bridge

from which

$$(R_x - jX_{C_x}) = \frac{(-jX_{C_2}R_4)(R_3 - jX_{C_3})}{-jX_{C_3}R_3}$$

$$= \frac{X_{C_2}R_4}{X_{C_3}R_3}(R_3 - jX_{C_3})$$

Equating the real parts gives

$$R_x = \frac{X_{C_2}R_4}{X_{C_3}} = \frac{(1/\omega C_2)R_4}{(1/\omega C_3)} = \frac{C_3R_4}{C_2}$$

Equating the imaginary parts gives

$$-X_{C_x} = \frac{-X_{C_2}R_4}{R_3}$$

i.e.,

$$\frac{1}{\omega C_x} = \frac{(1/\omega C_2)R_4}{R_3} = \frac{R_4}{\omega C_2 R_3}$$

from which

$$C_x = \frac{C_2R_3}{R_4}$$

Thus at balance the unknown components in the Schering bridge are given by

$$R_x = \frac{C_3R_4}{C_2} \quad \text{and} \quad C_x = \frac{C_2R_3}{R_4}$$

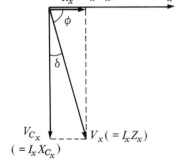

Figure 8 Phasor diagram for the unknown arm in the Schering bridge

The loss in a dielectric may be represented by either (a) a resistance in parallel with a capacitor, or (b) a lossless capacitor in series with a resistor.

If the dielectric is represented by an R–C series circuit, as shown by R_x and C_x in Fig. 7, the phasor diagram for the unknown arm is as shown in Fig. 8. Angle ϕ is given by

$$\phi = \arctan\frac{V_{C_x}}{V_{R_x}} = \arctan\frac{I_xX_{C_x}}{I_xR_x}$$

i.e.,

$$\phi = \arctan\left(\frac{1}{\omega C_xR_x}\right)$$

The power factor of the unknown arm is given by $\cos\phi$.

The angle $\delta(= 90° - \phi)$ is called the **loss angle** and is given by

$$\delta = \arctan\frac{V_{R_x}}{V_{C_x}} = \arctan \omega C_x R_x \text{ and } \phi = \arctan\left[\omega\left(\frac{C_2R_3}{R_4}\right)\left(\frac{C_3R_4}{C_2}\right)\right]$$

$$= \arctan(\omega R_3 C_3)$$

(see also chapter 11, page 277)

(g) The Wien bridge

This bridge is used to measure frequency in terms of known components

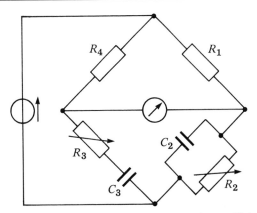

Figure 9 Wien bridge

(or, alternatively, to measure capacitance if the frequency is known). It may also be used as a frequency-stabilising network.

A typical circuit diagram of a Wien bridge is shown in Fig. 9, from which

$$Z_1 = R_1, Z_2 = \frac{1}{(1/R_2) + j\omega C_2} \quad \text{(see (ii), para (d), page 77)},$$

$$Z_3 = R_3 - jX_{C_3} \quad \text{and} \quad Z_4 = R_4.$$

At balance, $(Z_1)(Z_3) = (Z_2)(Z_4)$ from equation (3), i.e.,

$$(R_1)(R_3 - jX_{C_3}) = \left(\frac{1}{(1/R_2) + j\omega C_2}\right)(R_4).$$

Rearranging gives

$$\left(R_3 - \frac{j}{\omega C_3}\right)\left(\frac{1}{R_2} + j\omega C_2\right) = \frac{R_4}{R_1}$$

$$\frac{R_3}{R_2} + \frac{C_2}{C_3} - j\left(\frac{1}{\omega C_3 R_2}\right) + j\omega C_2 R_3 = \frac{R_4}{R_1}$$

Equating real parts gives

$$\frac{R_3}{R_2} + \frac{C_2}{C_3} = \frac{R_4}{R_1} \tag{4}$$

Equating imaginary parts gives

$$-\frac{1}{\omega C_3 R_2} + \omega C_2 R_3 = 0$$

i.e.,

$$\omega C_2 R_3 = \frac{1}{\omega C_3 R_2}$$

from which

$$\omega^2 = \frac{1}{C_2 C_3 R_2 R_3}$$

Since $\omega = 2\pi f$,

$$\textbf{frequency, } f = \frac{1}{2\pi\sqrt{(C_2 C_3 R_2 R_3)}} \tag{5}$$

Note that if $C_2 = C_3 = C$ and $R_2 = R_3 = R$,

$$\text{frequency, } f = \frac{1}{2\pi\sqrt{(C^2 R^2)}} = \frac{1}{2\pi CR}$$

Worked problems on a.c. bridges

Problem 1. The a.c. bridge shown in Fig. 10 is used to measure the capacitance C_x and resistance R_x. (a) Derive the balance equations of the bridge. (b) Given $R_3 = R_4$, $C_2 = 0.2\,\mu F$, $R_2 = 2.5\,k\Omega$ and the frequency of the supply is 1 kHz, determine the values of R_x and C_x at balance.

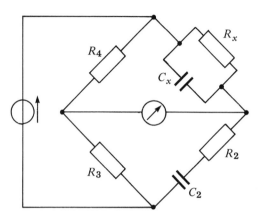

Figure 10

(a) Since C_x and R_x are the unknown values and are connected in parallel, it is easier to use the reciprocal impedance form for this branch $\left(\text{rather than } \dfrac{\text{product}}{\text{sum}} \right)$, i.e.,

$$\frac{1}{Z_x} = \frac{1}{R_x} + \frac{1}{-jX_{C_x}} = \frac{1}{R_x} + \frac{j}{X_{C_x}}$$

from which

$$Z_x = \frac{1}{(1/R_x) + j\omega C_x}$$

From Fig. 10, $Z_2 = R_2 - jX_{C_2}$, $Z_3 = R_3$ and $Z_4 = R_4$.
At balance, $(Z_x)(Z_3) = (Z_2)(Z_4)$

$$\left(\frac{1}{(1/R_x) + j\omega C_x} \right)(R_3) = (R_2 - jX_{C_2})(R_4)$$

$$\frac{R_3}{R_4(R_2 - jX_{C_2})} = \frac{1}{R_x} + j\omega C_x$$

Rationalising gives

$$\frac{R_3(R_2 + jX_{C_2})}{R_4(R_2^2 + X_{C_2}^2)} = \frac{1}{R_x} + j\omega C_x$$

Hence

$$\frac{1}{R_x} + j\omega C_x = \frac{R_3 R_2}{R_4(R_2^2 + (1/\omega^2 C_2^2)} + \frac{jR_3(1/\omega C_2)}{R_4(R_2^2 + (1/\omega^2 C_2^2))}$$

Equating the real parts gives

$$\frac{1}{R_x} = \frac{R_3 R_2}{R_4(R_2^2 + (1/\omega^2 C_2^2)}$$

i.e.,

$$R_x = \frac{R_4}{R_2 R_3}\left(\frac{R_2^2 \omega^2 C_2^2 + 1}{\omega^2 C_2^2}\right)$$

and

$$R_x = \frac{R_4(1 + \omega^2 C_2^2 R_2^2)}{R_2 R_3 \omega^2 C_2^2}$$

Equating the imaginary parts gives

$$\omega C_x = \frac{R_3(1/\omega C_2)}{R_4(R_2^2 + (1/\omega^2 C^2))}$$

$$= \frac{R_3}{\omega C_2 R_4((R_2^2 \omega^2 C_2^2 + 1)/\omega^2 C_2^2)}$$

i.e.,

$$\omega C_x = \frac{R_3 \omega^2 C_2^2}{\omega C_2 R_4(1 + \omega^2 C_2^2 R_2^2)}$$

and

$$C_x = \frac{R_3 C_2}{R_4(1 + \omega^2 C_2^2 R_2^2)}$$

(b) Substituting the given values gives

$$R_x = \frac{(1 + \omega^2 C_2^2 R_2^2)}{R_2 \omega^2 C_2^2} \quad \text{since } R_3 = R_4$$

i.e.,

$$R_x = \frac{1 + (2\pi 1000)^2 (0.2 \times 10^{-6})^2 (2.5 \times 10^3)^2}{(2.5 \times 10^3)(2\pi 1000)^2 (0.2 \times 10^{-6})^2}$$

$$= \frac{1 + 9.8696}{3.9478 \times 10^{-3}} \equiv 2.75\,\text{k}\Omega$$

$$C_x = \frac{C_2}{(1 + \omega^2 C_2^2 R_2^2)} \quad \text{since } R_3 = R_4$$

$$= \frac{(0.2 \times 10^{-6})}{1 + 9.8696}\,\mu\text{F} = 0.01840\,\mu\text{F} \text{ or } 18.40\,\text{nF}$$

Hence at balance R_x = 2.75 kW and C_x = 18.40 nF.

Problem 2. For the Wien bridge shown in Fig. 9, $R_2 = R_3 = 30\,\text{k}\Omega$, $R_4 = 1\,\text{k}\Omega$ and $C_2 = C_3 = 1\,\text{nF}$. Determine, when the bridge is balanced, (a) the value of resistance R_1, and (b) the frequency of the bridge.

(a) From equation (4),

$$\frac{R_3}{R_2} + \frac{C_2}{C_3} = \frac{R_4}{R_1}$$

i.e., $1 + 1 = 1000/R_1$, since $R_2 = R_3$ and $C_2 = C_3$, from which

$$\text{resistance } R_1 = \frac{1000}{2} = \textbf{500}\,\boldsymbol{\Omega}$$

(b) From equation (5),

$$\text{frequency, } f = \frac{1}{2\pi\sqrt{(C_2 C_3 R_2 R_3)}} = \frac{1}{2\pi\sqrt{[(10^{-9})^2(30 \times 10^3)^2]}}$$

$$= \frac{1}{2\pi(10^{-9})(30 \times 10^3)} \equiv \textbf{5.305 kHz}$$

Problem 3. A Schering bridge network is as shown in Fig. 7. Given $C_2 = 0.2\,\mu\text{F}$, $R_4 = 200\,\Omega$, $R_3 = 600\,\Omega$, $C_3 = 4000\,\text{pF}$ and the supply frequency is 1.5 kHz, determine, when the bridge is balanced, (a) the value of resistance R_x, (b) the value of capacitance C_x, (c) the phase angle of the unknown arm, (d) the power factor of the unknown arm and (e) its loss angle.

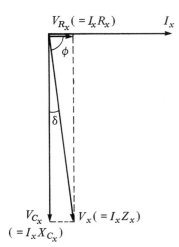

Figure 11

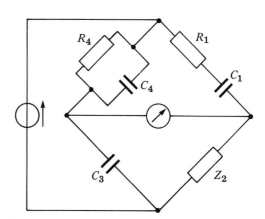

Figure 12

From para (f), the equation for R_x and C_x at balance are given by

$$R_x = \frac{R_4 C_3}{C_2} \quad \text{and} \quad C_x = \frac{C_2 R_3}{R_4}$$

(a) Resistance, $R_x = \dfrac{R_4 C_3}{C_2} = \dfrac{(200)(4000 \times 10^{-12})}{(0.2 \times 10^{-6})} = \textbf{4}\,\boldsymbol{\Omega}$

(b) Capacitance, $C_x = \dfrac{C_2 R_3}{R_4} = \dfrac{(0.2 \times 10^{-6})(600)}{(200)}\,\text{F} = \textbf{0.6}\,\boldsymbol{\mu}\textbf{F}$

(c) The phasor diagram for R_x and C_x in series is shown in Fig. 11.

Phase angle, $\phi = \arctan\dfrac{V_{C_x}}{V_{R_x}} = \arctan\dfrac{I_x X_{Cx}}{I_x R_x} = \arctan\dfrac{1}{\omega C_x R_x}$

i.e.,

$$\phi = \arctan\left(\frac{1}{(2\pi 1500)(0.6 \times 10^{-6})(4)}\right) = \arctan 44.21 = \textbf{88.7}^\circ \textbf{ lead.}$$

(d) Power factor of capacitor $= \cos\phi = \cos 88.7^\circ = \textbf{0.0227}$

(e) Loss angle, shown as δ in Fig. 11, is given by $\delta = 90^\circ - 88.7^\circ = \textbf{1.3}^\circ$.

Alternatively, loss angle $\delta = \arctan \omega C_x R_x$ (see para (f), page 80)

$$= \arctan\left(\frac{1}{44.21}\right) \quad \text{from (c) above,}$$

i.e.,

$$\delta = 1.3°$$

Problem 4. An a.c. bridge is shown in Fig. 12, where Z_2 is an unknown impedance. The components of the bridge have the following values at balance: $R_1 = 1.5\,\text{k}\Omega$, $C_1 = 0.8\,\mu\text{F}$, $C_3 = 0.4\,\mu\text{F}$, $R_4 = 3\,\text{k}\Omega$ and $C_4 = 0.08\,\mu\text{F}$. The supply frequency is 800 Hz. Determine, for the balance condition, the components comprising impedance Z_2 if they are series connected.

The equivalent impedance of R_4 in parallel with C_4 is given by

$$\frac{1}{Z_4} = \frac{1}{R_4} + \frac{1}{-jX_{C_4}} = \frac{1}{R_4} + \frac{j}{X_{C_4}},$$

from which

$$Z_4 = \frac{1}{(1/R_4) + (j/X_{C_4})}$$

$\left(\text{Alternatively } Z_4 \text{ may be obtained from } \dfrac{\text{product}}{\text{sum}}\right)$

From Fig. 12, $Z_1 = (R_1 - jX_{C_1})$ and $Z_3 = -jX_{C_3}$. At balance, $(Z_1)(Z_3) = (Z_2)(Z_4)$ from equation (3), i.e.,

$$(R_1 - jX_{C_1})(-jX_{C_3}) = (Z_2)\left(\frac{1}{(1/R_4) + (j/X_{C_4})}\right)$$

from which

$$Z_2 = (R_1 - jX_{C_1})(-jX_{C_3})\left(\frac{1}{R_4} + \frac{j}{X_{C_4}}\right)$$

$$= (-jR_1 X_{C_3} - X_{C_1} X_{C_3})\left(\frac{1}{R_4} + \frac{j}{X_{C_4}}\right)$$

$$= \frac{-jR_1 X_{C_3}}{R_4} + \frac{R_1 X_{C_3}}{X_{C_4}} - \frac{X_{C_1} X_{C_3}}{R_4} - j\frac{X_{C_1} X_{C_3}}{X_{C_4}}$$

$$= \left(\frac{R_1 X_{C_3}}{X_{C_4}} - \frac{X_{C_1} X_{C_3}}{R_4}\right) - j\left(\frac{R_1 X_{C_3}}{R_4} + \frac{X_{C_1} X_{C_3}}{X_{C_4}}\right)$$

i.e.,

$$Z_2 = \left(\frac{R_1 C_4}{C_3} - \frac{1}{\omega^2 C_1 C_3 R_4}\right) - j\left(\frac{R_1}{\omega C_3 R_4} + \frac{C_4}{\omega C_1 C_3}\right)$$

Substituting in given values gives

real part of impedance Z_2

$$= \left(\frac{(1500(0.08 \times 10^{-6})}{(0.4 \times 10^{-6})} - \frac{1}{(2\pi 800)^2 (0.8 \times 10^{-6})(0.4 \times 10^{-6})(3000)}\right)$$

$$= (300 - 41.23) = \mathbf{258.8\,\Omega}$$

imaginary part of impedance Z_2

$$= -\left(\frac{1500}{(2\pi800)(0.4 \times 10^{-6})(3000)} + \frac{0.08 \times 10^{-6}}{(2\pi800)(0.8 \times 10^{-6})(0.4 \times 10^{-6})}\right)$$

$$= -(248.68 + 49.74) = -298.4\,\Omega$$

Hence impedance $Z_2 = (258.8 - j298.4)\,\Omega$, i.e., resistance $= 258.8\,\Omega$ and capacitive reactance, $X_C = 298.4\,\Omega$. Thus

$$298.4 = \frac{1}{2\pi fC}$$

from which

$$\text{capacitance, } C = \frac{1}{2\pi(800)(298.4)} \equiv 0.667\,\mu F$$

Thus impedance Z_2 is comprised of a 258.8 Ω resistance in series with a 0.667 μF capacitor.

Further problems on a.c. bridges may be found in the following section (4), problems 1 to 15.

4. Further problems

1. A Maxwell–Wien bridge circuit $ABCD$ has the following arm impedances: AB, 250 Ω resistance; BC, 2 μF capacitor in parallel with a 10 kΩ resistor; CD, 400 Ω resistor; DA, unknown inductor having inductance L in series with resistance R. Determine the values of L and R if the bridge is balanced.
   $[L = 0.20\,\text{H}, R = 10\,\Omega]$

2. In a four-arm de Sauty a.c. bridge, arm 1 contains a 2 kΩ non-inductive resistor, arm 3 contains a loss-free 2.4 μF capacitor, and arm 4 contains a 5 kΩ non-inductive resistor. When the bridge is balanced, determine the value of the capacitor contained in arm 2.
   $[6\mu F]$

3. A four-arm bridge $ABCD$ consists of: AB—fixed resistor R_1; BC—variable resistor R_2 in series with a variable capacitor C_2; CD—fixed resistor R_3; DA—coil of unknown resistance R and inductance L. Determine the values of R and L if, at balance, $R_1 = 1\,\text{k}\Omega$, $R_2 = 2.5\,\text{k}\Omega$, $C_2 = 4000\,\text{pF}$, $R_3 = 1\,\text{k}\Omega$ and the supply frequency is 1.6 kHz.
   $[R = 4.00\,\Omega, L = 3.96\,\text{mH}]$

4. The bridge shown in Fig. 13 is used to measure capacitance C_x and resistance R_x. Derive the balance equations of the bridge and determine the values of C_x and R_x when $R_1 = R_4$, $C_2 = 0.1\,\mu F$, $R_2 = 2\,\text{k}\Omega$ and the supply frequency is 1 kHz.
   $[C_x = 38.77\,\text{nF}, R_x = 3.27\,\text{k}\Omega]$

5. In a Schering bridge network $ABCD$, the arms are made up as follows: AB—a standard capacitor C_1; BC—a capacitor C_2 in parallel with a resistor R_2; CD—a resistor R_3; DA—the capacitor under test, represented by a capacitor C_x in series with a resistor R_x. The detector is connected between B and D and the a.c. supply is connected between A and C. Derive the equations for R_x and C_x when the bridge is balanced. Evaluate R_x and C_x if, at balance, $C_1 = 1\,\text{nF}$, $R_2 = 100\,\Omega$, $R_3 = 1\,\text{k}\Omega$ and $C_2 = 10\,\text{nF}$.
   $[R_x = 10\,\text{k}\Omega, C_x = 100\,\text{pF}]$

6. The a.c. bridge shown in Fig. 14 is balanced when the values of the components are shown. Determine, at balance, the values of R_x and L_x.
   $[R_x = 2\,\text{k}\Omega, L_x = 0.2\,\text{H}]$

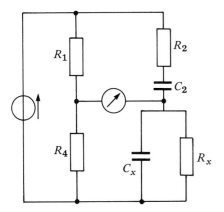

Figure 13

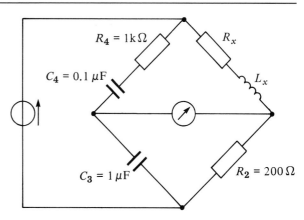

Figure 14

7. An a.c. bridge has, in arm AB, a pure capacitor of $0.4\,\mu\text{F}$; in arm BC, a pure resistor of $500\,\Omega$; in arm CD, a coil of $50\,\Omega$ resistance and $0.1\,\text{H}$ inductance; in arm DA, an unknown impedance comprising resistance R_x and capacitance C_x in series. If the frequency of the bridge at balance is $800\,\text{Hz}$, determine the values of R_x and C_x.

$$[R_x = 500\,\Omega, C_x = 4\,\mu\text{F}]$$

8. When the Wien bridge shown in Fig. 9 is balanced, the components have the following values: $R_2 = R_3 = 20\,\text{k}\Omega$, $R_4 = 500\,\Omega$, $C_2 = C_3 = 800\,\text{pF}$. Determine for the balance condition (a) the value of resistance R_1 and (b) the frequency of the bridge supply.

$$[\text{(a) } 250\,\Omega \quad \text{(b) } 9.95\,\text{kHz}]$$

9. An a.c. bridge $ABCD$ is balanced with the following component values: AB—a $2\,\text{k}\Omega$ resistor in parallel with a $0.06\,\mu\text{F}$ capacitor; BC—a $1\,\text{k}\Omega$ resistor in series with a $0.6\,\mu\text{F}$ capacitor; CD—an unknown impedance; DA—a $0.3\,\mu\text{F}$ capacitor. Determine the series components contained in arm CD if the supply frequency is $1\,\text{kHz}$.

$$[R = 129.6\,\Omega, C = 0.50\,\mu\text{F}]$$

10. The conditions at balance of a Schering bridge $ABCD$ used to measure the capacitance and loss angle of a paper capacitor are as follows: AB—a pure capacitance of $0.2\,\mu\text{F}$; BC—a pure capacitance of $3000\,\text{pF}$ in parallel with a $400\,\Omega$ resistance; CD—a pure resistance of $200\,\Omega$; DA—the capacitance under test which may be considered as a capacitance C_x in series with a resistance R_x. If the supply frequency is $1\,\text{kHz}$ determine (a) the value of R_x, (b) the value of C_x, (c) the power factor of the capacitor, and (d) its loss angle.

$$[\text{(a) } 3\,\Omega \quad \text{(b) } 0.4\,\mu\text{F} \quad \text{(c) } 0.0075 \quad \text{(d) } 0.432°]$$

11. At balance, an a.c. bridge $PQRS$ used to measure the inductance and resistance of an inductor has the following values: PQ—a non-inductive $400\,\Omega$ resistor; QR—the inductor with unknown inductance L_x in series with resistance R_x; RS—a $3\,\mu\text{F}$ capacitor in series with a non-inductive $250\,\Omega$ resistor; SP—a $15\,\mu\text{F}$ capacitor. A detector is connected between P and R and the a.c. supply is connected between Q and S. Derive the balance equations for R_x and L_x and determine their values.

$$[R_x = 2\,\text{k}\Omega, L_x = 1.5\,\text{H}]$$

12. A $1\,\text{kHz}$ a.c. bridge $ABCD$ has the following components in its four arms: AB—a pure capacitor of $0.2\,\mu\text{F}$; BC—a pure resistance of $500\,\Omega$; CD—an unknown

impedance; DA—a $400\,\Omega$ resistor in parallel with a $0.1\,\mu\text{F}$ capacitor. If the bridge is balanced, determine the series components comprising the impedance in arm CD.

$$[R = 59.41\,\Omega, L = 37.6\,\text{mH}]$$

13. An a.c. bridge $PQRS$ has the following components in its four arms: PQ—a $0.2\,\mu\text{F}$ capacitor in parallel with a resistance R_1; QR—a pure $2\,\text{k}\Omega$ resistor; RS—a coil of inductance $0.20\,\text{H}$ and resistance $200\,\Omega$; SP—an unknown resistor R_4.
    (a) Determine the values of R_1 and R_4 when the bridge is balanced.
    (b) If resistance R_1 in arm PQ is connected in series with the $0.2\,\mu\text{F}$ capacitor, determine the values of R_1 and R_4 when the bridge is balanced and the frequency is $1\,\text{kHz}$.

$$\left[\begin{array}{l} \text{(a) } R_1 = 5000\,\Omega, R_4 = 500\,\Omega \\ \text{(b) } R_1 = 126.7\,\Omega, R_4 = 512.7\,\Omega \end{array}\right]$$

14. An a.c. bridge $ABCD$ has in arm AB a standard lossless capacitor of $200\,\text{pF}$; arm BC, an unknown impedance, represented by a lossless capacitor C_x in series with a resistor R_x; arm CD, a pure $5\,\text{k}\Omega$ resistor; arm DA, a $6\,\text{k}\Omega$ resistor in parallel with a variable capacitor set at $250\,\text{pF}$. The frequency of the bridge supply is $1500\,\text{Hz}$. Determine for the condition when the bridge is balanced (a) the values of R_x and C_x, and (b) the loss angle.

$$[\text{(a) } R_x = 6.25\,\text{k}\Omega, C_x = 240\,\text{pF}; \quad \text{(b) } 0.81°]$$

15. An a.c. bridge $ABCD$ has the following constants: AB—a $1\,\text{k}\Omega$ resistance in parallel with a $0.2\,\mu\text{F}$ capacitor; BC—a $1.2\,\text{k}\Omega$ resistance; CD—a $750\,\Omega$ resistance; DA—a $0.8\,\mu\text{F}$ capacitor in series with an unknown resistance. Determine (a) the value of the unknown resistance in arm DA to produce balance, and (b) the frequency for which the bridge is in balance.

$$[\text{(a) } 375\,\Omega \quad \text{(b) } 649.7\,\text{Hz}]$$

6 Series and parallel resonance and Q-factor

1. Introduction

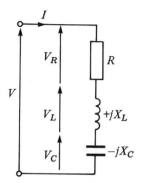

Figure 1 R–L–C series circuit

When the voltage V applied to an electrical network containing resistance, inductance and capacitance is in phase with the resulting current I, the circuit is said to be **resonant.** The phenomenon of **resonance** is of great value in all branches of radio, television and communications engineering, since it enables small portions of the communications frequency spectrum to be selected for amplificaton independently of the remainder.

At resonance, the equivalent network impedance Z is purely resistive since the supply voltage and current are in phase. The power factor of a resonant network is unity, (i.e., power factor $= \cos \phi = \cos 0 = 1$).

In electrical work there are two outstanding types of resonance—one associated with series circuits when the input impedance is a minimum, and the other associated with simple parallel networks, when the input impedance is a maximum.

2. Series resonance

Figure 1 shows a circuit comprising a coil of inductance L and resistance R connected in series with a capacitor C. The R–L–C series circuit has a total impedance Z given by $Z = R + j(X_L - X_C)$ ohms, or

$$Z = R + j\left(\omega L - \frac{1}{\omega C} \right) \text{ohms}$$

where $\omega = 2\pi f$. The circuit is at resonance when $(X_L - X_C) = 0$, i.e., when $X_L = X_C$ or $\omega L = 1/(\omega C)$. The phasor diagram for this condition is shown in Fig. 2, where $|V_L| = |V_C|$.

Since at resonance $\omega_r L = 1/(\omega_r C)$,

$$\omega_r^2 = \frac{1}{LC} \quad \text{and} \quad \omega_r = \frac{1}{\sqrt{(LC)}}$$

Thus resonant frequency,

$$\boxed{f_r = \frac{1}{2\pi \sqrt{(LC)}} \text{ hertz,}} \quad \text{since } \omega_r = 2\pi f_r$$

Figure 3 shows how inductive reactance X_L and capacitive reactance X_C vary with frequency. At the resonant frequency f_r, $|X_L| = |X_C|$. Since impedance $Z = R + j(X_L - X_C)$ and, at resonance, $(X_L - X_C) = 0$, then **impedance $Z = R$ at resonance**. This is the **minimum** value possible for the

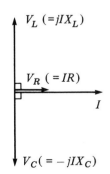

Figure 2 Phasor diagram $|V_L| = |V_C|$

Figure 3 Variation of X_L and X_C with frequency

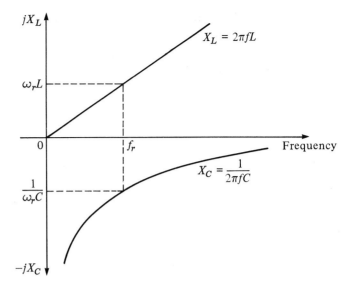

impedance as shown in the graph of the modulus of impedance, $|Z|$, against frequency in Fig. 4.

At frequencies less than f_r, $X_L < X_C$ and the circuit is capacitive; at frequencies greater than f_r, $X_L > X_C$ and the circuit is inductive.

Current $I = V/Z$. Since impedance Z is a minimum value at resonance, the **current I then has a maximum value**. At resonance, current $I = V/R$. A graph of current against frequency is shown in Fig. 4.

Figure 4 $|Z|$ and I plotted against frequency

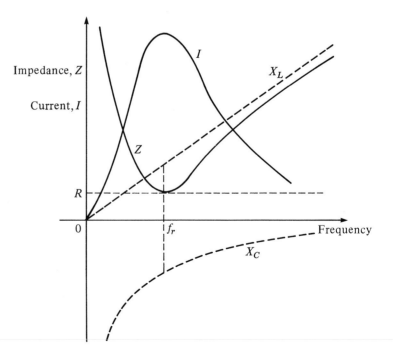

Worked problems on series resonance

Problem 1. A coil having a resistance of $10\,\Omega$ and an inductance of $75\,\text{mH}$ is connected in series with a $40\,\mu\text{F}$ capacitor across a $200\,\text{V}$ a.c. supply. Determine (a) at what frequency resonance occurs, and (b) the current flowing at resonance.

(a) Resonant frequency, $f_r = \dfrac{1}{2\pi\sqrt{(LC)}} = \dfrac{1}{2\pi\sqrt{[(75\times 10^{-3})(40\times 10^{-6})]}}$

i.e.,

$$f_r = 91.9\,\text{Hz}$$

(b) Current at resonance, $I = \dfrac{V}{R} = \dfrac{200}{10} = 20\,\text{A}.$

Problem 2. An R–L–C series circuit is comprised of a coil of inductance $10\,\text{mH}$ and resistance $8\,\Omega$ and a variable capacitor C. The supply frequency is $1\,\text{kHz}$. Determine the value of capacitor C for series resonance.

At resonance, $\omega_r L = 1/(\omega_r C)$, from which capacitance, $C = 1/(\omega_r^2 L)$. Hence

$$\text{capacitance } C = \frac{1}{(2\pi 1000)^2(10\times 10^{-3})} = 2.53\,\mu\text{F}$$

Problem 3. A coil having inductance L is connected in series with a variable capacitor C. The circuit possesses stray capacitance C_s which is assumed to be constant and effectively in parallel with the variable capacitor C. When the capacitor is set to $1000\,\text{pF}$ the resonant frequency of the circuit is $92.5\,\text{kHz}$, and when the capacitor is set to $500\,\text{pF}$ the resonant frequency is $127.8\,\text{kHz}$. Determine the values of (a) the stray capacitance C_s, and (b) the coil inductance L.

For a series R–L–C circuit the resonant frequency f_r is given by

$$f_r = \frac{1}{2\pi\sqrt{(LC)}}$$

The total capacitance of C in parallel with C_s is given by $(C + C_s)$. At $92.5\,\text{kHz}$, $C = 1000\,\text{pF}$. Hence

$$92.5\times 10^3 = \frac{1}{2\pi\sqrt{[L(1000 + C_s)10^{-12}]}} \qquad (1)$$

At $127.8\,\text{kHz}$, $C = 500\,\text{pF}$. Hence

$$127.8\times 10^3 = \frac{1}{2\pi\sqrt{[L(500 + C_s)10^{-12}]}} \qquad (2)$$

(a) Dividing equation (2) by equation (1) gives

$$\frac{127.8\times 10^3}{92.5\times 10^3} = \frac{1/(2\pi\sqrt{[L(500 + C_s)10^{-12}]})}{1/(2\pi\sqrt{[L(1000 + C_s)10^{-12}]})}$$

i.e.,

$$\frac{127.8}{92.5} = \frac{\sqrt{[L(1000 + C_s)10^{-12}]}}{\sqrt{[L(500 + C_s)10^{-12}]}} = \sqrt{\left(\frac{1000 + C_s}{500 + C_s}\right)}$$

where C_s is in picofarads, from which

$$\left(\frac{127.8}{92.5}\right)^2 = \frac{1000 + C_s}{500 + C_s}$$

i.e.,

$$1.909 = \frac{1000 + C_s}{500 + C_s}$$

Hence

$$1.909(500 + C_s) = 1000 + C_s$$
$$954.5 + 1.909\,C_s = 1000 + C_s$$
$$0.909\,C_s = 1000 - 954.5 = 45.5$$

Thus **stray capacitance $C_s = 45.5/0.909 = 50\,\text{pF}$**.

(b) Substituting $C_s = 50\,\text{pF}$ in equation (1) gives

$$92.5 \times 10^3 = \frac{1}{2\pi\sqrt{[L(1050 \times 10^{-12})]}}$$

Hence

$$(92.5 \times 10^3 \times 2\pi)^2 = \frac{1}{L(1050 \times 10^{-12})}$$

from which,

$$\textbf{inductance, } L = \frac{1}{(1050 \times 10^{-12})(92.5 \times 10^3 \times 2\pi)^2}\,\text{H} = \textbf{2.82\,mH}$$

Further problems on series resonance may be found in section 5, problems 1 to 6, page 110.

3. Q-factor and bandwidth

Q-factor is a figure of merit for a resonant device such as an L–C–R circuit. Such a circuit resonates by cyclic interchange of stored energy, accompanied by energy dissipation due to the resistance.

By definition, at resonance

$$Q = 2\pi\left(\frac{\text{maximum energy stored}}{\text{energy loss per cycle}}\right)$$

Since the energy loss per cycle is equal to the average power dissipated $\times$ periodic time,

$$Q = 2\pi\left(\frac{\text{maximum energy stored}}{\text{average power dissipated} \times \text{periodic time}}\right)$$

$$= 2\pi\left(\frac{\text{maximum energy stored}}{\text{average power dissipated} \times (1/f_r)}\right)$$

since the periodic time $T = 1/f_r$. Thus

$$Q = 2\pi f_r\left(\frac{\text{maximum energy stored}}{\text{average power dissipated}}\right)$$

i.e.,

$$Q = \omega_r\left(\frac{\text{maximum energy stored}}{\text{average power dissipated}}\right)$$

where ω_r is the angular frequency at resonance.

In an *L–C–R* circuit both of the reactive elements store energy during a quarter cycle of the alternating supply input and return it to the circuit source during the following quarter cycle. An inductor stores energy in its magnetic field, then transfers it to the electric field of the capacitor and then back to the magnetic field, and so on. Thus the inductive and capacitive elements transfer energy from one to the other successively with the source of supply ideally providing no additional energy at all. Practical reactors both store and dissipate energy.

Q-factor is an abbreviation for quality factor and refers to the "goodness" of a reactive component.

For an **inductor**,

$$Q = \omega_r \left(\frac{\text{maximum energy stored}}{\text{average power dissipated}} \right)$$

$$= \omega_r \left(\frac{\frac{1}{2}LI_m^2}{I^2R} \right) = \frac{\omega_r(\frac{1}{2}LI_m^2)}{(I_m/\sqrt{2})^2} = \frac{\omega_r L}{R} \tag{1}$$

For a **capacitor**,

$$Q = \frac{\omega_r(\frac{1}{2}CV_m^2)}{(I_m/\sqrt{2})^2 R} = \frac{\omega_r \frac{1}{2}C(I_m X_C)^2}{(I_m/\sqrt{2})^2 R} = \frac{\omega_r \frac{1}{2}CI_m^2 1/\omega C^2}{(I_m/\sqrt{2})^2 R}$$

i.e.,

$$Q = \frac{1}{\omega_r CR} \tag{2}$$

From expressions (1) and (2) it can be deduced that

$$Q = \frac{X_L}{R} = \frac{X_C}{R} = \frac{\text{reactance}}{\text{resistance}}$$

In fact, *Q*-factor can also be defined as

$$Q\text{-factor} = \frac{\text{reactive power}}{\text{active power}} = \frac{Q}{P}$$

where *Q* is the reactive power which is also the peak rate of energy storage, and *P* is the average energy dissipation rate. Hence

$$Q\text{-factor} = \frac{Q}{P} = \frac{I^2 X_L(\text{or } I^2 X_C)}{I^2 R} = \frac{X_L}{R} \left(\text{or } \frac{X_C}{R} \right)$$

i.e.,

$$Q = \frac{\textbf{reactance}}{\textbf{resistance}}$$

In an *R–L–C* series circuit the amount of energy stored at resonance is constant. When the capacitor voltage is a maximum, the inductor current is zero, and *vice versa*, i.e., $\frac{1}{2}LI_m^2 = \frac{1}{2}CV_m^2$.

Thus the *Q*-factor at resonance, Q_r, is given by

$$\boxed{Q_r = \frac{\omega_r L}{R} = \frac{1}{\omega_r CR}} \tag{3}$$

Figure 5 (a) High Q-factor
(b) Low Q-factor

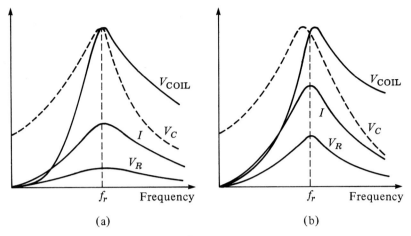

(a)   (b)

However, at resonance $\omega_r = 1/\sqrt{(LC)}$. Hence

$$Q_r = \frac{\omega_r L}{R} = \frac{1}{\sqrt{(LC)}}\left(\frac{L}{R}\right) = \frac{1}{R}\sqrt{\left(\frac{L}{C}\right)}$$

It should be noted that when Q-factor is referred to, it is nearly always assumed to mean "the Q-factor at resonance".

With reference to Figs. 1 and 2, at resonance, $V_L = V_C$.

$$V_L = IX_L = I\omega_r L = \frac{V}{R}\omega_r L = \left(\frac{\omega_r L}{R}\right)V = Q_r V$$

and

$$V_C = IX_C = \frac{1}{\omega_r C} = \frac{V/R}{\omega_r C} = \left(\frac{1}{\omega_r CR}\right)V = Q_r V$$

Hence, at resonance, $V_L = V_C = Q_r V$ or

$$\boxed{Q_r = \frac{V_L(\text{or } V_C)}{V}}$$

The voltages V_L and V_C at resonance may be much greater than that of the supply voltage V. For this reason Q is often called the **circuit magnification factor.** It represents a measure of the number of times V_L or V_C is greater than the supply voltage.

The Q-factor at resonance can have a value of several hundreds. Resonance is usually of interest only in circuits of Q-factor greater than about 10; circuits considerably below this value are effectively merely operating at unity power factor.

For a circuit with a high value of Q (say, exceeding 100), the maximum volt-drop across the coil, V_{COIL}, and the maximum volt-drop across the capacitor, V_C, coincide with the maximum circuit current at the resonant frequency f_r, as shown in Fig. 5(a). However, if a circuit of low Q (say, less than 10) is used, it may be shown experimentally that the maximum value of V_C occurs at a frequency less than f_r while the maximum value of V_{COIL} occurs at a frequency higher than f_r, as shown in Fig. 5(b). The maximum current, however, still occurs at the resonant frequency with low Q.

Bandwidth

Figure 6 shows how current I varies with frequency f in an R–L–C series circuit. At the resonant frequency f_r, current is a maximum value, shown as I_r. Also shown are the points A and B where the current is 0.707 of the maximum value at frequencies f_1 and f_2. The power delivered to the circuit is $I^2 R$. At $I = 0.707\,I_r$, the power is $(0.707\,I_r)^2 R = 0.5\,I_r^2 R$, i.e., half the power that occurs at frequency f_r. The points corresponding to f_1 and f_2 are called the **half-power points**. The distance between these points, i.e., $(f_2 - f_1)$, is called the **bandwidth**

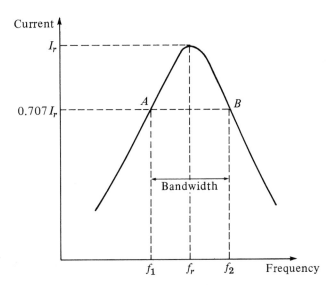

Figure 6 Bandwidth and half-power points f_1, f_2

When the ratio of two powers P_1 and P_2 is expressed in decibel units, the number of decibels N is given by

$$N = 10\lg\left(\frac{P_2}{P_1}\right) dB$$

Let the power at the half-power points be $(0.707\,I_r)^2 R = \dfrac{I_r^2 R}{2}$ and let the peak power be $I_r^2 R$. Then the ratio of the power in decibels is given by

$$10\lg\left\{\frac{I_r^2 R/2}{I_r^2 R}\right\} = 10\lg\tfrac{1}{2} = -\textbf{3 dB}$$

It is for this reason that the half-power points are often referred to as **"the −3 db points"**.

At the half-power frequencies, $I = 0.707\,I_r$, thus

$$\text{impedance } Z = \frac{V}{I} = \frac{V}{0.707\,I_r} = 1.414\left(\frac{V}{I_r}\right) = \sqrt{2}\,Z_r = \sqrt{2}\,R$$

(since $Z_r = R$ at resonance).

Figure 7 (a) Inductive impedance triangle (b) Capacitive impedance triangle

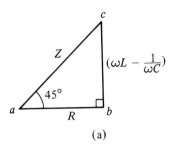

(a)

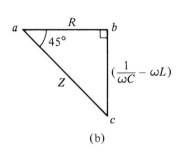

(b)

Since $Z = \sqrt{2}\,R$, an isosceles triangle is formed by the impedance triangles, as shown in Fig. 7, where $ab = bc$. From the impedance triangles it can be seen that the equivalent circuit reactance is equal to the circuit resistance at the half-power points.

At f_1, the lower half-power frequency $|X_C| > |X_L|$ (see Fig. 4). Thus

$$\frac{1}{2\pi f_1 C} - 2\pi f_1 L = R$$

from which

$$1 - 4\pi^2 f_1^2 LC = 2\pi f_1 CR$$

i.e.,

$$(4\pi^2 LC)f_1^2 + (2\pi CR)f_1 - 1 = 0$$

This is a quadratic equation in f_1. Using the quadratic formula gives

$$f_1 = \frac{-(2\pi CR) \pm \sqrt{[(2\pi CR)^2 - 4(4\pi^2 LC)(-1)]}}{2(4\pi^2 LC)}$$

$$= \frac{-(2\pi CR) \pm \sqrt{(4\pi^2 C^2 R^2 + 16\pi^2 LC)}}{8\pi^2 LC}$$

$$= \frac{-(2\pi CR) \pm \sqrt{[4\pi^2 C^2 (R^2 + (4L/C))]}}{8\pi^2 LC}$$

$$= \frac{-(2\pi CR) \pm 2\pi C\sqrt{(R^2 + (4L/C))}}{8\pi^2 LC}$$

Hence

$$f_1 = \frac{-R \pm \sqrt{(R^2 + (4L/C))}}{4\pi L} = \frac{-R + \sqrt{(R^2 + (4L/C))}}{4\pi L},$$

(since $\sqrt{(R^2 + (4L/C))} > R$ and f_1 cannot be negative).

At f_2, the upper half-power frequency $|X_L| > |X_C|$ (see Fig. 4). Thus

$$2\pi f_2 L - \frac{1}{2\pi f_2 C} = R$$

from which,

$$4\pi^2 f_2^2 LC - 1 = R(2\pi f_2 C)$$

i.e.,

$$(4\pi^2 LC)f_2^2 - (2\pi CR)f_2 - 1 = 0$$

This is a quadratic equation in f_2 and may be solved using the quadratic formula as for f_1, giving

$$f_2 = \frac{R + \sqrt{(R^2 + (4L/C))}}{4\pi L}$$

$$\text{Bandwidth} = (f_2 - f_1) = \left\{ \frac{R + \sqrt{(R^2 + (4L/C))}}{4\pi L} \right\}$$

$$- \left\{ \frac{-R + \sqrt{(R^2 + (4L/C))}}{4\pi L} \right\}$$

$$= \frac{2R}{4\pi L} = \frac{R}{2\pi L} = \frac{1}{2\pi L/R} = \frac{f_r}{2\pi f_r L/R} = \frac{f_r}{Q_r}$$

from equation (3). Hence for a series R–L–C circuit

$$\boxed{Q_r = \frac{f_r}{f_2 - f_1}}$$

(4)

An alternative equation involving f_r

At the lower half-power frequency f_1: $\dfrac{1}{\omega_1 C} - \omega_1 L = R.$

At the higher half-power frequency f_2: $\omega_2 L - \dfrac{1}{\omega_2 C} = R.$

Equating gives $\dfrac{1}{\omega_1 C} - \omega_1 L = \omega_2 L - \dfrac{1}{\omega_2 C}$

Multiplying throughout by C gives $\dfrac{1}{\omega_1} - \omega_1 LC = \omega_2 LC - \dfrac{1}{\omega_2}$

However, for series resonance, $\omega_r^2 = 1/(LC)$. Hence

$$\frac{1}{\omega_1} - \frac{\omega_1}{\omega_r^2} = \frac{\omega_2}{\omega_r^2} - \frac{1}{\omega_2}$$

i.e.,

$$\frac{1}{\omega_1} + \frac{1}{\omega_2} = \frac{\omega_2}{\omega_r^2} + \frac{\omega_1}{\omega_r^2} = \frac{\omega_1 + \omega_2}{\omega_r^2}$$

Therefore

$$\frac{\omega_2 + \omega_1}{\omega_1 \omega_2} = \frac{\omega_1 + \omega_2}{\omega_r^2}$$

from which

$$\omega_r^2 = \omega_1 \omega_2 \qquad \text{or} \qquad \omega_r = \sqrt{(\omega_1 \omega_2)}$$

Hence

$$2\pi f_r = \sqrt{[(2\pi f_1)(2\pi f_2)]}$$

and

$$f_r = \sqrt{(f_1 f_2)}$$

(5)

Selectivity is the ability of a circuit to respond more readily to signals of a particular frequency to which it is tuned than to signals of other frequencies. The response becomes progressively weaker as the frequency departs from the resonant frequency. Discrimination against other signals becomes more pronounced as circuit losses are reduced, i.e., as the Q-factor is increased. Thus $Q_r = f_r/(f_2 - f_1)$ is a measure of the circuit selectivity in terms of the points on each side of resonance where the circuit current has fallen to 0.707 of its maximum value reached at resonance. The higher the Q-factor, the narrower the bandwidth and the more selective is the circuit. Circuits having high Q-factors (say, in the order of 100 to 300) are therefore useful in communications engineering. A high Q-factor in a series power circuit has disadvantages in that it can lead to dangerously high voltages across the insulation and may result in electrical breakdown.

For example, suppose that the working voltage of a capacitor is stated as 1 kV and is used in a circuit having a supply voltage of 240 V. The maximum value of the supply will be $\sqrt{2}(240)$, i.e., 340 V. The working voltage of the capacitor would appear to be ample. However, if the Q-factor is, say, 10, the voltage across the capacitor will reach 2.4 kV. Since the capacitor is rated only at 1 kV, dielectric breakdown is more than likely to occur.

Low Q-factors, say, in the order of 5 to 25, may be found in power transformers using laminated iron cores.

A capacitor-start induction motor, as used in domestic appliances such as washing-machines and vacuum-cleaners, having a Q-factor as low as 1.5 at starting would result in a voltage across the capacitor 1.5 times that of the supply voltage; hence the cable joining the capacitor to the motor would require extra insulation.

Worked problems on Q-factor and bandwidth

Problem 1. A series circuit comprises a $10\,\Omega$ resistance, a $5\,\mu\text{F}$ capacitor and a variable inductance L. The supply voltage is $20\underline{/\,0°}$ volts at a frequency of 318.3 Hz. The inductance is adjusted until the p.d. across the $10\,\Omega$ resistance is a maximum. Determine for this condition (a) the value of inductance L, (b) the p.d. across each component and (c) the Q-factor.

(a) The maximum voltage across the resistance occurs at resonance when the current is a maximum. At resonance, $\omega_r L = 1/(\omega_r C)$, from which

$$\text{inductance } L = \frac{1}{\omega_r^2 C} = \frac{1}{(2\pi318.3)^2(5 \times 10^{-6})} = \textbf{0.050 H or 50 mH}$$

(b) Current at resonance $I_r = \dfrac{V}{R} = \dfrac{20\underline{/\,0°}}{10\underline{/\,0°}} = 2.0\underline{/\,0°}\,\text{A}$

p.d. across resistance, $V_R = I_r R = (2.0\underline{/\,0°})(10) = \textbf{20}\underline{/\,\textbf{0}°}\,\textbf{V}$

p.d. across inductance, $V_L = IX_L$

$$X_L = 2\pi(318.3)(0.050) = 100\,\Omega$$

Hence $V_L = (2.0\underline{/\,0°})(100\underline{/\,90°}) = \textbf{200}\underline{/\,\textbf{90}°}\,\textbf{V}.$

p.d. across capacitor, $V_C = IX_C = (2.0\underline{/\,0°})(100\underline{/\,-90°}) = \textbf{200}\underline{/\,\textbf{-90}°}\,\textbf{V}.$

(c) Q-factor at resonance, $Q_r = \dfrac{V_L(\text{or } V_C)}{V} = \dfrac{200}{20} = \textbf{10}$

$$\left(\text{Alternatively, } Q_r = \frac{\omega_r L}{R} = \frac{100}{10} = \textbf{10}\right.$$

or

$$Q_r = \frac{1}{\omega_r C R} = \frac{1}{2\pi(318.3)(5 \times 10^{-6})(10)} = \textbf{10}$$

or

$$\left. Q_r = \frac{1}{R}\sqrt{\left(\frac{L}{C}\right)} = \frac{1}{10}\sqrt{\left(\frac{0.050}{5 \times 10^{-6}}\right)} = \textbf{10}\right)$$

Problem 2. A filter in the form of a series $L-R-C$ circuit is designed to operate at a resonant frequency of 10 kHz. Included within the filter is a 10 mH inductance and 5 Ω resistance. Determine the bandwidth of the filter.

Q-factor at resonance is given by

$$Q_r = \frac{\omega_r L}{R} = \frac{(2\pi 10\,000)(10 \times 10^{-3})}{5} = 125.66$$

Since $Q_r = f_r/(f_2 - f_1)$,

$$\textbf{bandwidth, } (f_2 - f_1) = \frac{f_r}{Q_r} = \frac{10000}{125.66} = \textbf{79.6 Hz}$$

Problem 3. An $R-L-C$ series circuit has a resonant frequency of 1.2 kHz and a Q-factor at resonance of 30. If the impedance of the circuit at resonance is 50 Ω, determine the values of (a) the inductance, and (b) the capacitance. Find also (c) the bandwidth, (d) the lower and upper half-power frequencies and (e) the value of the circuit impedance at the half-power frequencies.

(a) At resonance the circuit impedance, $Z = R$, i.e., $R = 50\,\Omega$. Q-factor at resonance, $Q_r = \omega_r L/R$. Hence

$$\text{inductance, } L = \frac{Q_r R}{\omega_r} = \frac{(30)(50)}{(2\pi 1200)} = \textbf{0.199 H} \quad \text{or} \quad \textbf{199 mH}$$

(b) At resonance $\omega_r L = 1/(\omega_r C)$. Hence

$$\text{capacitance, } C = \frac{1}{\omega_r^2 L} = \frac{1}{(2\pi 1200)^2 (0.199)} = \textbf{0.088 } \mu\textbf{F} \quad \text{or} \quad \textbf{88 nF}$$

(c) Q-factor at resonance is also given by $Q_r = f_r/(f_2 - f_1)$, from which

$$\textbf{bandwidth, } (f_2 - f_1) = \frac{f_r}{Q_r} = \frac{1200}{30} = \textbf{40 Hz}$$

(d) From equation (5), resonant frequency, $f_r = \sqrt{(f_1 f_2)}$, i.e., $1200 = \sqrt{(f_1 f_2)}$ from which,

$$f_1 f_2 = (1200)^2 = 1.44 \times 10^6 \tag{6}$$

From part (c),

$$f_2 - f_1 = 40 \tag{7}$$

From equation (6), $f_1 = (1.44 \times 10^6)/f_2$. Substituting in equation (7) gives

$$f_2 - \frac{1.44 \times 10^6}{f_2} = 40$$

Multiplying throughout by f_2 gives $f_2^2 - 1.44 \times 10^6 = 40 f_2$,

i.e., $\qquad\qquad\qquad f_2^2 - 40 f_2 - 1.44 \times 10^6 = 0$

This is a quadratic equation in f_2. Using the quadratic formula gives

$$f_2 = \frac{40 \pm \sqrt{[(40)^2 - 4(-1.44 \times 10^6)]}}{2} = \frac{40 \pm 2400}{2} = \frac{40 + 2400}{2}$$

(since f_2 cannot be negative). **Hence the upper half-power frequency, $f_2 = $ 1220 Hz.**

From equation (7), **the lower half-power frequency,**

$$f_1 = f_2 - 40 = 1220 - 40 = \textbf{1180 Hz}$$

Note that the upper and lower half-power frequency values are symmetrically placed about the resonance frequency. This is usually the case when the Q-factor has a high value (say, > 10).

(e) At the half-power frequencies, current $I = 0.707 I_r$. Hence

$$\text{impedance}, Z = \frac{V}{I} = \frac{V}{0.707 I_r} = 1.414\left(\frac{V}{I_r}\right) = \sqrt{2}\, Z_r = \sqrt{2}\, R$$

Thus impedance at the half-power frequencies, $Z = \sqrt{2}\, R = \sqrt{2(50)} = 70.71\,\Omega$

Problem 4. A series R–L–C circuit is connected to a 0.2 V supply and the current is at its maximum value of 4 mA when the supply frequency is adjusted to 3 kHz. The Q-factor of the circuit under these conditions is 100. Determine the value of (a) the circuit resistance, (b) the circuit inductance, (c) the circuit capacitance, and (d) the voltage across the capacitor.

Since the current is at its maximum, the circuit is at resonance and the resonant frequency is 3 kHz.

(a) At resonance,

$$\text{impedance}, Z = R = \frac{V}{I} = \frac{0.2}{4 \times 10^{-3}} = 50\,\Omega$$

Hence the circuit resistance is $50\,\Omega$

(b) Q-factor at resonance is given by $Q_r = \omega_r L / R$, from which

$$\textbf{inductance } L = \frac{Q_r R}{\omega_r} = \frac{(100)(50)}{2\pi 3000} = \textbf{0.265 H}$$

(c) Q-factor at resonance is also given by $Q_r = 1/(\omega_r C R)$, from which

$$\textbf{capacitance}, C = \frac{1}{\omega_r R Q_r} = \frac{1}{(2\pi 3000)(50)100} = \textbf{0.0106}\,\mu\textbf{F} \quad \text{or} \quad \textbf{10.6 nF}$$

(d) Q-factor at resonance in a series circuit represents the voltage magnification, i.e., $Q_r = V_C / V$, from which, $V_C = Q_r V = (100)(0.2) = 20$ V. **Hence the voltage across the capacitor is 20 V.**

$$\left(\text{Alternatively}, V_C = I X_C = \frac{I}{\omega_r C} = \frac{4 \times 10^{-3}}{(2\pi 3000)(0.0106 \times 10^{-6})} = 20\,\text{V} \right)$$

Problem 5. A coil of inductance 351.8 mH and resistance $8.84\,\Omega$ is connected in series with a 20 μF capacitor. Determine (a) the resonant frequency, (b) the Q-factor at resonance, (c) the bandwidth, and (d) the lower and upper -3 dB frequencies.

(a) Resonant frequency, $f_r = \dfrac{1}{2\pi\sqrt{(LC)}} = \dfrac{1}{2\pi\sqrt{[(0.3518)(20 \times 10^{-6})]}} = \textbf{60.0 Hz.}$

(b) Q-factor at resonance, $Q_r = \dfrac{1}{R}\sqrt{\left(\dfrac{L}{C}\right)} = \dfrac{1}{8.84}\sqrt{\left(\dfrac{0.3518}{20 \times 10^{-6}}\right)} = \textbf{15}$

$$\left[\text{Alternatively}, Q_r = \frac{\omega_r L}{R} = \frac{2\pi(60.0)(0.3518)}{8.84} = \textbf{15} \right.$$

or

$$Q_r = \frac{1}{\omega_r CR} = \frac{1}{(2\pi 60.0)(20 \times 10^{-6})(8.84)} = 15$$

(c) Bandwidth, $(f_2 - f_1) = \dfrac{f_r}{Q_r} = \dfrac{60.0}{15} = 4\,\text{Hz.}$

(d) With a Q-factor of 15 it may be assumed that the lower and upper -3dB frequencies, f_1 and f_2, are symmetrically placed about the resonant frequency of 60.0 Hz. **Hence the lower -3dB frequency, $f_1 = 58\,\text{Hz}$ and the upper -3dB frequency, $f_2 = 62\,\text{Hz.}$**
(This may be checked by using $(f_2 - f_1) = 4$ and $f_r = \sqrt{(f_1 f_2)}$.)

Further problems on Q-factor and bandwidth may be found in section 5, problems 7 to 15, page 111.

4. Parallel resonance

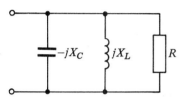

Figure 8 Parallel R–L–C circuit

A parallel network containing resistance R, pure inductance L and pure capacitance C connected in parallel is shown in Fig. 8. Since the inductance and capacitance are considered as pure components, this circuit is something of an 'ideal' circuit. However, it may be used to highlight some important points regarding resonance which are applicable to any parallel circuit.

From Fig. 8,

the admittance of the resistive branch, $G = \dfrac{1}{R}$

the admittance of the inductive branch, $B_L = \dfrac{1}{jX_L} = \dfrac{-j}{\omega L}$

the admittance of the capacitive branch, $B_C = \dfrac{1}{-jX_C} = \dfrac{j}{1/\omega C} = j\omega C.$

Total circuit admittance, $Y = G + j(B_C - B_L)$, i.e.,

$$Y = \frac{1}{R} + j\left(\omega C - \frac{1}{\omega L}\right)$$

The circuit is at resonance when the imaginary part is zero, i.e., when $\omega C - (1/\omega L) = 0$. Hence at resonance $\omega_r C = 1/(\omega_r L)$ and $\omega_r^2 = 1/(LC)$, from which $\omega_r = 1/\sqrt{(LC)}$ and the resonant frequency

$$f_r = \frac{1}{2\pi\sqrt{(LC)}}\ \textbf{hertz}$$

the same expression as for a series R–L–C circuit.

Figure 9 shows typical graphs of B_C, B_L, G and Y against frequency f for the circuit shown in Fig. 8. At resonance, $B_C = B_L$ and admittance $Y = G = 1/R$. This represents the condition of **minimum admittance** for the circuit and thus **maximum impedance.**

Since current $I = V/Z = VY$, the current is also at a **minimum** value at resonance in a parallel network.

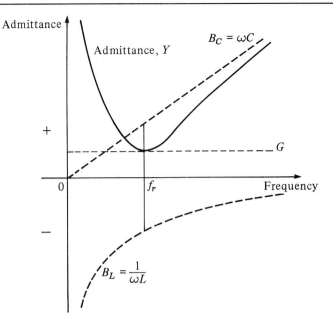

Figure 9 $|Y|$ plotted against frequency

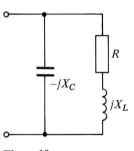

Figure 10

From the ideal circuit of Fig. 8 we have therefore established the following facts which apply to any parallel circuit. At resonance:

(i) admittance Y is a minimum
(ii) impedance Z is a maximum
(iii) current I is a minimum
(iv) an expression for the resonant frequency f_r may be obtained by making the "imaginary" part of the complex expression for admittance equal to zero.

A more practical network, containing a coil of inductance L and resistance R in parallel with a pure capacitance C, is shown in Fig. 10.

$$\text{Admittance of coil, } Y_{\text{COIL}} = \frac{1}{R + jX_L} = \frac{R - jX_L}{R^2 + X_L^2} = \frac{R}{R^2 + \omega^2 L^2} - \frac{j\omega L}{R^2 + \omega^2 L^2}$$

$$\text{Admittance of capacitor, } Y_C = \frac{1}{-jX_C} = \frac{j}{X_C} = j\omega C.$$

$$\text{Total circuit admittance, } Y = Y_{\text{COIL}} + Y_C = \frac{R}{R^2 + \omega^2 L^2} - \frac{j\omega L}{R^2 + \omega^2 L^2} + j\omega C$$

$$(8)$$

At resonance, the total circuit admittance Y is real $(Y = R/(R^2 + \omega^2 L^2))$, i.e., the imaginary part is zero. Hence, at resonance

$$\frac{-\omega_r L}{R^2 + \omega_r^2 L^2} + \omega_r C = 0$$

Therefore

$$\frac{\omega_r L}{R^2 + \omega_r^2 L^2} = \omega_r C \quad \text{and} \quad \frac{L}{C} = R^2 + \omega_r^2 L^2$$

Thus

$$\omega_r^2 L^2 = \frac{L}{C} - R^2$$

and

$$\omega_r^2 = \frac{L}{CL^2} - \frac{R^2}{L^2} = \frac{1}{LC} - \frac{R^2}{L^2} \qquad (9)$$

Hence

$$\omega_r = \sqrt{\left(\frac{1}{LC} - \frac{R^2}{L^2}\right)}$$

and

$$\text{resonant frequency, } f_r = \frac{1}{2\pi}\sqrt{\left(\frac{1}{LC} - \frac{R^2}{L^2}\right)} \qquad (10)$$

(Note that when $R^2/L^2 \ll 1/(LC)$ then $f_r = 1/(2\pi\sqrt{(LC)})$, as for the series R–L–C circuit.)

Dynamic resistance

Since the current at resonance is in phase with the voltage, the impedance of the network acts as a resistance. This resistance is known as the **dynamic resistance, R_D**. Impedance at resonance, $R_D = V/I_r$, where I_r is the current at resonance.

$$I_r = VY_r = V\left(\frac{R}{R^2 + \omega_r^2 L^2}\right)$$

from equation (8) with the j terms equal to zero. Hence

$$R_D = \frac{V}{VR/(R^2 + \omega_r^2 L^2)} = \frac{R^2 + \omega_r^2 L^2}{R} = \frac{R^2 + L^2((1/LC) - (R^2/L^2))}{R}$$

$$= \frac{R^2 + (L/C) - R^2}{R} = \frac{L}{CR}$$

Hence

$$\text{dynamic resistance, } R_D = \frac{L}{CR} \qquad (11)$$

A more general network comprising a coil of inductance L and resistance R_L in parallel with a capacitance C and resistance R_C in series is shown in Fig. 11.

Admittance of inductive branch,

$$Y_L = \frac{1}{R_L + jX_L} = \frac{R_L - jX_L}{R_L^2 + X_L^2} = \frac{R_L}{R_L^2 + X_L^2} - \frac{jX_L}{R_L^2 + X_L^2}$$

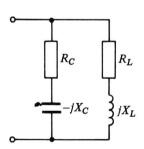

Figure 11

Admittance of capacitive branch,

$$Y_C = \frac{1}{R_C - jX_C} = \frac{R_C + jX_C}{R_C^2 + X_C^2} = \frac{R_C}{R_C^2 + X_C^2} + \frac{jX_C}{R_C^2 + X_C^2}$$

Total network admittance,

$$Y = Y_L + Y_C = \frac{R_L}{R_L^2 + X_L^2} - \frac{jX_L}{R_L^2 + X_L^2} + \frac{R_C}{R_C^2 + X_C^2} + \frac{jX_C}{R_C^2 + X_C^2}$$

At resonance the admittance is a minimum, i.e., when the imaginary part of Y is zero. Hence, at resonance,

$$\frac{-X_L}{R_L^2 + X_L^2} + \frac{X_C}{R_C^2 + X_C^2} = 0$$

i.e.,

$$\frac{\omega_r L}{R_L^2 + \omega_r^2 L^2} = \frac{1/(\omega_r C)}{R_C^2 + (1/\omega_r^2 C^2)} \qquad (12)$$

Rearranging gives

$$\omega_r L \left(R_C^2 + \frac{1}{\omega_r^2 C^2} \right) = \frac{1}{\omega_r C} (R_L^2 + \omega_r^2 L^2)$$

$$\omega_r L R_C^2 + \frac{L}{\omega_r C^2} = \frac{R_L^2}{\omega_r C} + \frac{\omega_r L^2}{C}$$

Multiplying throughout by $\omega_r C^2$ gives

$$\omega_r^2 C^2 L R_C^2 + L = R_L^2 C + \omega_r^2 L^2 C$$
$$\omega_r^2 (C^2 L R_C^2 - L^2 C) = R_L^2 C - L$$
$$\omega_r^2 CL(CR_C^2 - L) = R_L^2 C - L$$

Hence

$$\omega_r^2 = \frac{(CR_L^2 - L)}{LC(CR_C^2 - L)}$$

i.e.,

$$\omega_r = \frac{1}{\sqrt{(LC)}} \sqrt{\left(\frac{R_L^2 - (L/C)}{R_C^2 - (L/C)} \right)}$$

Hence

resonant frequency, $f_r = \dfrac{1}{2\pi\sqrt{(LC)}} \sqrt{\left(\dfrac{R_L^2 - (L/C)}{R_C^2 - (L/C)} \right)}$

It is clear from equation (12) that parallel resonance may be achieved in such a circuit in several ways—by varying either the frequency f, the inductance L, the capacitance C, the resistance R_L or the resistance R_C.

Q-factor in a parallel network

The Q-factor in the series R–L–C circuit is a measure of the voltage magnification. In a parallel circuit, currents higher than the supply current can circulate within the parallel branches of a parallel resonant network, the current leaving the capacitor and establishing the magnetic field of the inductance, this then collapsing and recharging the capacitor, and so on. The Q-factor of a parallel resonant circuit is the ratio of the current circulating in the parallel branches of the circuit to the supply current, i.e., in a parallel circuit, Q-factor is a measure of the **current magnification**.

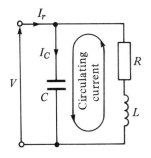

Figure 12

Circulating currents may be several hundreds of times greater than the supply current at resonance. For the parallel network of Fig. 12, the Q-factor at resonance is given by

$$Q_r = \frac{\text{circulating current}}{\text{current at resonance}} = \frac{\text{capacitor current}}{\text{current at resonance}} = \frac{I_C}{I_r}$$

Current in capacitor, $I_C = V/X_C = V\omega_r C$. Current at resonance,

$$I_r = \frac{V}{R_D} = \frac{V}{L/(CR)} = \frac{VCR}{L}$$

Hence

$$Q_r = \frac{V\omega_r C}{VCR/L} = \frac{\omega_r L}{R}$$

the same expression as for series resonance.

The difference between the resonant frequency of a series circuit and that of a parallel circuit can be quite small. The resonant frequency of a coil in parallel with a capacitor is shown in equation (10); however, around the closed loop comprising the coil and capacitor the energy would naturally resonate at a frequency given by that for a series $L–C–R$ circuit, as shown in section 2. This latter frequency is termed the **natural frequency, f_n,** and the frequency of resonance seen at the terminals of Fig. 12 is often called the **forced resonant frequency, f_r.** (For a series circuit, the forced and natural frequencies coincide.)

From the coil-capacitor loop of Fig. 12,

$$f_n = \frac{1}{2\pi\sqrt{(LC)}}$$

and the forced resonant frequency,

$$f_r = \frac{1}{2\pi}\sqrt{\left(\frac{1}{LC} - \frac{R^2}{L^2}\right)}$$

Thus

$$\frac{f_r}{f_n} = \frac{(1/2\pi)\sqrt{[(1/LC) - (R^2/L^2)]}}{1/(2\pi\sqrt{(LC)})} = \sqrt{\left(1 - \frac{R^2 C}{L}\right)}$$

i.e.,

$$f_r = f_n\sqrt{\left(1 - \frac{1}{Q^2}\right)}$$

since, from section 3,

$$Q = \frac{1}{R}\sqrt{\left(\frac{L}{C}\right)}$$

Thus it is seen that even with small values of Q the difference between f_r and f_n tends to be very small. A high value of Q makes the parallel resonant frequency tend to the same value as that of the series resonant frequency.

The expressions already obtained for bandwidth and resonant frequency, i.e., $Q_r = f_r/(f_2 - f_1)$ and $f_r = \sqrt{(f_1 f_2)}$, also apply to parallel circuits.

By similar reasoning to that of the series R–L–C circuit it may be shown that at the half-power frequencies the admittance is $\sqrt{2}$ times its minimum value at resonance and, since $Z = 1/Y$, the value of impedance at the half-power frequencies is $1/\sqrt{2}$ or 0.707 times its maximum value at resonance.

Worked problems on parallel resonance

Problem 1. A coil of inductance 5 mH and resistance 10 Ω is connected in parallel with a 250 nF capacitor across a 50 V variable-frequency supply. Determine (a) the resonant frequency, (b) the dynamic resistance, (c) the current at resonance, and (d) the circuit Q-factor at resonance.

(a) Resonance frequency, $f_r = \dfrac{1}{2\pi}\sqrt{\left(\dfrac{1}{LC} - \dfrac{R^2}{L^2}\right)}$, from equation (10),

$$= \frac{1}{2\pi}\sqrt{\left(\frac{1}{5\times 10^{-3}\times 250\times 10^{-9}} - \frac{10^2}{(5\times 10^{-3})^2}\right)}$$

$$= \frac{1}{2\pi}\sqrt{(800\times 10^6 - 4\times 10^6)} = \frac{1}{2\pi}\sqrt{(796\times 10^6)}$$

$$= \textbf{4490 Hz}$$

(b) From equation (11), dynamic resistance, $R_D = \dfrac{L}{CR} = \dfrac{5\times 10^{-3}}{(250\times 10^{-9})(10)} = \textbf{2000 Ω}$

(c) Current at resonance, $I_r = \dfrac{V}{R_D} = \dfrac{50}{2000} = \textbf{25 mA}$

(d) Q-factor at resonance,

$$Q_r = \frac{\omega_r L}{R} = \frac{(2\pi 4490)(5\times 10^{-3})}{10} = \textbf{14.1}$$

Problem 2. In the parallel network of Fig. 13, inductance, $L = 100$ mH and capacitance, $C = 40\,\mu$F. Determine the resonant frequency for the network if (a) $R_L = 0$ and (b) $R_L = 30\,\Omega$.

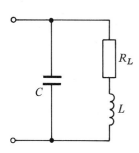

Figure 13

Total circuit admittance, $Y = \dfrac{1}{R_L + jX_L} + \dfrac{1}{-jX_C}$

$$= \frac{R_L - jX_L}{R_L^2 + X_L^2} + \frac{j}{X_C}$$

$$= \frac{R_L}{R_L^2 + X_L^2} - \frac{jX_L}{R_L^2 + X_L^2} + \frac{j}{X_C}$$

The network is at resonance when the admittance is at a minimum value, i.e., when the imaginary part is zero. Hence, at resonance,

$$\frac{-X_L}{R_L^2 + X_L^2} + \frac{1}{X_C} = 0$$

or

$$\omega_r C = \frac{\omega_r L}{R_L^2 + \omega_r^2 L^2} \qquad (13)$$

(a) When $R_L = 0$,

$$\omega_r C = \frac{\omega_r L}{\omega_r^2 L^2}$$

from which

$$\omega_r^2 = \frac{1}{LC} \quad \text{and} \quad \omega_r = \frac{1}{\sqrt{(LC)}}$$

Hence resonant frequency,

$$f_r = \frac{1}{2\pi\sqrt{(LC)}} = \frac{1}{2\pi\sqrt{(100 \times 10^{-3} \times 40 \times 10^{-6})}} = \textbf{79.6 Hz}$$

(b) When $R_L = 30\,\Omega$,

$$\omega_r C = \frac{\omega_r L}{30^2 + \omega_r^2 L^2} \qquad \text{from equation (13) above,}$$

from which

$$30^2 + \omega_r^2 L^2 = \frac{L}{C}$$

i.e.,

$$\omega_r^2 (100 \times 10^{-3})^2 = \frac{100 \times 10^{-3}}{40 \times 10^{-6}} - 900$$

i.e., $\omega_r^2\,(0.01) = 2500 - 900 = 1600$.

Thus, $\omega_r^2 = 1600/0.01 = 160\,000$ and $\omega_r = \sqrt{160\,000} = 400$ rad/s. Hence

$$\text{resonant frequency,} \quad f_r = \frac{400}{2\pi} = \textbf{63.7 Hz}$$

Hence, as the resistance of a coil increases, the resonant frequency decreases in the circuit of Fig. 13.

Problem 3. A coil of inductance 120 mH and resistance 150 Ω is connected in parallel with a variable capacitor across a 20 V, 4 kHz supply. Determine for the condition when the supply current is a minimum, (a) the capacitance of the capacitor, (b) the dynamic resistance, (c) the supply current, (d) the Q-factor, (e) the bandwidth, (f) the upper and lower -3 dB frequencies, and (g) the value of the circuit impedance at the -3 dB frequencies.

(a) The supply current is a minimum when the parallel network is at resonance.

$$\text{Resonant frequency,} \quad f_r = \frac{1}{2\pi}\sqrt{\left(\frac{1}{LC} - \frac{R^2}{L^2}\right)} \quad \text{from equation (10),}$$

from which

$$(2\pi f_r)^2 = \frac{1}{LC} - \frac{R^2}{L^2}$$

Hence

$$\frac{1}{LC} = (2\pi f_r)^2 + \frac{R^2}{L^2}$$

and

$$\text{capacitance } C = \frac{1}{L\{(2\pi f_r)^2 + (R^2/L^2)\}}$$

$$= \frac{1}{120 \times 10^{-3}\{(2\pi 4000)^2 + (150^2/(120 \times 10^{-3})^2)\}}$$

$$= \frac{1}{0.12(631.65 \times 10^6 + 1.5625 \times 10^6)}$$

$$= 0.01316\,\mu\text{F} \quad \text{or} \quad \textbf{13.16\,nF}$$

(b) Dynamic resistance, $R_D = \dfrac{L}{CR} = \dfrac{120 \times 10^{-3}}{(13.16 \times 10^{-9})(150)} = \textbf{60.8\,k}\Omega$

(c) Supply current at resonance, $I_r = \dfrac{V}{R_D} = \dfrac{20}{60.8 \times 10^3} = \textbf{0.329\,mA or 329}\,\mu\textbf{A.}$

(d) Q-factor at resonance, $Q_r = \dfrac{\omega_r L}{R} = \dfrac{(2\pi 4000)(120 \times 10^{-3})}{150} = \textbf{20.1}$

$\left(\begin{array}{l} \text{Note that the expressions} \end{array}\right.$

$$Q_r = \frac{1}{\omega_r CR} \quad \text{or} \quad Q_r = \frac{1}{R}\sqrt{\left(\frac{L}{C}\right)}$$

used for the R–L–C series circuit may also be used in parallel circuits when the resistance of the coil is much smaller than the inductive reactance of the coil. In this case $R = 150\,\Omega$ and $X_L = 2\pi(4000)(120 \times 10^{-3}) = 3016\,\Omega$. Hence, alternatively,

$$Q_r = \frac{1}{\omega_r CR} = \frac{1}{(2\pi 4000)(13.16 \times 10^{-9})(150)} = \textbf{20.2}$$

or

$$Q_r = \frac{1}{R}\sqrt{\left(\frac{L}{C}\right)} = \frac{1}{150}\sqrt{\left(\frac{120 \times 10^{-3}}{13.16 \times 10^{-9}}\right)} = \textbf{20.1} \Bigg)$$

(e) If the lower and upper -3 dB frequencies are f_1 and f_2 respectively then the bandwidth is $(f_2 - f_1)$. Q-factor at resonance is given by $Q_r = f_r/(f_2 - f_1)$, from which

$$\text{bandwidth, } (f_2 - f_1) = \frac{f_r}{Q_r} = \frac{4000}{20.1} = \textbf{199\,Hz}$$

(f) Resonant frequency, $f_r = \sqrt{(f_1 f_2)}$, from which

$$f_1 f_2 = f_r^2 = (4000)^2 = 16 \times 10^6 \qquad (14)$$

Also, from part (e),

$$f_2 - f_1 = 199 \qquad (15)$$

From equation (14),

$$f_1 = \frac{16 \times 10^6}{f_2}$$

Substituting in equation (15) gives

$$f_2 - \frac{16 \times 10^6}{f_2} = 199$$

i.e., $f_2^2 - 16 \times 10^6 = 199\,f_2$ and $f_2^2 - 199\,f_2 - 16 \times 10^6 = 0$. Solving this quadratic equation gives

$$f_2 = \frac{199 \pm \sqrt{[(199)^2 - 4(-16 \times 10^6)]}}{2} = \frac{199 \pm 8002.5}{2}$$

i.e., **the upper 3 dB frequency, $f_2 = 4100$ Hz** (neglecting the negative answer).
From equation (14),

$$\text{the lower } -3\,\text{dB frequency, } f_1 = \frac{16 \times 10^6}{f_2} = \frac{16 \times 10^6}{4100} = \textbf{3900 Hz}$$

(Note that f_1 and f_2 are equally displaced about the resonant frequency, f_r.)
(g) The value of the circuit impedance, Z, at the -3 dB frequencies is given by

$$Z = \frac{1}{\sqrt{2}} Z_r$$

where Z_r is the impedance at resonance.
The impedance at resonance $Z_r = R_D$, the dynamic resistance. Hence

$$\textbf{impedance at the } -\textbf{3 dB frequencies} = \frac{1}{\sqrt{2}}(60.8 \times 10^3) = \textbf{43.0 k}\boldsymbol{\Omega}$$

Figure 14 shows impedance plotted against frequency for the circuit in the region of the resonant frequency.

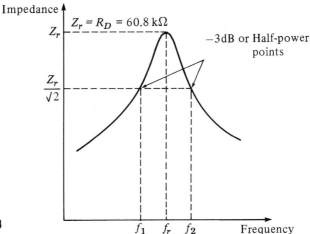

Figure 14

Problem 4. A two branch parallel network is shown in Fig. 15. Determine the resonant frequency of the network.

It is shown on pages 103–104 that, for the circuit of Fig. 15,

$$\text{resonant frequency, } f_r = \frac{1}{2\pi\sqrt{(LC)}} \sqrt{\left(\frac{R_L^2 - (L/C)}{R_C^2 - (L/C)} \right)}$$

where $R_L = 5\,\Omega$, $R_C = 3\,\Omega$, $L = 2$ mH and $C = 25\,\mu$F. Thus

$$f_r = \frac{1}{2\pi\sqrt{[(2 \times 10^3)(25 \times 10^{-6})]}} \sqrt{\left\{ \frac{5^2 - ((2 \times 10^{-3})/(25 \times 10^{-6}))}{3^2 - ((2 \times 10^{-3})/(25 \times 10^{-6}))} \right\}}$$

$$= \frac{1}{2\pi\sqrt{(5 \times 10^{-8})}} \sqrt{\left(\frac{25 - 80}{9 - 80} \right)} = \frac{10^4}{2\pi\sqrt{5}} \sqrt{\left(\frac{-55}{-71} \right)} = \textbf{627 Hz}$$

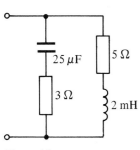

Figure 15

Figure 16

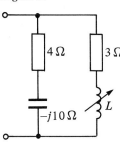

Problem 5. Determine for the parallel network shown in Fig. 16 the values of inductance L for which the network is resonant at a frequency of 1 kHz.

The total network admittance, Y, is given by

$$Y = \frac{1}{3+jX_L} + \frac{1}{4-j10} = \frac{3-jX_L}{3^2+X_L^2} + \frac{4+j10}{4^2+10^2}$$

$$= \frac{3}{3^2+X_L^2} - \frac{jX_L}{3^2+X_L^2} + \frac{4}{116} + \frac{j10}{116} = \left(\frac{3}{3^2+X_L^2} + \frac{4}{116}\right) + j\left(\frac{10}{116} - \frac{X_L}{3^2+X_L^2}\right)$$

Resonance occurs when the admittance is a minimum, i.e., when the imaginary part of Y is zero. Hence, at resonance,

$$\frac{10}{116} - \frac{X_L}{3^2+X_L^2} = 0 \quad \text{i.e.,} \quad \frac{10}{116} = \frac{X_L}{9+X_L^2}$$

Therefore

$$10\,(9+X_L^2) = 116\,X_L, \qquad 10\,X_L^2 - 116\,X_L + 90 = 0$$

from which

$$X_L^2 - 11.6\,X_L + 9 = 0$$

Solving the quadratic equation gives

$$X_L = \frac{11.6 \pm \sqrt{[(-11.6)^2 - 4(9)]}}{2} = \frac{11.6 \pm 9.93}{2}$$

i.e., $X_L = 10.765\,\Omega$ or $0.835\,\Omega$. Hence $10.765 = 2\pi f_r L_1$, from which

$$\text{inductance } L_1 = \frac{10.765}{2\pi(1000)} = 1.71\,\text{mH}$$

and $0.835 = 2\pi f_r L_2$, from which

$$\text{inductance, } L_2 = \frac{0.835}{2\pi(1000)} = 0.13\,\text{mH}$$

Thus the conditions for the circuit of Fig. 16 to be resonant are that inductance L is either 1.71 mH or 0.13 mH.

Further problems on parallel resonance may be found in the following section (5), problems 16 to 28, page 112.

5. Further problems

Series resonance

1. A coil having an inductance of 50 mH and resistance 8.0 Ω is connected in series with a 25 μF capacitor across a 100 V a.c. supply. Determine (a) the resonant frequency of the circuit, and (b) the current flowing at resonance.
   [(a) 142.4 Hz   (b) 12.5 A]
2. The current at resonance in a series R–L–C circuit is 0.12 mA. The circuit has an inductance of 0.05 H and the supply voltage is 24 mV at a frequency of 40 kHz. Determine (a) the circuit resistance, and (b) the circuit capacitance.
   [(a) 200 Ω (b) 316.6 pF]
3. An R–L–C series circuit is comprised of a variable capacitor and a coil of resistance 10 Ω and inductance 15 mH. If the supply frequency is 2 kHz and the

current flowing is 5 A, determine for series resonance (a) the value of capacitance required, (b) the supply p.d., and (c) the p.d. across the capacitor.
[(a) 0.422 μF   (b) 50 V   (c) 943 V]

4. A coil of inductance 2.0 mH and resistance 4.0 Ω is connected in series with a 0.3 μF capacitor. The circuit is connected to a 5.0 V, variable frequency supply. Calculate (a) the frequency at which resonance occurs, (b) the voltage across the capacitance at resonance, and (c) the voltage across the coil at resonance.
[(a) 6.50 kHz,   (b) 102.1 V,   (c) 102.2 V]

5. A series R–L–C circuit having an inductance of 0.40 H has an instantaneous voltage, $v = 60 \sin (4000t - (\pi/6))$ volts and an instantaneous current, $i = 2.0 \sin 4000t$ amperes. Determine (a) the values of the circuit resistance and capacitance, and (b) the frequency at which the circuit will be resonant.
[(a) 26.0 Ω; 154.8 nF   (b) 639.6 Hz]

6. A variable capacitor C is connected in series with a coil having inductance L. The circuit possesses stray capacitance C_s which is assumed to be constant and effectively in parallel with the variable capacitor C. When the capacitor is set to 2.0 nF the resonant frequency of the circuit is 86.85 kHz, and when the capacitor is set to 1.0 nF the resonant frequency is 120 kHz. Determine the values of (a) the stray circuit capacitance C_s, and (b) the coil inductance L.
[(a) 100 pF   (b) 1.60 mH]

Q-factor and bandwidth

7. A series R–L–C circuit comprises a 5 μF capacitor, a 4 Ω resistor and a variable inductance L. The supply voltage is 10$\underline{/\ 0°}$ V at a frequency of 159.1 Hz. The inductance is adjusted until the p.d. across the 4 Ω resistance is a maximum. Determine for this condition (a) the value of inductance, (b) the p.d. across each component, and (c) the Q-factor of the circuit.

$$\left[\begin{matrix}\text{(a) 200 mH} \quad \text{(b) } V_R = 10\underline{/\ 0°}\text{ V}; V_L = 500\underline{/\ 90°}\text{ V}; V_C = 500\underline{/\ -90°}\text{ V} \\ \text{(c) 50}\end{matrix}\right]$$

8. A coil of resistance 10.05 Ω and inductance 400 mH is connected in series with a 0.396 μF capacitor. Determine (a) the resonant frequency, (b) the resonant Q-factor, (c) the bandwidth, and (d) the lower and upper half-power frequencies.
[(a) 400 Hz   (b) 100   (c) 4 Hz   (d) 398 Hz and 402 Hz]

9. An R–L–C series circuit has a resonant frequency of 2 kHz and a Q-factor at resonance of 40. If the impedance of the circuit at resonance is 30 Ω determine the values of (a) the inductance and (b) the capacitance. Find also (c) the bandwidth, (d) the lower and upper − 3 dB frequencies, and (e) the impedance at the − 3 dB frequencies.

$$\left[\begin{matrix}\text{(a) 95.5 mH} \quad \text{(b) 66.3 nF} \quad \text{(c) 50 Hz} \\ \text{(d) 1975 Hz and 2025 Hz} \quad \text{(e) 42.43 } \Omega\end{matrix}\right]$$

10. A filter in the form of a series L–C–R circuit is designed to operate at a resonant frequency of 20 kHz and incorporates a 20 mH inductor and 30 Ω resistance. Determine the bandwidth of the filter.
[238.7 Hz]

11. An R–L–C series circuit has a maximum current of 2 mA flowing in it when the frequency of the 0.1 V supply is 4 kHz. The Q-factor of the circuit under these conditions is 90. Determine (a) the voltage across the capacitor, and (b) the values of the circuit resistance, inductance and capacitance.
[(a) 9 V   (b) 50 Ω; 0.179 H; 8.84 nF]

12. Calculate the inductance of a coil which must be connected in series with a

4000 pF capacitor to give a resonant frequency of 200 kHz. If the coil has a resistance of 12 Ω, determine the circuit Q-factor.

[158.3 μH; 16.58]

13. A circuit consists of a coil of inductance 200 μH and resistance 8.0 Ω in series with a lossless 500 pF capacitor. Determine (a) the −3 dB bandwidth of the circuit, and (b) the resonant Q-factor.

[(a) 6366 Hz   (b) 79.06]

14. A capacitor of capacitance 20 nF is connected in series with a coil having resistance R and inductance L. The resonant frequency of the circuit is 10 kHz and the Q-factor is 80. Determine (a) the resistance R of the coil, (b) the inductance L of the coil, (c) the bandwidth, and (d) the upper and lower −3 dB frequencies.

[(a) 9.95 Ω   (b) 12.67 mH   (c) 125 Hz   (d) 10.06 kHz; 9.94 kHz]

15. A coil of inductance 200 μH and resistance 50.27 Ω and a variable capacitor are connected in series to a 5 mV supply of frequency 2 MHz. Determine (a) the value of capacitance to tune the circuit to resonance, (b) the supply current at resonance, (c) the p.d. across the capacitor at resonance, (d) the bandwidth, and (e) the half-power frequencies.

$$\begin{bmatrix} \text{(a) } 31.65 \text{ pF} & \text{(b) } 99.46 \,\mu\text{A} & \text{(c) } 250 \text{ mV} \\ \text{(d) } 40 \text{ kHz} & \text{(e) } 2.02 \text{ MHz}; 1.98 \text{ MHz} & \end{bmatrix}$$

Parallel resonance

16. A coil of resistance 20 Ω and inductance 100 mH is connected in parallel with a 50 μF capacitor across a 30 V variable-frequency supply. Determine (a) the resonant frequency of the circuit, (b) the dynamic resistance, (c) the current at resonance, and (d) the circuit Q-factor at resonance.

[(a) 63.66 Hz   (b) 100 Ω   (c) 0.30 A   (d) 2]

17. (i) Explain the meaning of the dynamic resistance of a parallel resonant network.

   (ii) A 25 V, 2.5 kHz supply is connected to a network comprising a variable capacitor in parallel with a coil of resistance 250 Ω and inductance 80 mH. Determine for the condition when the supply current is a minimum (a) the capacitance of the capacitor, (b) the dynamic resistance, (c) the supply current, (d) the Q-factor, (e) the bandwidth, (f) the upper and lower half-power frequencies and (g) the value of the circuit impedance at the −3 dB frequencies.

$$\begin{bmatrix} \text{(a) } 48.73 \text{ nF} & \text{(b) } 6.57 \text{ k}\Omega & \text{(c) } 3.81 \text{ mA} & \text{(d) } 5.03 \\ \text{(e) } 497.3 \text{ Hz} & \text{(f) } 2761 \text{ Hz}; 2264 \text{ Hz} & \text{(g) } 4.64 \text{ k}\Omega & \end{bmatrix}$$

18. A 0.1 μF capacitor and a pure inductance of 0.02 H are connected in parallel across a 12 V variable-frequency supply. Determine (a) the resonant frequency of the circuit, and (b) the current circulating in the capacitance and inductance at resonance.

[(a) 3.56 kHz   (b) 26.84 mA]

19. A 20 μF capacitor is connected in parallel with a coil of inductance 40 mH and unknown resistance R across a 100 V, 50 Hz supply. If the circuit has an overall power factor of unity, determine (a) the value of R, (b) the current in the coil and (c) the supply current.

[(a) 42.92 Ω   (b) 2.24 A   (c) 2.15 A]

20. A coil of resistance 300 Ω and inductance 100 mH and a 4000 pF capacitor are connected (i) in series and (ii) in parallel. Find for each connection (a) the

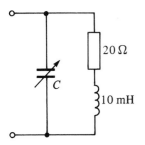

Figure 17

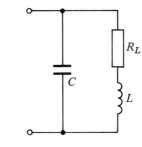

Figure 18

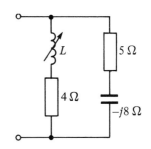

Figure 19

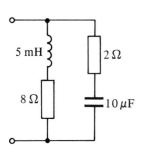

Figure 20

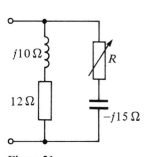

Figure 21

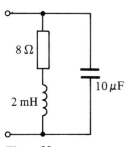

Figure 22

resonant frequency, (b) the Q-factor, and (c) the impedance at resonance.

$$\begin{bmatrix} \text{(i) (a) } 7958\,\text{Hz} & \text{(b) } 16.67 & \text{(c) } 300\,\Omega \\ \text{(ii) (a) } 7943\,\text{Hz} & \text{(b) } 16.64 & \text{(c) } 83.33\,\text{k}\Omega \end{bmatrix}$$

21. A network comprises a coil of resistance $100\,\Omega$ and inductance $0.8\,\text{H}$ and a capacitor having capacitance $30\,\mu\text{F}$. Determine the resonant frequency of the network when the capacitor is connected (a) in series with, and (b) in parallel with the coil.

[(a) 32.5 Hz   (b) 25.7 Hz]

22. Determine the value of capacitor C shown in Fig. 17 for which the resonant frequency of the network is 1 kHz.

[2.30 μF]

23. In the parallel network shown in Fig. 18, inductance L is $40\,\text{mH}$ and capacitance C is $5\,\mu\text{F}$. Determine the resonant frequency of the circuit if (a) $R_L=0$ and (b) $R_L=40\,\Omega$.

[(a) 355.9 Hz   (b) 318.3 Hz]

24. A capacitor of reactance $5\,\Omega$ is connected in series with a $10\,\Omega$ resistor. The whole circuit is then connected in parallel with a coil of inductive reactance $20\,\Omega$ and a variable resistor. Determine the value of this resistance for which the parallel network is resonant.

[10Ω]

25. Determine, for the parallel network shown in Fig. 19, the values of inductance L for which the circuit is resonant at a frequency of 600 Hz.

[2.50 mH or 0.45 mH]

26. Find the resonant frequency of the two-branch parallel network shown in Fig. 20.

[667 Hz]

27. Determine the value of the variable resistance R in Fig. 21 for which the parallel network is resonant.

[11.87Ω]

28. For the parallel network shown in Fig. 22, determine the resonant frequency. Find also the value of resistance to be connected in series with the $10\,\mu\text{F}$ capacitor to change the resonant frequency to 1 kHz.

[928 Hz; 5.27Ω]

7 Introduction to network analysis

1. Introduction

Voltage sources in series-parallel networks cause currents to flow in each branch of the circuit and corresponding volt-drops occur across the circuit components. A.C. circuit (or network) analysis involves the determination of the currents in the branches and/or the voltages across components.

The laws which determine the currents and voltage drops in a.c. networks are:

(a) **current, $I = V/Z$,** where Z is the complex impedance and V the voltage across the impedance;

(b) **the laws for impedances in series and parallel,** i.e., total impedance, $Z_T = Z_1 + Z_2 + Z_3 + \cdots + Z_n$ for n impedances connected in series, and

$$\frac{1}{Z_T} = \frac{1}{Z_1} + \frac{1}{Z_2} + \frac{1}{Z_3} + \cdots + \frac{1}{Z_n}$$

for n impedances connected in parallel; and

(c) **Kirchhoff's laws,** which may be stated as:

(i) "At any point in an electrical circuit the phasor sum of the currents flowing towards that junction is equal to the phasor sum of the currents flowing away from the junction."

(ii) "In any closed loop in a network, the phasor sum of the voltage drops (i.e., the products of current and impedance) taken around the loop is equal to the phasor sum of the emf's acting in that loop."

In any circuit the currents and voltages at any point may be determined by applying Kirchhoff's laws, or by extensions of Kirchhoff's laws, called **mesh-current analysis** and **nodal analysis** (see sections 3 to 5).

However, for more complicated circuits, a number of circuit theorems have been developed as alternatives to use of Kirchhoff's laws to solve problems involving both d.c. and a.c. electrical networks. These include:

(a) the superposition theorem (see section 6)

(b) Thévenin's theorem (see chapter 8)

(c) Norton's theorem (see chapter 8),

(d) the maximum power transfer theorems (see chapter 9).

In addition to these theorems, and often used as a preliminary to using circuit theorems, star–delta (or T–π) and delta-star (or π–T) transformations provide a method for simplifying certain circuits (see chapter 9).

In a.c. circuit analysis involving Kirchhoff's laws or circuit theorems, the use of complex numbers is essential.

The above laws and theorems apply to **linear circuits**, i.e., circuits containing impedances whose values are independent of the direction and magnitude of the current flowing in them.

2. Solution of simultaneous equations using determinants

When Kirchhoff's laws are applied to electrical circuits, simultaneous equations result which require solution. If two loops are involved, two simultaneous equations containing two unknowns need to be solved; if three loops are involved, three simultaneous equations containing three unknowns need to be solved, and so on. The elimination and substitution methods of solving simultaneous equations may be used to solve such equations. However a more convenient method is to use **determinants.**

Two unknowns

When solving linear simultaneous equations in two unknowns using determinants:
(i) the equations are initially written in the form

$$a_1 x + b_1 y + c_1 = 0$$
$$a_2 x + b_2 y + c_2 = 0$$

(ii) the solution is given by:

$$\frac{x}{D_x} = \frac{-y}{D_y} = \frac{1}{D}$$

where

$$D_x = \begin{vmatrix} b_1 & c_1 \\ b_2 & c_2 \end{vmatrix},$$

i.e., the determinant of the coefficients left when the x-column is "covered up",

$$D_y = \begin{vmatrix} a_1 & c_1 \\ a_2 & c_2 \end{vmatrix},$$

i.e., the determinant of the coefficients left when the y-column is "covered up", and

$$D = \begin{vmatrix} a_1 & b_1 \\ a_2 & b_2 \end{vmatrix},$$

i.e., the determinant of the coefficients left when the constants-column is "covered up".
A "2×2" determinant

$$\begin{vmatrix} a & b \\ c & d \end{vmatrix}$$

is evaluated as $ad - bc$.

Three unknowns

When solving linear simultaneous equations in three unknowns using determinants:

(i) the equations are initially written in the form

$$a_1 x + b_1 y + c_1 z + d_1 = 0$$
$$a_2 x + b_2 y + c_2 z + d_2 = 0$$
$$a_3 x + b_3 y + c_3 z + d_3 = 0$$

(ii) the solution is given by

$$\frac{x}{D_x} = \frac{-y}{D_y} = \frac{z}{D_z} = \frac{-1}{D}$$

where

$$D_x = \begin{vmatrix} b_1 & c_1 & d_1 \\ b_2 & c_2 & d_2 \\ b_3 & c_3 & d_3 \end{vmatrix},$$

i.e., the determinant of the coefficients left when the x-column is "covered up",

$$D_y = \begin{vmatrix} a_1 & c_1 & d_1 \\ a_2 & c_2 & d_2 \\ a_3 & c_3 & d_3 \end{vmatrix},$$

i.e. the determinant of the coefficients left when the y-column is "covered up",

$$D_z = \begin{vmatrix} a_1 & b_1 & d_1 \\ a_2 & b_2 & d_2 \\ a_3 & b_3 & d_3 \end{vmatrix},$$

i.e., the determinant of the coefficients left when the z-column is "covered up", and

$$D = \begin{vmatrix} a_1 & b_1 & c_1 \\ a_2 & b_2 & c_2 \\ a_3 & b_3 & c_3 \end{vmatrix},$$

i.e., the determinant of the coefficients left when the constants-column is "covered up".

A "3 × 3" determinant

$$\begin{vmatrix} a & b & c \\ d & e & f \\ g & h & j \end{vmatrix}$$

is evaluated as

$$a \begin{vmatrix} e & f \\ h & j \end{vmatrix} - b \begin{vmatrix} d & f \\ g & j \end{vmatrix} + c \begin{vmatrix} d & e \\ g & h \end{vmatrix}$$

using the top row.

3. Network analysis using Kirchhoff's laws

Kirchhoff's laws may be applied to d.c. and a.c. circuits. To demonstrate the method of analysis, consider the d.c. network shown in Fig. 1. If the current flowing in each branch is required, the following three-step procedure may be used.

(i) Label branch currents and their directions on the circuit diagram. The directions chosen are arbitrary but, as a starting-point, a useful guide is to assume that current flows from the positive terminals of the voltage sources. This is shown in Fig. 2 where the three branch currents are expressed in terms of I_1 and I_2 only, since the current through resistance R, by Kirchhoff's current law, is $(I_1 + I_2)$.

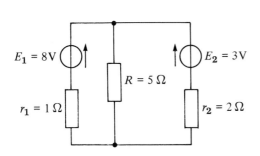

Figure 1

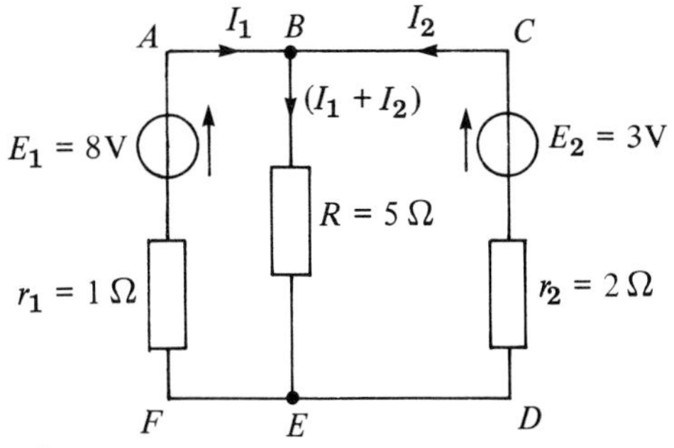

Figure 2

(ii) Divide the circuit into loops—two in this case (see Fig. 2)—and then apply Kirchhoff's voltage law to each loop in turn. From loop $ABEF$, and moving in a clockwise direction (the choice of loop direction is arbitrary), $E_1 = I_1 r + (I_1 + I_2)R$ (note that the two voltage drops are positive since the loop direction is the same as the current directions involved in the volt drops). Hence $8 = I_1 + 5(I_1 + I_2)$ or

$$6I_1 + 5I_2 = 8 \tag{1}$$

From loop $BCDE$, and moving in an anticlockwise direction as shown in Fig. 2 (note that the direction does not have to be the same as that used for the first loop), $E_2 = I_2 r_2 + (I_1 + I_2)R$, i.e., $3 = 2I_2 + 5(I_1 + I_2)$ or

$$5I_1 + 7I_2 = 3 \tag{2}$$

(iii) Solve simultaneous equations (1) and (2) for I_1 and I_2.

Multiplying equation (1) by 7 gives $\quad 42I_1 + 35I_2 = 56 \quad$ (3)

Multiplying equation (2) by 5 gives $\quad 25I_1 + 35I_2 = 15 \quad$ (4)

Equation (3) − equation (4) gives $\quad 17I_1 \qquad = 41$

from which current $I_1 = 41/17 = 2.412\,\text{A} = \textbf{2.41 A}$, correct to two dec. places.

From equation (1): $6(2.412) + 5I_2 = 8$, from which

$$\textbf{current } I_2 = \frac{8 - 6(2.412)}{5} = -1.294\,\text{A}$$

$$= -\textbf{1.29 A}, \text{ correct to two dec. places.}$$

The minus sign indicates that current I_2 flows in the opposite direction to that shown in Fig. 2.

The current flowing through resistance R is

$(I_1 + I_2) = 2.412 + (-1.294) = 1.118$ A $= \textbf{1.12 A,}$ correct to two dec. places

(A third loop may be selected in Fig. 2 (just as a check), moving clockwise around the outside of the network. Then $E_1 - E_2 = I_1 r_1 - I_2 r_2$, i.e., $8 - 3 = I_1 - 2I_2$. Thus $5 = 2.412 - 2(-1.294) = 5$.)

An alternative method of solving equations (1) and (2) is shown below using determinants. Since

$$6I_1 + 5I_2 - 8 = 0 \tag{1}$$

$$5I_1 + 7I_2 - 3 = 0 \tag{2}$$

then

$$\frac{I_1}{\begin{vmatrix} 5 & -8 \\ 7 & -3 \end{vmatrix}} = \frac{-I_2}{\begin{vmatrix} 6 & -8 \\ 5 & -3 \end{vmatrix}} = \frac{1}{\begin{vmatrix} 6 & 5 \\ 5 & 7 \end{vmatrix}}$$

i.e.,

$$\frac{I_1}{-15 + 56} = \frac{-I_2}{-18 + 40} = \frac{1}{42 - 25}$$

$$\frac{I_1}{41} = \frac{-I_2}{22} = \frac{1}{17}$$

from which $I_1 = 41/17 = \textbf{2.41 A}$ and $I_2 = -22/17 = -\textbf{1.29 A,}$ as obtained previously.

The above procedure is shown for a simple d.c. circuit having two unknown values of current. The procedure however applies equally well to a.c. networks and/or to circuits where three unknown currents are involved. This is illustrated in the following worked problems.

Worked problems on network analysis using Kirchhoff's laws

Problem 1. Use Kirchhoff's laws to find the current flowing in each branch of the network shown in Fig. 3.

(i) The branch currents and their directions are labelled as shown in Fig. 4.

Figure 3

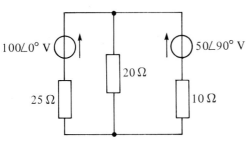

Figure 4

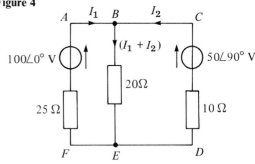

(ii) Two loops are chosen. From loop $ABEF$, and moving clockwise,

$$25I_1 + 20(I_1 + I_2) = 100\underline{/\ 0°}$$

i.e.,

$$45I_1 + 20I_2 = 100 \tag{1}$$

From loop $BCDE$, and moving anticlockwise,

$$10I_2 + 20(I_1 + I_2) = 50\underline{/\ 90°}$$

i.e.,

$$20I_1 + 30I_2 = j50 \tag{2}$$

(iii) $3 \times$ equation (1) gives $135I_1 + 60I_2 = 300$ (3)

$2 \times$ equation (2) gives $40I_1 + 60I_2 = j100$ (4)

Equation (3) − equation (4) gives $95I_1 = 300 - j100$, from which

$$\textbf{current } I_1 = \frac{300 - j100}{95} = \textbf{3.329}\underline{/-\textbf{18.43°}}\textbf{ A or (3.158} - \textbf{\textit{j}1.052)A}$$

Substituting in equation (1) gives $45(3.158 - j1.052) + 20I_2 = 100$, from which

$$I_2 = \frac{100 - 45(3.158 - j1.052)}{20}$$

$$= (-\textbf{2.106} + \textbf{\textit{j}2.367})\textbf{ A or 3.168}\underline{/\ \textbf{131.66°}}\textbf{ A}$$

Thus $I_1 + I_2 = (3.158 - j1.052) + (-2.106 + j2.367) = \textbf{(1.052} + \textbf{\textit{j}1.315) A}$ or **1.684**$\underline{/\ \textbf{51.34°}}$ **A.**

Problem 2. Determine the current flowing in the $2\,\Omega$ resistor of the circuit shown in Fig. 5 by using Kirchhoff's laws. Find also the power dissipated in the $3\,\Omega$ resistance.

(i) Currents and their directions are assigned as shown in Fig. 6.

(ii) Three loops are chosen since three unknown currents are required. The choice of loop directions is arbitrary. From loop $ABCDE$, and moving anticlockwise,

$$5I_1 + 6I_2 + 4(I_2 - I_3) = 8$$

i.e.,

$$5I_1 + 10I_2 - 4I_3 = 8 \tag{1}$$

From loop $EDGF$, and moving clockwise,

$$6I_2 + 2I_3 - 1(I_1 - I_2) = 0$$

i.e.,

$$-I_1 + 7I_2 + 2I_3 = 0 \tag{2}$$

Figure 5

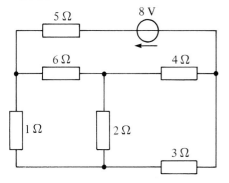

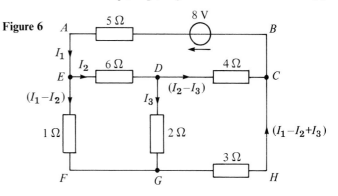

Figure 6

From loop *DCHG*, and moving anticlockwise,

$$2I_3 + 3(I_1 - I_2 + I_3) - 4(I_2 - I_3) = 0$$

i.e.,

$$3I_1 - 7I_2 + 9I_3 = 0 \tag{3}$$

(iii) Thus

$$5I_1 + 10I_2 - 4I_3 - 8 = 0$$
$$-I_1 + 7I_2 + 2I_3 - 0 = 0$$
$$3I_1 - 7I_2 + 9I_3 - 0 = 0$$

Hence, using determinants,

$$\frac{I_1}{\begin{vmatrix} 10 & -4 & -8 \\ 7 & 2 & 0 \\ -7 & 9 & 0 \end{vmatrix}} = \frac{-I_2}{\begin{vmatrix} 5 & -4 & -8 \\ -1 & 2 & 0 \\ 3 & 9 & 0 \end{vmatrix}} = \frac{I_3}{\begin{vmatrix} 5 & 10 & -8 \\ -1 & 7 & 0 \\ 3 & -7 & 0 \end{vmatrix}} = \frac{-1}{\begin{vmatrix} 5 & 10 & -4 \\ -1 & 7 & 2 \\ 3 & -7 & 9 \end{vmatrix}}$$

Thus

$$\frac{I_1}{-8\begin{vmatrix} 7 & 2 \\ -7 & 9 \end{vmatrix}} = \frac{-I_2}{-8\begin{vmatrix} -1 & 2 \\ 3 & 9 \end{vmatrix}} = \frac{I_3}{-8\begin{vmatrix} -1 & 7 \\ 3 & -7 \end{vmatrix}}$$

$$= \frac{-1}{5\begin{vmatrix} 7 & 2 \\ -7 & 9 \end{vmatrix} - 10\begin{vmatrix} -1 & 2 \\ 3 & 9 \end{vmatrix} - 4\begin{vmatrix} -1 & 7 \\ 3 & -7 \end{vmatrix}}$$

$$\frac{I_1}{-8(63+14)} = \frac{-I_2}{-8(-9-6)} = \frac{I_3}{-8(7-21)} = \frac{-1}{5(63+14) - 10(-9-6) - 4(7-21)}$$

$$\frac{I_1}{-616} = \frac{-I_2}{120} = \frac{I_3}{112} = \frac{-1}{591}$$

Hence

$$I_1 = \frac{616}{591} = 1.042 \text{ A}, \qquad I_2 = \frac{120}{591} = 0.203 \text{ A}$$

and

$$I_3 = \frac{-112}{591} = -0.190 \text{ A}$$

Thus the current flowing in the $2\,\Omega$ resistance **is 0.190 A** in the opposite direction to that shown in Fig. 2.

Current in the $3\,\Omega$ resistance $= I_1 - I_2 + I_3 = 1.042 - 0.203 + (-0.190) = 0.649$ A. Hence power dissipated in the $3\,\Omega$ resistance, $I^2(3) = (0.649)^2(3) = \textbf{1.26 W}$.

Problem 3. For the a.c. network shown in Fig. 7, determine the current flowing in each branch using Kirchhoff's laws.

(i) Currents I_1 and I_2 with their directions are shown in Fig. 8.
(ii) Two loops are chosen with their directions both clockwise. From loop *ABEF*,

$$(5 + j0) = I_1(3 + j4) + (I_1 - I_2)(6 + j8)$$

i.e.,

$$5 = (9 + j12)I_1 - (6 + j8)I_2 \tag{1}$$

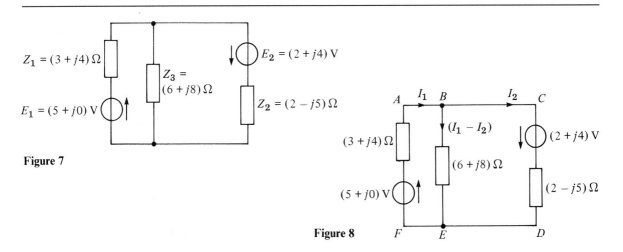

Figure 7

Figure 8

From loop $BCDE$,

$$(2+j4) = I_2(2-j5) - (I_1 - I_2)(6+j8)$$

i.e.,

$$(2+j4) = -(6+j8)I_1 + (8+j3)I_2 \qquad (2)$$

(iii) Multiplying equation (1) by $(8+j3)$ gives

$$5(8+j3) = (8+j3)(9+j12)I_1 - (8+j3)(6+j8)I_2 \qquad (3)$$

Multiplying equation (2) by $(6+j8)$ gives

$$(6+j8)(2+j4) = -(6+j8)(6+j8)I_1 + (6+j8)(8+j3)I_2 \qquad (4)$$

Adding equations (3) and (4) gives

$$5(8+j3) + (6+j8)(2+j4) = [(8+j3)(9+j12) - (6+j8)(6+j8)]I_1$$

i.e.,

$$(20+j55) = (64+j27)I_1$$

from which

$$I_1 = \frac{20+j55}{64+j27} = \frac{58.52 \angle 70.02°}{69.46 \angle 22.87°} = \mathbf{0.842 \angle 47.15°\ A}$$

$$= (0.573 + j0.617)\,\text{A} = \mathbf{(0.57 + j0.62)\,A}, \text{ correct to two dec. places}$$

From equation (1),

$$5 = (9+j12)(0.573+j0.617) - (6+j8)I_2$$
$$5 = (-2.247 + j12.429) - (6+j8)I_2$$

from which

$$I_2 = \frac{-2.247 + j12.429 - 5}{(6+j8)} = \frac{14.39 \angle 120.25°}{10 \angle 53.13°}$$

$$= \mathbf{1.439 \angle 67.12°\ A} = \mathbf{(0.559 + j1.326)\,A}$$

$$= \mathbf{(0.56 + j1.33)\,A}, \text{ correct to two dec. places}$$

The current in the $(6+j8)\,\Omega$ impedance,

$$I_1 - I_2 = (0.573 + j0.617) - (0.559 + j1.326)$$

$$= \mathbf{(0.01 - j0.71)\,A} \text{ or } \mathbf{0.71 \angle -88.87°\ A}$$

An alternative method of solving equations (1) and (2) is shown below, using determinants.

$$(9+j12)I_1 - (6+j8)I_2 - 5 = 0 \tag{1}$$

$$-(6+j8)I_1 + (8+j3)I_2 - (2+j4) = 0 \tag{2}$$

Thus

$$\frac{I_1}{\begin{vmatrix} -(6+j8) & -5 \\ (8+j3) & -(2+j4) \end{vmatrix}} = \frac{-I_2}{\begin{vmatrix} (9+j12) & -5 \\ -(6+j8) & -(2+j4) \end{vmatrix}} = \frac{1}{\begin{vmatrix} (9+j12) & -(6+j8) \\ -(6+j8) & (8+j3) \end{vmatrix}}$$

$$\frac{I_1}{(-20+j40)+(40+j15)} = \frac{-I_2}{(30-j60)-(30+j40)} = \frac{1}{(36+j123)-(-28+j96)}$$

$$\frac{I_1}{20+j55} = \frac{-I_2}{-j100} = \frac{1}{64+j27}$$

Hence

$$I_1 = \frac{20+j55}{64+j27} = \frac{58.52\angle70.02°}{69.46\angle22.87°} = \textbf{0.84}\angle\textbf{47.15° A}$$

and

$$I_2 = \frac{100\angle90°}{69.46\angle22.87°} = \textbf{1.44}\angle\textbf{67.13° A}$$

The current flowing in the $(6+j8)\,\Omega$ impedance is given by

$$I_1 - I_2 = 0.842\angle47.15° - 1.440\angle67.13°\text{ A}$$
$$= \textbf{(0.013}-\textbf{j0.709) A}\text{ or }\textbf{0.71}\angle\textbf{-88.96° A}$$

Problem 4. For the network shown in Fig. 9, use Kirchhoff's laws to determine the magnitude of the current in the $(4+j3)\,\Omega$ impedance.

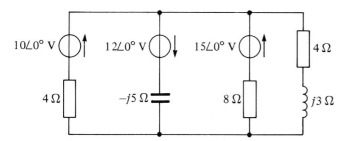

Figure 9

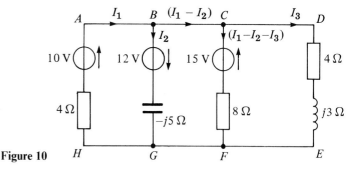

Figure 10

(i) Currents I_1, I_2 and I_3 with their directions are shown in Fig. 10. The current in the $(4 + j3)\,\Omega$ impedance is specified by one symbol only (i.e., I_3), which means that the three equations formed need to be solved for only one unknown current.

(ii) Three loops are chosen. From loop $ABGH$, and moving clockwise,

$$4I_1 - j5I_2 = 10 + 12 \tag{1}$$

From loop $BCFG$, and moving anticlockwise,

$$-j5I_2 - 8(I_1 - I_2 - I_3) = 15 + 12 \tag{2}$$

From loop $CDEF$, and moving clockwise,

$$-8(I_1 - I_2 - I_3) + (4 + j3)(I_3) = 15 \tag{3}$$

Hence

$$4I_1 - j5I_2 + 0I_3 - 22 = 0$$
$$-8I_1 + (8 - j5)I_2 + 8I_3 - 27 = 0$$
$$-8I_1 + 8I_2 + (12 + j3)I_3 - 15 = 0$$

Solving for I_3 using determinants gives

$$\frac{I_3}{\begin{vmatrix} 4 & -j5 & -22 \\ -8 & (8-j5) & -27 \\ -8 & 8 & -15 \end{vmatrix}} = \frac{-1}{\begin{vmatrix} 4 & -j5 & 0 \\ -8 & (8-j5) & 8 \\ -8 & 8 & (12+j3) \end{vmatrix}}$$

Thus

$$\frac{I_3}{4\begin{vmatrix}(8-j5) & -27 \\ 8 & -15\end{vmatrix} + j5\begin{vmatrix}-8 & -27 \\ -8 & -15\end{vmatrix} - 22\begin{vmatrix}-8 & (8-j5) \\ -8 & 8\end{vmatrix}}$$

$$= \frac{-1}{4\begin{vmatrix}(8-j5) & 8 \\ 8 & (12+j3)\end{vmatrix} + j5\begin{vmatrix}-8 & 8 \\ -8 & (12+j3)\end{vmatrix}}$$

$$\frac{I_3}{384 + j700} = \frac{-1}{308 - j304}$$

and

$$I_3 = \frac{-(384 + j700)}{(308 - j304)} = \frac{798.41\angle -118.75°}{432.76\angle -44.63°} = 1.85\angle -74.12\ \text{A}$$

Hence the magnitude of the current flowing in the $(4 + j3)\,\Omega$ impedance is 1.85 A.

Further problems on network analysis using Kirchhoff's laws may be found in section 7, problems 1 to 11, page 145.

4. Mesh-current analysis

Mesh-current analysis is merely an extension of the use of Kirchhoff's laws. Figure 11 shows a network whose circulating currents I_1, I_2 and I_3 have been assigned to closed loops in the circuit rather than to branches. Currents I_1, I_2 and I_3 are called mesh-currents or loop-currents.

In mesh-current analysis the loop-currents are all arranged to circulate in the same direction (in Fig. 11, shown as clockwise direction). Kirchhoff's

Figure 11

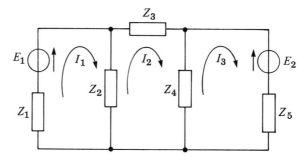

second law is applied to each of the loops in turn, which in the circuit of Fig. 11 produces three equations in three unknowns which may be solved for I_1, I_2 and I_3. The three equations produced from Fig. 11 are

$$I_1(Z_1 + Z_2) - I_2 Z_2 = E_1$$
$$I_2(Z_2 + Z_3 + Z_4) - I_1 Z_2 - I_3 Z_4 = 0$$
$$I_3(Z_4 + Z_5) - I_2 Z_4 = -E_2$$

The branch currents are determined by taking the phasor sum of the mesh currents common to that branch. For example, the current flowing in impedance Z_2 of Fig. 11 is given by $(I_1 - I_2)$ phasorially. The method of mesh-current analysis, called **Maxwell's theorem**, is demonstrated in the following worked problems.

Worked problems on mesh-current analysis

Problem 1. Use mesh-current analysis to determine the current flowing in (a) the 5 Ω resistance, and (b) the 1 Ω resistance of the d.c. circuit shown in Fig. 12.

Figure 12

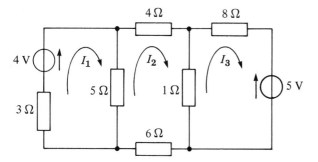

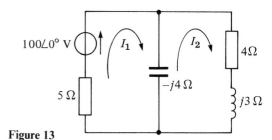

Figure 13

The mesh currents I_1, I_2 and I_3 are shown in Fig. 12.

For loop 1, $(3+5)I_1 - 5I_2 = 4$ using Kirchhoff's second law   (1)
For loop 2, $(4+1+6+5)I_2 - 5I_1 - (1)I_3 = 0$ (2)
For loop 3, $(1+8)I_3 - (1)I_2 = -5$ (3)

Thus

$$8I_1 - 5I_2 \qquad\qquad - 4 = 0 \tag{1'}$$

$$-5I_1 + 16I_2 - I_3 \qquad = 0 \tag{2'}$$

$$-I_2 + 9I_3 + 5 = 0 \tag{3'}$$

Using determinants,

$$\frac{I_1}{\begin{vmatrix} -5 & 0 & -4 \\ 16 & -1 & 0 \\ -1 & 9 & 5 \end{vmatrix}} = \frac{-I_2}{\begin{vmatrix} 8 & 0 & -4 \\ -5 & -1 & 0 \\ 0 & 9 & 5 \end{vmatrix}} = \frac{I_3}{\begin{vmatrix} 8 & -5 & -4 \\ -5 & 16 & 0 \\ 0 & -1 & 5 \end{vmatrix}} = \frac{-1}{\begin{vmatrix} 8 & -5 & 0 \\ -5 & 16 & -1 \\ 0 & -1 & 9 \end{vmatrix}}$$

$$\frac{I_1}{-5\begin{vmatrix} -1 & 0 \\ 9 & 5 \end{vmatrix} - 4\begin{vmatrix} 16 & -1 \\ -1 & 9 \end{vmatrix}} = \frac{-I_2}{8\begin{vmatrix} -1 & 0 \\ 9 & 5 \end{vmatrix} - 4\begin{vmatrix} -5 & -1 \\ 0 & 9 \end{vmatrix}}$$

$$= \frac{I_3}{-4\begin{vmatrix} -5 & 16 \\ 0 & -1 \end{vmatrix} + 5\begin{vmatrix} 8 & -5 \\ -5 & 16 \end{vmatrix}} = \frac{-1}{8\begin{vmatrix} 16 & -1 \\ -1 & 9 \end{vmatrix} + 5\begin{vmatrix} -5 & -1 \\ 0 & 9 \end{vmatrix}}$$

$$\frac{I_1}{-5(-5)-4(143)} = \frac{-I_2}{8(-5)-4(-45)} = \frac{I_3}{-4(5)+5(103)} = \frac{-1}{8(143)+5(-45)}$$

$$\frac{I_1}{-547} = \frac{-I_2}{140} = \frac{I_3}{495} = \frac{-1}{919}$$

Hence

$$I_1 = \frac{547}{919} = 0.595 \text{ A}, \qquad I_2 = \frac{140}{919} = 0.152 \text{ A}$$

and

$$I_3 = -\frac{495}{919} = -0.539 \text{ A}$$

Thus **current in the 5 Ω resistance** $= I_1 - I_2 = 0.595 - 0.152 = \mathbf{0.44 \ A}$,
current in the 1 Ω resistance $= I_2 - I_3 = 0.152 - (-0.539) = \mathbf{0.69 \ A}$.

Problem 2. For the a.c. network shown in Fig. 13 determine, using mesh-current analysis, (a) the mesh currents I_1 and I_2, (b) the current flowing in the capacitor, and (c) the active power delivered by the $100 \underline{/ \ 0^\circ}$ V voltage source.

(a) For the first loop $(5-j4)I_1 - (-j4I_2) = 100\underline{/ \ 0^\circ}$ (1)
For the second loop $(4+j3-j4)I_2 - (-j4I_1) = 0$ (2)
Rewriting equations (1) and (2) gives

$$(5-j4)I_1 + j4I_2 - 100 = 0 \qquad (1')$$

$$j4I_1 + (4-j)I_2 = 0 \qquad (2')$$

Thus, using determinants,

$$\frac{I_1}{\begin{vmatrix} j4 & -100 \\ (4-j) & 0 \end{vmatrix}} = \frac{-I_2}{\begin{vmatrix} (5-j4) & -100 \\ j4 & 0 \end{vmatrix}} = \frac{1}{\begin{vmatrix} (5-j4) & j4 \\ j4 & (4-j) \end{vmatrix}}$$

$$\frac{I_1}{(400-j100)} = \frac{-I_2}{j400} = \frac{1}{(32-j21)}$$

Hence

$$I_1 = \frac{(400-j100)}{(32-j21)} = \frac{412.31\underline{/ \ -14.04^\circ}}{38.28\underline{/ \ -33.27^\circ}} = 10.77\underline{/ \ 19.23^\circ} \text{ A}$$

$$= \mathbf{10.8\underline{/ \ 19.2^\circ} \ A}, \text{ correct to one dec. place.}$$

$$I_2 = \frac{400\underline{/-90^\circ}}{38.28\underline{/-33.27^\circ}} = 10.45\underline{/-56.73^\circ}\,\text{A}$$

$$= \mathbf{10.5\underline{/-56.7^\circ}\,A}, \text{ correct to one dec. place.}$$

(b) Current flowing in capacitor $= I_1 - I_2 = 10.77\underline{/19.23^\circ} - 10.45\underline{/-56.73^\circ}$
$$= 4.44 + j12.28 = 13.1\underline{/70.12^\circ}\,\text{A},$$

i.e., **the current in capacitor is 13.1 A.**

(c) Source power $P = VI\cos\phi = (100)(10.77)\cos 19.23^\circ = \mathbf{1016.9\,W}$
$$= 1020\,\text{W}, \text{ correct to three sig. figures.}$$

(Check: power in $5\,\Omega$ resistor $= I_1^2(5) = (10.77)^2(5) = 579.97\,\text{W}$
and power in $4\,\Omega$ resistor $= I_2^2(4) = (10.45)^2(4) = 436.81\,\text{W}$.
Thus total power dissipated $= 579.97 + 436.81 = 1016.8\,\text{W} = 1020\,\text{W}$, correct to three sig. figures.)

Problem 3. A balanced star-connected 3-phase load is shown in Fig. 14. Determine the value of the line currents I_R, I_Y and I_B using mesh-current analysis.

Two mesh currents I_1 and I_2 are chosen as shown in Fig. 14.
From loop 1, $\quad I_1(3+j4) + I_1(3+j4) - I_2(3+j4) = 415\underline{/120^\circ}$

i.e., $\quad (6+j8)I_1 - (3+j4)I_2 - 415\underline{/120^\circ} = 0$ \hfill (1)

From loop 2, $\quad I_2(3+j4) - I_1(3+j4) + I_2(3+j4) = 415\underline{/0^\circ}$

i.e., $\quad -(3+j4)I_1 + (6+j8)I_2 - 415\underline{/0^\circ} = 0$ \hfill (2)

Solving equations (1) and (2) using determinants gives

$$\frac{I_1}{\begin{vmatrix} -(3+j4) & -415\underline{/120^\circ} \\ (6+j8) & -415\underline{/0^\circ} \end{vmatrix}} = \frac{-I_2}{\begin{vmatrix} (6+j8) & -415\underline{/120^\circ} \\ -(3+j4) & -415\underline{/0^\circ} \end{vmatrix}} = \frac{1}{\begin{vmatrix} (6+j8) & -(3+j4) \\ -(3+j4) & (6+j8) \end{vmatrix}}$$

$$\frac{I_1}{2075\underline{/53.13^\circ} + 4150\underline{/173.13^\circ}} = \frac{-I_2}{-4150\underline{/53.13^\circ} - 2075\underline{/173.13^\circ}}$$

$$= \frac{1}{100\underline{/106.26^\circ} - 25\underline{/106.26^\circ}}$$

$$\frac{I_1}{3594\underline{/143.13^\circ}} = \frac{I_2}{3594\underline{/83.13^\circ}} = \frac{1}{75\underline{/106.26^\circ}}$$

Hence

$$I_1 = \frac{3594\underline{/143.13^\circ}}{75\underline{/106.26^\circ}} = 47.9\underline{/36.87^\circ}\,\text{A}$$

$$I_2 = \frac{3594\underline{/83.13^\circ}}{75\underline{/106.26^\circ}} = 47.9\underline{/-23.13^\circ}\,\text{A}$$

Thus

line current $I_R = I_1 = \mathbf{47.9\underline{/36.87^\circ}\,A}$

$$I_B = -I_2 = -(47.9\underline{/-23.13^\circ}\,\text{A}) = \mathbf{47.9\underline{/156.87^\circ}\,A}$$

and

$$I_Y = I_2 - I_1 = 47.9\underline{/-23.13^\circ} - 47.9\underline{/36.87^\circ}$$
$$= \mathbf{47.9\underline{/-83.13^\circ}\,A}$$

Further problems on mesh-current analysis may be found in section 7, problems 12 to 21, page 147.

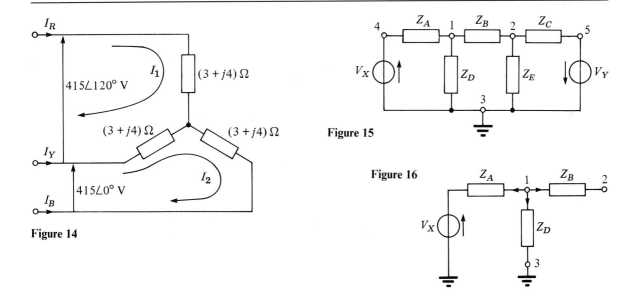

Figure 14

Figure 15

Figure 16

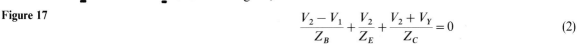

5. Nodal analysis

A **node** of a network is defined as a point where two or more branches are joined. If three or more branches join at a node, then that node is called a **principal node** or **junction**. In Fig. 15, points 1, 2, 3, 4 and 5 are nodes, and points 1, 2 and 3 are principal nodes.

A node voltage is the voltage of a particular node with respect to a node called the reference node. If in Fig. 15, for example, node 3 is chosen as the reference node then V_{13} is assumed to mean the voltage at node 1 with respect to node 3 (as distinct from V_{31}). Similarly, V_{23} would be assumed to mean the voltage at node 2 with respect to node 3, and so on. However, since the node voltage is always determined with respect to a particular chosen reference node, the notation V_1 for V_{13} and V_2 for V_{23} would always be used in this instance.

The object of nodal analysis is to determine the values of voltages at all the principal nodes with respect to the reference node, e.g., to find voltages V_1 and V_2 in Fig. 15. When such voltages are determined, the currents flowing in each branch can be found.

Kirchhoff's current law is applied to nodes 1 and 2 in turn in Fig. 15 and two equations in unknowns V_1 and V_2 are obtained which may be simultaneously solved using determinants.

The branches leading to node 1 are shown separately in Fig. 16. Let us assume that all branch currents are leaving the node as shown. Since the sum of currents at a junction is zero,

$$\frac{V_1 - V_X}{Z_A} + \frac{V_1}{Z_D} + \frac{V_1 - V_2}{Z_B} = 0 \qquad (1)$$

Similarly, for node 2, assuming all branch currents are leaving the node as shown in Fig. 17,

$$\frac{V_2 - V_1}{Z_B} + \frac{V_2}{Z_E} + \frac{V_2 + V_Y}{Z_C} = 0 \qquad (2)$$

Figure 17

In equations (1) and (2), the currents are all assumed to be leaving the node. In fact, any selection in the direction of the branch currents may be made—the resulting equations will be identical. (For example, if for node 1 the current flowing in Z_B is considered as flowing towards node 1 instead of away, then the equation for node 1 becomes

$$\frac{V_1 - V_X}{Z_A} + \frac{V_1}{Z_D} = \frac{V_2 - V_1}{Z_B},$$

which if rearranged is seen to be exactly the same as equation (1).)

Rearranging equations (1) and (2) gives

$$\left(\frac{1}{Z_A} + \frac{1}{Z_B} + \frac{1}{Z_D}\right)V_1 - \left(\frac{1}{Z_B}\right)V_2 - \left(\frac{1}{Z_A}\right)V_X = 0 \qquad (3)$$

$$-\left(\frac{1}{Z_B}\right)V_1 + \left(\frac{1}{Z_B} + \frac{1}{Z_C} + \frac{1}{Z_E}\right)V_2 + \left(\frac{1}{Z_C}\right)V_Y = 0 \qquad (4)$$

Equations (3) and (4) may be rewritten in terms of admittances (where admittance $Y = 1/Z$):

$$(Y_A + Y_B + Y_D)V_1 - Y_B V_2 - Y_A V_X = 0 \qquad (5)$$

$$- Y_B V_1 + (Y_B + Y_C + Y_E)V_2 + Y_C V_Y = 0 \qquad (6)$$

Equations (5) and (6) may be solved for V_1 and V_2 by using determinants. Thus

$$\frac{V_1}{\begin{vmatrix} -Y_B & -Y_A \\ (Y_B + Y_C + Y_E) & Y_C \end{vmatrix}} = \frac{-V_2}{\begin{vmatrix} (Y_A + Y_B + Y_D) & -Y_A \\ -Y_B & Y_C \end{vmatrix}}$$

$$= \frac{1}{\begin{vmatrix} (Y_A + Y_B + Y_D) & -Y_B \\ -Y_B & (Y_B + Y_C + Y_E) \end{vmatrix}}$$

Current equations, and hence voltage equations, may be written at each principal node of a network with the exception of a reference node. The number of equations necessary to produce a solution for a circuit is, in fact, always one less than the number of principal nodes.

Whether mesh-current analysis or nodal analysis is used to determine currents in circuits depends on the number of loops and nodes the circuit contains. Basically, the method that requires the least number of equations is used. The method of nodal analysis is demonstrated in the following worked problems.

Worked problems on nodal analysis

Problem 1. For the network shown in Fig. 18, determine the voltage V_{AB} by using nodal analysis.

Figure 18 contains two principal nodes (at 1 and B) and thus only one nodal equation is required. B is taken as the reference node and the equation for node 1 is obtained as follows. Applying Kirchhoff's current law to node 1 gives

$$I_X + I_Y = I$$

i.e.,

$$\frac{V_1}{16} + \frac{V_1}{(4+j3)} = 20\underline{/0^\circ}$$

Thus

$$V_1\left(\frac{1}{16} + \frac{1}{4+j3}\right) = 20$$

$$V_1\left(0.0625 + \frac{4-j3}{4^2+3^2}\right) = 20$$

$$V_1(0.0625 + 0.16 - j0.12) = 20$$
$$V_1(0.2225 - j0.12) = 20$$

from which

$$V_1 = \frac{20}{(0.2225 - j0.12)} = \frac{20}{0.2528\underline{/-28.34^\circ}}$$

i.e.,

$$\text{voltage } V_1 = 79.1\underline{/28.34^\circ}\text{ V}$$

The current through the $(4+j3)\,\Omega$ branch, $I_Y = V_1/(4+j3)$. Hence the voltage drop between points A and B, i.e., across the $4\,\Omega$ resistance, is given by

$$V_{AB} = (I_Y)(4) = \frac{V_1(4)}{(4+j3)} = \frac{79.1\underline{/28.34^\circ}}{5\underline{/36.87^\circ}}(4) = \mathbf{63.3\underline{/-8.53^\circ}\text{ V}}$$

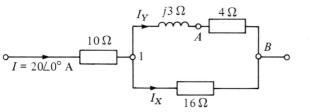

Figure 18   **Figure 19**

Problem 2. Determine the value of voltage V_{XY} shown in the circuit of Fig. 19.

The circuit contains no principal nodes. However, if point Y is chosen as the reference node then an equation may be written for node X assuming that current leaves point X by both branches. Thus

$$\frac{V_X - 8\underline{/0^\circ}}{(5+4)} + \frac{V_X - 8\underline{/90^\circ}}{(3+j6)} = 0$$

from which

$$V_X\left(\frac{1}{9} + \frac{1}{3+j6}\right) = \frac{8}{9} + \frac{j8}{3+j6}$$

$$V_X(0.1778 - j0.1333) = 0.8889 + \frac{j8(3-j6)}{45}$$

$$V_X(0.2222\underline{/-36.86^\circ}) = 1.9556 + j0.5333 = 2.027\underline{/15.25^\circ}$$

Since point Y is the reference node,

$$\text{voltage } V_X = V_{XY} = \frac{2.027 \angle 15.25°}{0.2222 \angle -36.86°} = 9.12 \angle 52.11° \text{ V}$$

Problem 3. Use nodal analysis to determine the current flowing in each branch of the network shown in Fig. 20.

This is the same problem as problem 1 of section 3, page 118, which was solved using Kirchhoff's laws. A comparison of methods can be made.

There are only two principal nodes in Fig. 20 so only one nodal equation is required. Node 2 is taken as the reference node.

The equation at node 1 is $I_1 + I_2 + I_3 = 0$

i.e.,

$$\frac{V_1 - 100 \angle 0°}{25} + \frac{V_1}{20} + \frac{V_1 - 50 \angle 90°}{10} = 0$$

i.e.,

$$\left(\frac{1}{25} + \frac{1}{20} + \frac{1}{10}\right) V_1 - \frac{100 \angle 0°}{25} - \frac{50 \angle 90°}{10} = 0$$

$$0.19 V_1 = 4 + j5$$

Thus the voltage at node 1,

$$V_1 = \frac{4 + j5}{0.19} = 33.70 \angle 51.34° \text{ V} \quad \text{or} \quad (21.05 + j26.32) \text{ V}$$

Hence the current in the 25 Ω resistance,

$$I_1 = \frac{V_1 - 100 \angle 0°}{25} = \frac{21.05 + j26.32 - 100}{25}$$

$$= \frac{-78.95 + j26.32}{25}$$

$$= 3.33 \angle 161.56° \text{ A flowing away from node 1}$$

(or $3.33 \angle (161.56° - 180°)$ A $= 3.33 \angle -18.44°$ A flowing toward node 1).

The current in the 20 Ω resistance,

$$I_2 = \frac{V_1}{20} = \frac{33.70 \angle 51.34°}{20} = 1.69 \angle 51.34° \text{ A,}$$

flowing from node 1 to node 2.

The current in the 10 Ω resistor,

$$I_3 = \frac{V_1 - 50 \angle 90°}{10} = \frac{21.05 + j26.32 - j50}{10} = \frac{21.05 - j23.68}{10}$$

$$= 3.17 \angle -48.36° \text{ A away from node 1}$$

(or $3.17 \angle (-48.36° - 180°) = 3.17 \angle -228.36°$ A $= 3.17 \angle 131.64°$ A toward node 1.)

Problem 4. In the network of Fig. 21 use nodal analysis to determine (a) the voltage at nodes 1 and 2, (b) the current in the $j4$ Ω inductance, (c) the current in the 5 Ω resistance, and (d) the magnitude of the active power dissipated in the 2.5 Ω resistance.

Figure 20

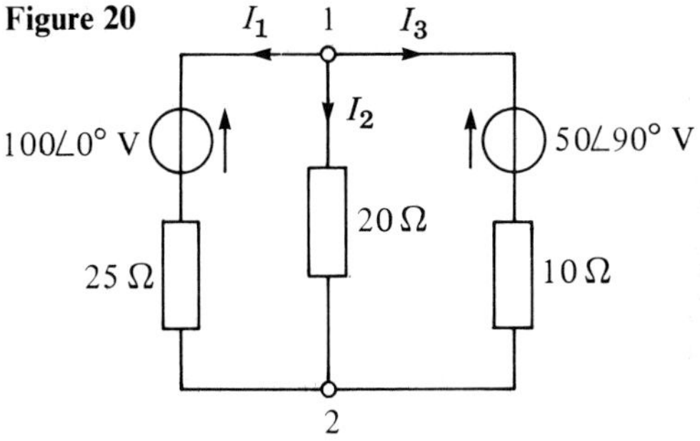

Figure 21

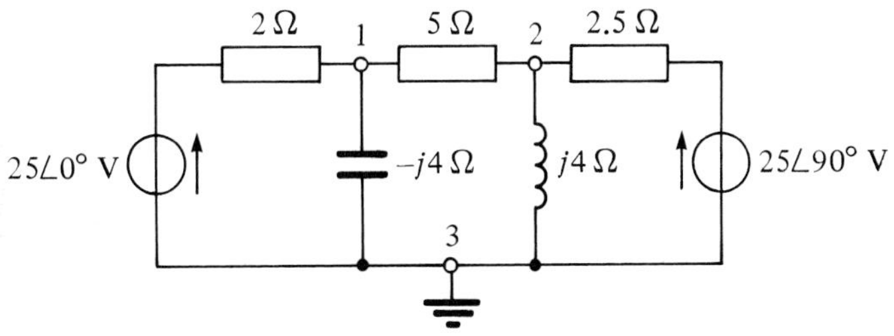

(a) At node 1,

$$\frac{V_1 - 25\angle\,0°}{2} + \frac{V_1}{-j4} + \frac{V_1 - V_2}{5} = 0$$

Rearranging gives

$$\left(\frac{1}{2} + \frac{1}{-j4} + \frac{1}{5}\right)V_1 - \left(\frac{1}{5}\right)V_2 - \frac{25\angle\,0°}{2} = 0$$

i.e.,

$$(0.7 + j0.25)V_1 - 0.2V_2 - 12.5 = 0 \tag{1}$$

At node 2,

$$\frac{V_2 - 25\angle\,90°}{2.5} + \frac{V_2}{j4} + \frac{V_2 - V_1}{5} = 0$$

Rearranging gives

$$-\left(\frac{1}{5}\right)V_1 + \left(\frac{1}{2.5} + \frac{1}{j4} + \frac{1}{5}\right)V_2 - \frac{25\angle\,90°}{2.5} = 0$$

i.e.,

$$-0.2V_1 + (0.6 - j0.25)V_2 - j10 = 0 \tag{2}$$

Thus two simultaneous equations have been formed with two unknowns, V_1 and V_2. Using determinants, if

$$(0.7 + j0.25)V_1 - 0.2V_2 - 12.5 = 0 \tag{1}$$

and

$$-0.2V_1 + (0.6 - j0.25)V_1 - j10 = 0 \tag{2}$$

then

$$\frac{V_1}{\begin{vmatrix} -0.2 & -12.5 \\ (0.6-j0.25) & -j10 \end{vmatrix}} = \frac{-V_2}{\begin{vmatrix} (0.7+j0.25) & -12.5 \\ -0.2 & -j10 \end{vmatrix}} = \frac{1}{\begin{vmatrix} (0.7+j0.25) & -0.2 \\ -0.2 & (0.6-j0.25) \end{vmatrix}}$$

i.e.,

$$\frac{V_1}{(j2 + 7.5 - j3.125)} = \frac{-V_2}{(-j7 + 2.5 - 2.5)} = \frac{1}{(0.42 - j0.175 + j0.15 + 0.0625 - 0.04)}$$

and

$$\frac{V_1}{7.584\angle\,-8.53°} = \frac{-V_2}{-7\angle\,90°} = \frac{1}{0.443\angle\,-3.23°}$$

Thus

$$\text{Voltage, } V_1 = \frac{7.584 \angle -8.53°}{0.443 \angle -3.23°} = 17.12 \angle -5.30° \text{ V}$$

$$= 17.1 \angle -5.3° \text{ V, correct to one dec. place}$$

and

$$\text{Voltage, } V_2 = \frac{7 \angle 90°}{0.443 \angle -3.23°} = 15.80 \angle 93.23° \text{ V}$$

$$= 15.8 \angle 93.2° \text{ V, correct to one dec. place}$$

(b) The current in the $j4\,\Omega$ inductance is given by,

$$\frac{V_2}{j4} = \frac{15.80 \angle 93.23°}{4 \angle 90°}$$

$$= 3.95 \angle 3.23° \text{ A flowing toward node 2.}$$

(c) The current in the $5\,\Omega$ resistance is given by

$$I_5 = \frac{V_1 - V_2}{5} = \frac{17.12 \angle -5.30° - 15.80 \angle 93.23°}{5}$$

i.e.,

$$I_5 = \frac{(17.05 - j1.58) - (-0.89 + j15.77)}{5} = \frac{17.94 - j17.35}{5}$$

$$= \frac{24.96 \angle -44.04°}{5} = 4.99 \angle -44.04° \text{ A flowing from node 1 to node 2.}$$

(d) The active power dissipated in the 2.5 Ω resistor is given by $P_{2.5} = (I_{2.5})^2 (2.5)$ i.e.,

$$P_{2.5} = \left(\frac{V_2 - 25 \angle 90°}{2.5} \right)^2 (2.5) = \frac{(-0.89 + j15.77 - j25)^2}{2.5}$$

$$= \frac{(9.273 \angle -95.51°)^2}{2.5} = \frac{85.99 \angle -191.02°}{2.5} \quad \text{by de Moivre's theorem}$$

$$= 34.4 \angle 169° \text{ W}$$

Thus the magnitude of the active power dissipated in the 2.5 Ω resistance is 34.4 W.

Problem 5. In the network shown in Fig. 22 determine the voltage V_{XY} using nodal analysis.

Node 3 is taken as the reference node.

At node 1, $25 \angle 0° = \dfrac{V_1}{4 + j3} + \dfrac{V_1 - V_2}{5}$

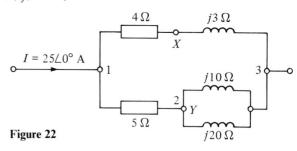

Figure 22

i.e.,
$$\left(\frac{4-j3}{25}+\frac{1}{5}\right)V_1 \ -\frac{1}{5}V_2-25=0$$

or
$$(0.379\angle-18.43°)V_1-0.2V_2-25=0 \tag{1}$$

At node 2, $\dfrac{V_2}{j10}+\dfrac{V_2}{j20}+\dfrac{V_2-V_1}{5}=0$

i.e.,
$$-0.2V_1+\left(\frac{1}{j10}+\frac{1}{j20}+\frac{1}{5}\right)V_2=0$$

or
$$-0.2V_1+(-j0.1-j0.05+0.2)V_2=0$$

i.e.,
$$-0.2V_1+(0.25\angle-36.87°)V_2=0 \tag{2}$$

Simultaneous equations (1) and (2) may be solved for V_1 and V_2 by using determinants. Thus,

$$\frac{V_1}{\begin{vmatrix} -0.2 & -25 \\ 0.25\angle-36.87° & 0 \end{vmatrix}}=\frac{-V_2}{\begin{vmatrix} 0.379\angle-18.43° & -25 \\ -0.2 & 0 \end{vmatrix}}$$

$$=\frac{1}{\begin{vmatrix} 0.379\angle-18.43° & -0.2 \\ -0.2 & 0.25\angle-36.87° \end{vmatrix}}$$

i.e.,

$$\frac{V_1}{6.25\angle-36.87°}=\frac{-V_2}{-5}=\frac{1}{0.09475\angle-55.30°-0.04}$$

$$=\frac{1}{0.079\angle-79.85°}$$

Hence,
$$\text{voltage } V_1=\frac{6.25\angle-36.87°}{0.079\angle-79.85°}=\textbf{79.11}\angle\,\textbf{42.98°}\textbf{ V}$$

and
$$\text{voltage } V_2=\frac{5}{0.079\angle-79.85°}=\textbf{63.29}\angle\,\textbf{79.85°}\textbf{ V}$$

The current flowing in the $(4+j3)\Omega$ branch is $V_1/(4+j3)$. Hence the voltage between point X and node 3 is

$$\frac{V_1}{(4+j3)}(j3)=\frac{(79.11\angle42.98°)(3\angle90°)}{5\angle36.87°}=47.47\angle96.11°\text{ V}$$

Thus the voltage
$$V_{XY}=V_X-V_Y=V_X-V_2=47.47\angle96.11°-63.29\angle79.85°$$
$$=-16.21-j15.10=\textbf{22.15}\angle\,\textbf{−137°}\textbf{ V}$$

Problem 6. Use nodal analysis to determine the voltages at nodes 2 and 3 in Fig. 23 and hence determine the current flowing in the $2\,\Omega$ resistor and the power dissipated in the $3\,\Omega$ resistor.

Figure 23

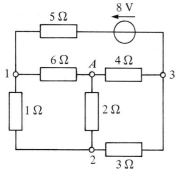

This is the same problem as problem 2 of section 3, page 119, which was solved using Kirchhoff's laws. In Fig. 23, the reference node is shown at point A.

At node 1,

$$\frac{V_1 - V_2}{1} + \frac{V_1}{6} + \frac{V_1 - 8 - V_3}{5} = 0$$

i.e.,

$$1.367V_1 - V_2 - 0.2V_3 - 1.6 = 0 \tag{1}$$

At node 2,

$$\frac{V_2}{2} + \frac{V_2 - V_1}{1} + \frac{V_2 - V_3}{3} = 0$$

i.e.,

$$-V_1 + 1.833V_2 - 0.333V_3 + 0 = 0 \tag{2}$$

At node 3,

$$\frac{V_3}{4} + \frac{V_3 - V_2}{3} + \frac{V_3 + 8 - V_1}{5} = 0$$

i.e.,

$$-0.2V_1 - 0.333V_2 + 0.783V_3 + 1.6 = 0 \tag{3}$$

Equations (1) to (3) can be solved for V_1, V_2 and V_3 by using determinants. Hence

$$\frac{V_1}{\begin{vmatrix} -1 & -0.2 & -1.6 \\ 1.833 & -0.333 & 0 \\ -0.333 & 0.783 & 1.6 \end{vmatrix}} = \frac{-V_2}{\begin{vmatrix} 1.367 & -0.2 & -1.6 \\ -1 & -0.333 & 0 \\ -0.2 & 0.783 & 1.6 \end{vmatrix}}$$

$$= \frac{V_3}{\begin{vmatrix} 1.367 & -1 & -1.6 \\ -1 & 1.833 & 0 \\ -0.2 & -0.333 & 1.6 \end{vmatrix}} = \frac{-1}{\begin{vmatrix} 1.367 & -1 & -0.2 \\ -1 & 1.833 & -0.333 \\ -0.2 & -0.333 & 0.783 \end{vmatrix}}$$

Solving for V_2 gives

$$\frac{-V_2}{-1.6(-0.8496) + 1.6(-0.6552)} = \frac{-1}{1.367(1.3244) + 1(-0.8496) - 0.2(0.6996)}$$

$$\frac{-V_2}{0.31104} = \frac{-1}{0.82093}$$

voltage $V_2 = 0.3789$ V

Thus the current in the $2\,\Omega$ resistor $= \dfrac{V_2}{2} = \dfrac{0.3789}{2} = \mathbf{0.19\,A}$, flowing from node 2 to node 4.

Solving for V_3 gives

$$\frac{V_3}{-1.6(0.6996) + 1.6(1.5057)} = \frac{-1}{0.82093}$$

$$\frac{V_3}{1.2898} = \frac{-1}{0.82093}$$

voltage $V_3 = \dfrac{-1.2898}{0.82093} = -1.57$ V

Power in the 3 Ω resistor $= (I_3)^2(3) = \left(\dfrac{V_2 - V_3}{3}\right)^2 (3) = \dfrac{(0.3789 - (-1.571))^2}{3}$

$$= \mathbf{1.27\ W}$$

Problem 7. For the network shown in Fig. 24 determine (a) the voltage at node 1, (b) the current in each branch, (c) the active power output of the $40\underline{/\ 0°}$ V source, and (d) the active power in each of the circuit resistors.

(a) Node 2 is shown in Fig. 24 as the reference node.
    At node 1,

Figure 24

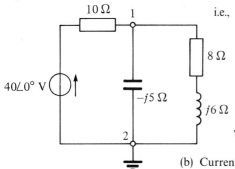

$$\dfrac{V_1 - 40\underline{/\ 0°}}{10} + \dfrac{V_1}{-j5} + \dfrac{V_1}{8 + j6} = 0$$

i.e.,

$$\left(\dfrac{1}{10} + \dfrac{1}{-j5} + \dfrac{1}{8 + j6}\right)V_1 = \dfrac{40\underline{/\ 0°}}{10}$$

$$\left(0.1 + j0.2 + \dfrac{8 - j6}{8^2 + 6^2}\right)V_1 = 4$$

$$(0.18 + j0.14)V_1 = 4$$

voltage, $V_1 = \dfrac{4}{(0.18 + j0.14)} = \dfrac{4}{0.228\underline{/\ 37.87°}} = \mathbf{17.54\underline{/\ -37.87°}\ V}$

(b) Current in the branch containing the $40\underline{/\ 0°}$ V source

$$= \dfrac{V_1 - 40\underline{/\ 0°}}{10} = \dfrac{17.54\underline{/\ -37.87°} - 40}{10} = \dfrac{-26.15 - j10.77}{10}$$

$$= 2.828\underline{/\ -157.62°}\ \text{A, flowing away from node 1,}$$
$$\text{or } \mathbf{2.828\underline{/\ 22.38°}\ A, \text{flowing toward node 1.}}$$

Current in the capacitive branch

$$= \dfrac{V_1}{-j5} = \dfrac{17.54\underline{/\ -37.87°}}{5\underline{/\ -90°}}$$

$$= \mathbf{3.508\underline{/\ 52.13°}\ A,} \text{ flowing away from node 1.}$$

Current in the inductive branch

$$= \dfrac{V_1}{8 + j6} = \dfrac{17.54\underline{/\ -37.87°}}{10\underline{/\ 36.87°}}$$

$$= \mathbf{1.754\underline{/\ -74.74°}\ A,} \text{ flowing away from node 1.}$$

(c) Active power output of the $40\underline{/\ 0°}$ V source $= VI \cos \phi$
$$= (40)(2.828)\cos 22.38°$$
$$= \mathbf{104.6\ W.}$$

(d) Active power in the 10 Ω resistor $= (I_{10})^2(10) = (2.828)^2(10) = \mathbf{80.0\ W.}$
    Active power in the 8 Ω resistor $= (I_8)^2(8) = (1.754)^2(8) = \mathbf{24.6\ W.}$
    (Note that since no active power is dissipated in the capacitor or inductance, the total active power delivered by the source, i.e. 104.6 W, equals the sum of the active powers dissipated by the two resistors, i.e., 80.0 + 24.6 = 104.6 W.)

Further problems on nodal analysis may be found in section 7, problems 22 to 28, page 149.

6. The superposition theorem

The **superposition theorem** states:

"In any network made up of linear impedances and containing more than one source of emf, the resultant current flowing in any branch is the phasor sum of the currents that would flow in that branch if each source were considered separately, all other sources being replaced at that time by their respective internal impedances."

The superposition theorem may be applied to d.c. and a.c. networks. A d.c. network is shown in Fig. 25 and will serve to demonstrate the principle of application of the superposition theorem.

Figure 25

Figure 26

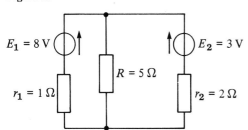

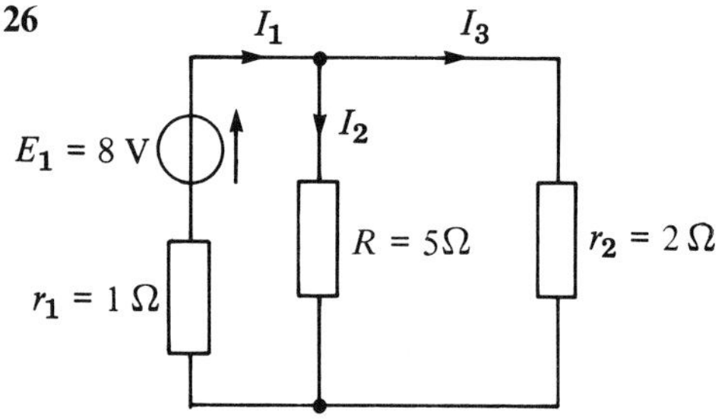

To find the current flowing in each branch of the circuit, the following six-step procedure can be adopted:

(i) Redraw the original network with one of the sources, say E_2, removed and replaced by r_2 only, as shown in Fig. 26.

(ii) Label the current in each branch and its direction as shown in Fig. 26, and then determine its value. The choice of current direction for I_1 depends on the source polarity which, by convention, is taken as flowing from the positive terminal as shown.

R in parallel with r_2 gives an equivalent resistance of

$$(5 \times 2)/(5+2) = 10/7 = 1.429 \, \Omega$$

as shown in the equivalent network of Fig. 27. From Fig. 27,

$$\text{current } I_1 = \frac{E_1}{(r_1 + 1.429)} = \frac{8}{2.429} = 3.294 \, \text{A}$$

From Fig. 26,

$$\text{current } I_2 = \left(\frac{r_2}{R+r_2}\right)(I_1) = \left(\frac{2}{5+2}\right)(3.294) = 0.941 \, \text{A}$$

and

$$\text{current } I_3 = \left(\frac{5}{5+2}\right)(3.294) = 2.353 \, \text{A}$$

(iii) Redraw the original network with source E_1 removed and replaced by r_1 only, as shown in Fig. 28.

(iv) Label the currents in each branch and their directions as shown in Fig. 28, and determine their values.

R and r_1 in parallel gives an equivalent resistance of

$$(5 \times 1)/(5+1) = 5/6\,\Omega \quad \text{or} \quad 0.833\,\Omega,$$

as shown in the equivalent network of Fig. 29. From Fig. 29,

$$\text{current } I_4 = \frac{E_2}{r_2 + 0.833} = \frac{3}{2.833} = 1.059 \text{ A}$$

From Fig. 28,

$$\text{current } I_5 = \left(\frac{1}{1+5}\right)(1.059) = 0.177 \text{ A}$$

and

$$\text{current } I_6 = \left(\frac{5}{1+5}\right)(1.059) = 0.8825 \text{ A}$$

(v) Superimpose Fig. 26 on Fig. 28, as shown in Fig. 30.
(vi) Determine the algebraic sum of the currents flowing in each branch. (Note that in an a.c. circuit it is the phasor sum of the currents that is required.)

From Fig. 30, the resultant current flowing through the 8 V source is given by $I_1 - I_6 = 3.294 - 0.8825 = \mathbf{2.41\ A}$ (discharging, i.e., flowing

Figure 27

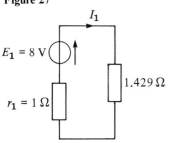

Figure 28

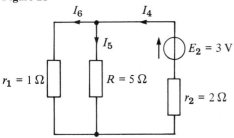

Figure 29

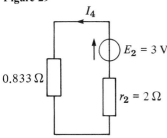

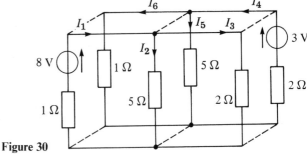

Figure 30

from the positive terminal of the source). The resultant current flowing in the 3 V source is given by $I_3 - I_4 = 2.353 - 1.059 = \mathbf{1.29\ A}$ (charging, i.e., flowing into the positive terminal of the source). The resultant current flowing in the 5 Ω resistance is given by $I_2 + I_5 = 0.941 + 0.177 = \mathbf{1.12\ A}$.

The values of current are the same as those obtained on pages 117–118 by using Kirchhoff's laws.

The following worked problems demonstrate further the use of the superposition theorem in analysing a.c. as well as d.c. networks. The theorem is straightforward to apply, but is lengthy. Thévenin's and Norton's theorems (described in chapter 8) produce results more quickly.

Worked problems involving the superposition theorem

Problem 1. A.C. sources of $100\underline{/\,0°}$ V and internal resistance 25 Ω, and $50\underline{/\,90°}$ V

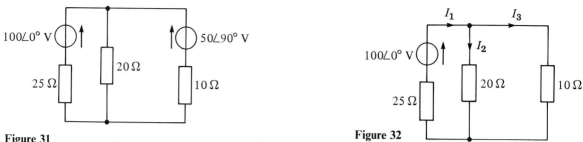

Figure 31

Figure 32

and internal resistance $10\,\Omega$, are connected in parallel across a $20\,\Omega$ load. Determine, using the superposition theorem, the current in the $20\,\Omega$ load and the current in each voltage source.

(This is the same problem as problem 1 on page 118 and a comparison of methods may be made.)

The circuit diagram is shown in Fig. 31. Following the above procedure:

(i) The network is redrawn with the $50\underline{/\,90°}$ V source removed as shown in Fig. 32.

(ii) Currents I_1, I_2 and I_3 are labelled as shown in Fig. 32.

$$I_1 = \frac{100\underline{/\,0°}}{25 + (10 \times 20)/(10 + 20)} = \frac{100\underline{/\,0°}}{25 + 6.667} = 3.158\underline{/\,0°}\,\text{A}$$

$$I_2 = \left(\frac{10}{10 + 20}\right)(3.158\underline{/\,0°}) = 1.053\underline{/\,0°}\,\text{A}$$

$$I_3 = \left(\frac{20}{10 + 20}\right)(3.158\underline{/\,0°}) = 2.105\underline{/\,0°}\,\text{A}$$

(iii) The network is redrawn with the $100\underline{/\,0°}$ V source removed as shown in Fig. 33.

(iv) Currents I_4, I_5 and I_6 are labelled as shown in Fig. 33.

$$I_4 = \frac{50\underline{/\,90°}}{10 + (25 \times 20/25 + 20)} = \frac{50\underline{/\,90°}}{10 + 11.111} = 2.368\underline{/\,90°}\,\text{A or } j2.368\,\text{A}$$

$$I_5 = \left(\frac{25}{20 + 25}\right)(j2.368) = j1.316\,\text{A}$$

$$I_6 = \left(\frac{20}{20 + 25}\right)(j2.368) = j1.052\,\text{A}$$

(v) Figure 34 shows Fig. 33 superimposed on Fig. 32, giving the currents shown.

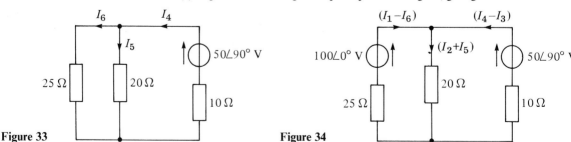

Figure 33

Figure 34

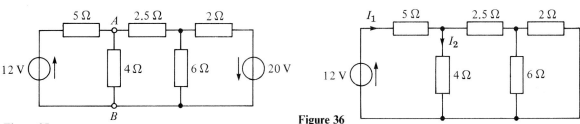

Figure 35

Figure 36

(vi) Current in the 20 Ω load, $I_2 + I_5 = (1.053 + j1.316)$ A or **1.69$\angle$ 51.33° A.**
Current in the 100$\angle$ 0° V source, $I_1 - I_6 = (3.158 - j1.052)$ A   or
3.33$\angle$ −18.42° A.
Current in the 50$\angle$ 90° V source, $I_4 - I_3 = (j2.368 - 2.105)$ or **3.17$\angle$ 131.64° A.**

Problem 2. Use the superposition theorem to determine the current in the 4 Ω resistor of the network shown in Fig. 35.

(i) Removing the 20 V source gives the network shown in Fig. 36.
(ii) Currents I_1 and I_2 are shown labelled in Fig. 36. It is unnecessary to determine the currents in all the branches since only the current in the 4 Ω resistance is required.
From Fig. 36, 6 Ω in parallel with 2 Ω gives $(6 \times 2)/(6 + 2) = 1.5$ Ω, as shown in Fig. 37. 2.5 Ω in series with 1.5 Ω gives 4 Ω, 4 Ω in parallel with 4 Ω gives 2 Ω, and 2 Ω in series with 5 Ω gives 7 Ω. Thus

$$\text{current } I_1 = \frac{12}{7} = 1.714 \text{ A}$$

and

$$\text{current } I_2 = \left(\frac{4}{4 + 4}\right)1.714 = 0.857 \text{ A}$$

Figure 37

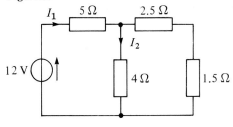

Figure 38

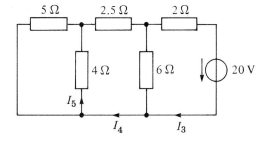

(iii) Removing the 12 V source from the original network gives the network shown in Fig. 38.
(iv) Currents I_3, I_4 and I_5 are shown labelled in Fig. 38.
From Fig. 38, 5 Ω in parallel with 4 Ω gives $(5 \times 4)/(5 + 4) = 20/9 = 2.222$ Ω, as shown in Fig. 39, 2.222 Ω in series with 2.5 Ω gives 4.722 Ω, 4.722 Ω in parallel with 6 Ω gives $(4.722 \times 6)/(4.722 + 6) = 2.642$ Ω, 2.642 Ω in series with 2 Ω gives 4.642 Ω. Hence

$$\text{current } I_3 = \frac{20}{4.642} = 4.308 \text{ A}$$

$$\text{current } I_4 = \left(\frac{6}{6 + 4.722}\right)(4.308) = 2.411 \text{ A, from Fig. 39.}$$

Figure 39

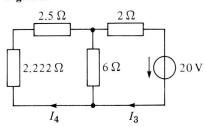

Figure 40

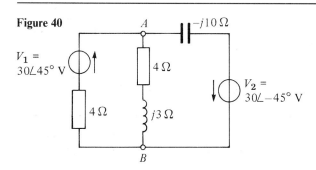

Figure 41

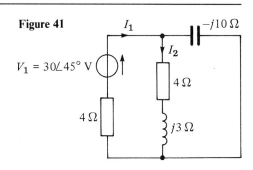

current $I_5 = \left(\dfrac{5}{4+5}\right)(2.411) = 1.339$ A, from Fig. 38.

(v) Superimposing Fig. 38 on Fig. 36 shows that the current flowing in the $4\,\Omega$ resistor is given by $I_5 - I_2$.

(vi) $I_5 - I_2 = 1.339 - 0.857 = \mathbf{0.48}$ **A, flowing from B toward A** (see Fig. 35).

Problem 3. Use the superposition theorem to obtain the current flowing in the $(4+j3)\,\Omega$ impedance of Fig. 40.

(i) The network is redrawn with V_2 removed, as shown in Fig. 41.

(ii) Current I_1 and I_2 are shown in Fig. 41. From Fig. 41, $(4+j3)\,\Omega$ in parallel with $-j10\,\Omega$ gives an equivalent impedance of

$$\frac{(4+j3)(-j10)}{(4+j3-j10)} = \frac{30-j40}{4-j7} = \frac{50\angle -53.13°}{8.062\angle -60.26°}$$
$$= 6.202\angle 7.13° \text{ or } (6.154+j0.770)\,\Omega$$

Total impedance of Fig. 41 is

$$6.154 + j0.770 + 4 = (10.154 + j0.770)\,\Omega \text{ or } 10.183\angle 4.34°\,\Omega$$

Hence

$$\text{current } I_1 = \frac{30\angle 45°}{10.183\angle 4.34°} = 2.946\angle 40.66° \text{ A}$$

and

$$\text{current } I_2 = \left(\frac{-j10}{4-j7}\right)(2.946\angle 40.66°) = \frac{(10\angle -90°)(2.946\angle 40.66°)}{8.062\angle -60.26°}$$
$$= 3.654\angle 10.92° \text{ A} \text{ or } (3.588+j0.692)\,\text{A}$$

Figure 42

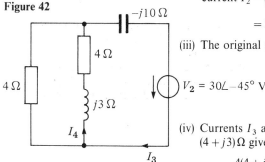

(iii) The original network is redrawn with V_1 removed, as shown in Fig. 42.

(iv) Currents I_3 and I_4 are shown in Fig. 42. From Fig. 42, $4\,\Omega$ in parallel with $(4+j3)\,\Omega$ gives an equivalent impedance of

$$\frac{4(4+j3)}{4+4+j3} = \frac{16+j12}{8+j3} = \frac{20\angle 36.87°}{8.544\angle 20.56°} = 2.341\angle 16.31°\,\Omega$$

$$\text{or } (2.247 + j0.657)\,\Omega$$

Total impedance of Fig. 42 is

$$2.247 + j0.657 - j10 = (2.247 - j9.343)\,\Omega \quad \text{or} \quad 9.609\underline{/-76.48°}\,\Omega$$

Hence

$$\text{current } I_3 = \frac{30\underline{/-45°}}{9.609\underline{/-76.48°}} = 3.122\underline{/31.48°}\,\text{A}$$

and

$$\text{current } I_4 = \left(\frac{4}{8+j3}\right)(3.122\underline{/31.48°}) = \frac{(4\underline{/0°})(3.122\underline{/31.48°})}{8.544\underline{/20.56°}}$$
$$= 1.462\underline{/10.92°}\,\text{A} \quad \text{or} \quad (1.436 + j0.277)\,\text{A}.$$

(v) If the network of Fig. 42 is superimposed on the network of Fig. 41, it can be seen that the current in the $(4+j3)\,\Omega$ is given by $I_2 - I_4$.

(vi) $I_2 - I_4 = (3.588 + j0.692) - (1.436 + j0.277) = \mathbf{(2.152 + j0.415)\,A}$ or **2.192$\underline{/10.92°}$ A, flowing from A to B** in Fig. 40.

Problem 4. For the a.c. network shown in Fig. 43 determine, using the superposition theorem, (a) the current in each branch, (b) the magnitude of the voltage across the $(6+j8)\,\Omega$ impedance, and (c) the total active power delivered to the network.

Figure 43

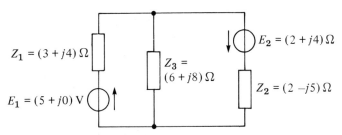

Figure 44

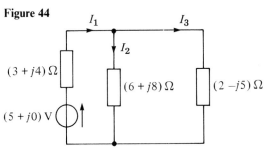

Figure 45

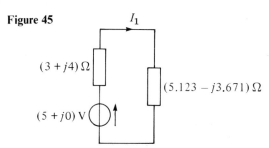

(a) (i) The original network is redrawn with E_2 removed, as shown in Fig. 44.

   (ii) Currents I_1, I_2 and I_3 are labelled as shown in Fig. 44. From Fig. 44, $(6+j8)\,\Omega$ in parallel with $(2-j5)\,\Omega$ gives an equivalent impedance of

$$\frac{(6+j8)(2-j5)}{(6+j8)+(2-j5)} = (5.123 - j3.671)\,\Omega$$

From the equivalent network of Fig. 45,

$$\text{current } I_1 = \frac{5+j0}{(3+j4)+(5.123-j3.671)} = (0.614 - j0.025)\,\text{A}$$

Figure 46

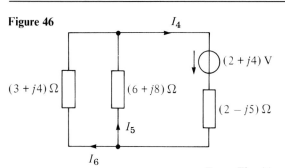

Figure 47

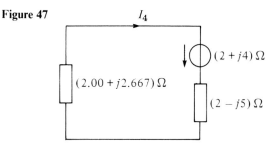

From Fig. 44

$$\text{current } I_2 = \left[\frac{(2-j5)}{(6+j8)+(2-j5)}\right](0.614-j0.025) = (-0.00731-j0.388)\,\text{A}$$

and

$$\text{current } I_3 = \left[\frac{(6+j8)}{(6+j8)+(2-j5)}\right](0.614-j0.025) = (0.622+j0.363)\,\text{A}$$

(iii) The original network is redrawn with E_1 removed, as shown in Fig 46.
(iv) Currents I_4, I_5 and I_6 are shown labelled in Fig. 46 with I_4 flowing away from the positive terminal of the $(2+j4)\,\text{V}$ source.

From Fig. 46, $(3+j4)\,\Omega$ in parallel with $(6+j8)\,\Omega$ gives an equivalent impedance of

$$\frac{(3+j4)(6+j8)}{(3+j4)+(6+j8)} = (2.00+j2.667)\,\Omega$$

From the equivalent network of Fig. 47,

$$\text{current } I_4 = \frac{(2+j4)}{(2.00+j2.667)+(2-j5)} = (-0.062+j0.964)\,\text{A}$$

From Fig. 46

$$\text{current } I_5 = \left[\frac{(3+j4)}{(3+j4)+(6+j8)}\right](-0.062+j0.964) = (-0.0207+j0.321)\,\text{A}$$

and

$$\text{current } I_6 = \left[\frac{6+j8}{(3+j4)+(6+j8)}\right](-0.062+j0.964) = (-0.041+j0.643)\,\text{A}$$

(v) If Fig. 46 is superimposed on Fig. 44, the resultant currents are as shown in Fig. 48.

Resultant current flowing from $(5+j0)\,\text{V}$ source is given by

$$I_1 + I_6 = (0.614-j0.025)+(-0.041+j0.643)$$
$$= \textbf{(0.573+j0.618)\,A}\quad\text{or}\quad\textbf{0.843}\underline{/\,\textbf{47.16°}}\,\textbf{A}$$

Resultant current flowing from $(2+j4)\,\text{V}$ source is given by

$$I_3 + I_4 = (0.622+j0.363)+(-0.062+j0.964)$$
$$= \textbf{(0.560+j1.327)\,A}\quad\text{or}\quad\textbf{1.440}\underline{/\,\textbf{67.12°}}\,\textbf{A}$$

Resultant current flowing through the $(6+j8)\,\Omega$ impedance is given by

$$I_2 - I_5 = (-0.00731-j0.388)-(-0.0207+j0.321)$$
$$= \textbf{(0.0134-j0.709)\,A}\quad\text{or}\quad\textbf{0.709}\underline{/-\textbf{88.92°}}\,\textbf{A}$$

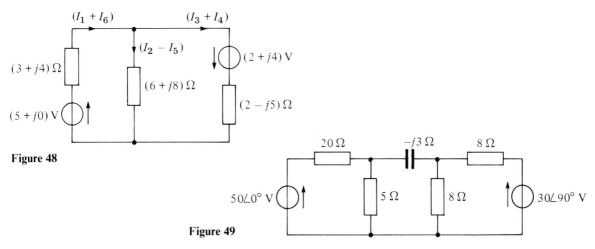

Figure 48

Figure 49

(b) Voltage across $(6 + j8)\,\Omega$ impedance is given by

$$(I_2 - I_5)(6 + j8) = (0.709\angle -88.92°)(10\angle 53.13°)$$
$$= 7.09\angle -35.79° \text{ V},$$

i.e., the magnitude of the voltage across the $(6 + j8)\,\Omega$ impedance is **7.09 V.**

(c) Total active power P delivered to the circuit is given by

$$P = E_1(I_1 + I_6)\cos\phi_1 + E_2(I_3 + I_4)\cos\phi_2$$

where ϕ_1 is the phase angle between E_1 and $(I_1 + I_6)$ and ϕ_2 is the phase angle between E_2 and $(I_3 + I_4)$, i.e.,

$$P = (5)(0.843)\cos(47.16° - 0°) + (\sqrt{(2^2 + 4^2)})(1.440)\cos(67.12° - \arctan\tfrac{4}{2})$$
$$= 2.866 + 6.427 = 9.293 \text{ W} = \textbf{9.3 W}, \text{ correct to one dec. place.}$$

(This value may be checked since total active power dissipated is given by

$$P = (I_1 + I_6)^2(3) + (I_2 - I_5)^2(6) + (I_3 + I_4)^2(2)$$
$$= (0.843)^2(3) + (0.709)^2(6) + (1.440)^2(2)$$
$$= 2.132 + 3.016 + 4.147 = 9.295 \text{ W} = \textbf{9.3 W}, \text{ correct to one dec. place.})$$

Problem 5. Use the superposition theorem to determine, for the network shown in Fig. 49, (a) the magnitude of the current flowing in the capacitor, (b) the p.d. across the $5\,\Omega$ resistance, (c) the active power dissipated in the $20\,\Omega$ resistance and (d) the total active power taken from the supply.

(i) The network is redrawn with the $30\angle 90°$ V source removed, as shown in Fig. 50.

Figure 50

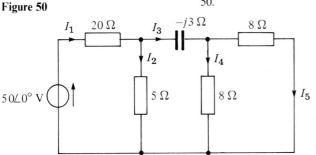

(ii) Currents I_1 to I_5 are shown labelled in Fig. 50. From Fig. 50, two $8\,\Omega$ resistors in parallel give an equivalent resistance of $4\,\Omega$. Hence

$$I_1 = \frac{50\underline{/\,0^\circ}}{20 + (5(4 - j3)/(5 + 4 - j3))} = 2.220\underline{/\,2.12^\circ}\,\text{A}$$

$$I_2 = \frac{(4 - j3)}{(5 + 4 - j3)}I_1 = 1.170\underline{/\,-16.32^\circ}\,\text{A}$$

$$I_3 = \left(\frac{5}{5 + 4 - j3}\right)I_1 = 1.170\underline{/\,20.55^\circ}\,\text{A}$$

$$I_4 = \left(\frac{8}{8 + 8}\right)I_3 = 0.585\underline{/\,20.55^\circ}\,\text{A} = I_5$$

(iii) The original network is redrawn with the $50\underline{/\,0^\circ}$ V source removed, as shown in Fig. 51.

(iv) Currents I_6 to I_{10} are shown labelled in Fig. 51. From Fig. 51, $20\,\Omega$ in parallel with $5\,\Omega$ gives an equivalent resistance of $(20 \times 5)/(20 + 5) = 4\,\Omega$. Hence

$$I_6 = \frac{30\underline{/\,90^\circ}}{8 + (8(4 - j3)/(8 + 4 - j3))} = 2.715\underline{/\,96.52^\circ}\,\text{A}$$

$$I_7 = \frac{(4 - j3)}{(8 + 4 - j3)}I_6 = 1.097\underline{/\,73.69^\circ}\,\text{A}$$

$$I_8 = \left(\frac{8}{8 + 4 - j3}\right)I_6 = 1.756\underline{/\,110.56^\circ}\,\text{A}$$

and

$$I_9 = \left(\frac{20}{20 + 5}\right)I_8 = 1.405\underline{/\,110.56^\circ}\,\text{A}$$

$$I_{10} = \left(\frac{5}{20 + 5}\right)I_8 = 0.351\underline{/\,110.56^\circ}\,\text{A}$$

(a) The current flowing in the capacitor is given by

$$(I_3 - I_8) = 1.170\underline{/\,20.55^\circ} - 1.756\underline{/\,110.56^\circ} = (1.712 - j1.233)\,\text{A} \text{ or}$$
$$2.11\underline{/\,-35.76^\circ}\,\text{A}$$

i.e., **the magnitude of the current in the capacitor is 2.11 A.**

(b) The p.d. across the $5\,\Omega$ resistance is given by $(I_2 + I_9)(5)$.

$$(I_2 + I_9) = 1.170\underline{/\,-16.32^\circ} + 1.405\underline{/\,110.56^\circ} = (0.629 + j0.987)\,\text{A} \text{ or}$$
$$1.17\underline{/\,57.49^\circ}\,\text{A}$$

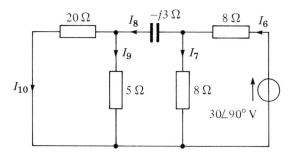

Figure 51

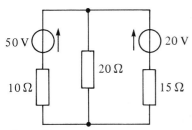

50 V   20 V

20 Ω

10 Ω   15 Ω

Figure 52

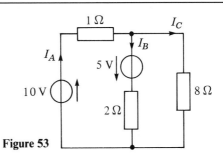

1 Ω I_C

I_A I_B

5 V

10 V

2 Ω   8 Ω

Figure 53

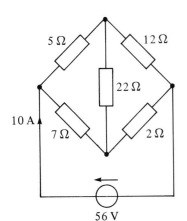

5 Ω   12 Ω

22 Ω

10 A

7 Ω   2 Ω

56 V

Figure 54

Hence the magnitude of the p.d. across the 5 Ω resistance is $(1.17)(5) = \mathbf{5.85\,V}$.

(c) Active power dissipated in the 20 Ω resistance is given by $(I_1 - I_{10})^2 \, (20)$.

$$(I_1 - I_{10}) = 2.220 \underline{/\,2.12°} - 0.351 \underline{/\,110.56°} = (2.342 - j0.247)\,\text{A or}$$
$$2.355 \underline{/\,-6.02°}\,\text{A}$$

Hence active power dissipated in the 20 Ω resistance is $(2.355)^2(20) = \mathbf{111\,W}$.

(d) Active power developed by $50 \underline{/\,0°}$ V source,

$$P_1 = V(I_1 - I_{10})\cos\phi_1 = (50)(2.355)\cos(6.02° - 0°) = 117.1\,\text{W}$$

Active power developed by $30 \underline{/\,90°}$ V source, $P_2 = 30\,(I_6 - I_5)\cos\phi_2$.

$$(I_6 - I_5) = 2.715 \underline{/\,96.52°} - 0.585 \underline{/\,20.55°} = (-0.856 + j2.492)\,\text{A or}$$
$$2.635 \underline{/\,108.96°}\,\text{A}$$

Hence $P_2 = (30)(2.635)\cos(108.96° - 90°) = 74.8\,\text{W}$. Total power developed, $P = P_1 + P_2 = 117.1 + 74.8 = \mathbf{191.9\,W}$
(This value may be checked by summing the I^2R powers dissipated in the four resistors.)

Further problems on the superposition theorem may be found in section 7 following, problems 29 to 36, page 150.

7. Further problems

Kirchhoff's laws

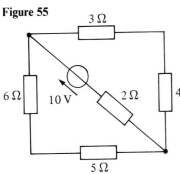

Figure 55

3 Ω

6 Ω   10 V   2 Ω   4 Ω

5 Ω

1. For the network shown in Fig. 52, determine the current flowing in each branch.
$$\begin{bmatrix} \text{50 V source discharges at 2.08 A} \\ \text{20 V source charges at 0.62 A} \\ \text{Current through 20 Ω resistor is 1.46 A} \end{bmatrix}$$

2. Determine the value of currents I_A, I_B and I_C for the network shown in Fig. 53.
$$[I_A = 5.39\,\text{A}, I_B = 4.81\,\text{A}, I_C = 0.58\,\text{A}]$$

3. For the bridge shown in Fig. 54, determine the current flowing in (a) the 5 Ω resistance, (b) the 22 Ω resistance, and (c) the 2 Ω resistance.
$$[\text{(a) } 4\,\text{A} \qquad \text{(b) } 1\,\text{A} \qquad \text{(c) } 7\,\text{A}]$$

4. For the circuit shown in Fig. 55, determine (a) the current flowing in the 10 V source, (b) the p.d. across the 6 Ω resistance, and (c) the active power dissipated in the 4 Ω resistance.
$$[\text{(a) } 1.59\,\text{A} \qquad \text{(b) } 3.72\,\text{V} \qquad \text{(c) } 3.79\,\text{W}]$$

5. Use Kirchhoff's laws to determine the current flowing in each branch of the network shown in Fig. 56.
$$\begin{bmatrix} 40 \underline{/\,90°}\ \text{V source discharges at } 4.40 \underline{/\,74.48°}\ \text{A} \\ 20 \underline{/\,0°}\ \text{V source discharges at } 2.94 \underline{/\,53.13°}\ \text{A} \\ \text{Current in 10 Ω resistance is } 1.97 \underline{/\,107.35°}\ \text{A} \\ \text{(downward).} \end{bmatrix}$$

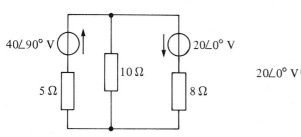

Figure 56

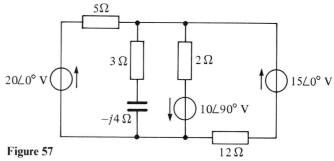

Figure 57

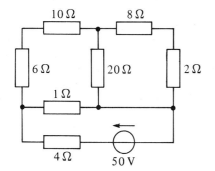

Figure 58

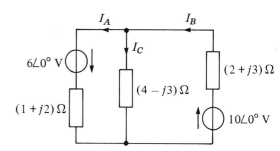

Figure 59

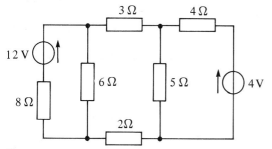

Figure 60

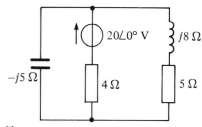

Figure 61

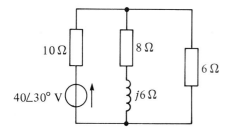

Figure 62

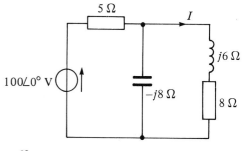

Figure 63

6. For the network shown in Fig. 57, use Kirchhoff's laws to determine the current flowing in the capacitive branch.   [5.87 A]

7. Use Kirchhoff's laws to determine, for the network shown in Fig. 58, the current flowing in (a) the 20 Ω resistance, and (b) the 4 Ω resistance. Determine also (c) the p.d. across the 8 Ω resistance, and (d) the active power dissipated in the 10 Ω resistance.
   [(a) 0.14 A   (b) 10.1 A   (c) 2.27 V   (d) 1.82 W]

8. Determine the value of currents I_A, I_B and I_C shown in the network of Fig. 59, using Kirchhoff's laws.
$$\left[\begin{array}{l} I_A = 2.80\underline{/-59.59°}\ A,\ I_B = 2.71\underline{/-58.78°}\ A, \\ \qquad\qquad I_C = 0.096\underline{/\ 97.01}\ A \end{array}\right]$$

9. Use Kirchhoff's laws to determine the currents flowing in (a) the 3 Ω resistance, (b) the 6 Ω resistance and (c) the 4 V source of the network shown in Fig. 60. Determine also the active power dissipated in the 5 Ω resistance.
   [(a) 0.27 A   (b) 0.70 A   (c) 0.29 A discharging   (d) 1.60 W]

10. Determine the current flowing in each branch of the network shown in Fig. 61. Determine also the active power output of the 20$\underline{/\ 0°}$ V source.
$$\left[\begin{array}{l} 3.07\ A\ \text{in the capacitive branch,} \\ 1.90\ A\ \text{through the source,} \\ 1.63\ A\ \text{in the inductive branch; } 27.70\ W \end{array}\right]$$

11. Determine the magnitude of the p.d. across the $(8 + j6)\,\Omega$ impedance shown in Fig. 62 by using Kirchhoff's laws.
                    [11.37 V]

Mesh-current analysis

12. Repeat problems 1 to 11 using mesh-current analysis.

13. For the network shown in Fig. 63, use mesh-current analysis to determine the value of current I and the active power output of the voltage source.
                    [6.96$\underline{/-50°}$ A; 645 W]

Figure 64

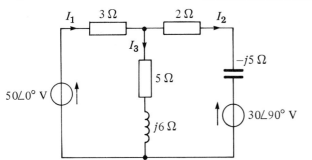

Figure 65

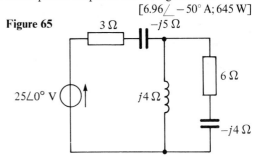

14. Use mesh-current analysis to determine currents I_1, I_2 and I_3 for the network shown in Fig. 64.
$$\left[\begin{array}{l} I_1 = 8.73\underline{/-1.37°}\ A,\ I_2 = 7.02\underline{/\ 17.25°}\ A, \\ \qquad\qquad I_3 = 3.05\underline{/-48.68°}\ A \end{array}\right]$$

15. Determine the current in each branch of the network shown in Fig. 65 and the active power supplied by the source.
$$\left[\begin{array}{l} 25\underline{/\ 0°}\ \text{V source discharges at } 4.35\underline{/\ 10°}\ \text{A;} \\ \text{Current in capacitive branch is } 2.90\underline{/\ 100°}\ \text{A downward;} \\ \text{Current in inductive branch is } 5.22\underline{/-23.68°}\ \text{A downward;} \\ P = 107\ \text{W.} \end{array}\right]$$

16. For the network shown in Fig. 66, determine the current flowing in the $(4 + j3)\,\Omega$ impedance.
                    [0]

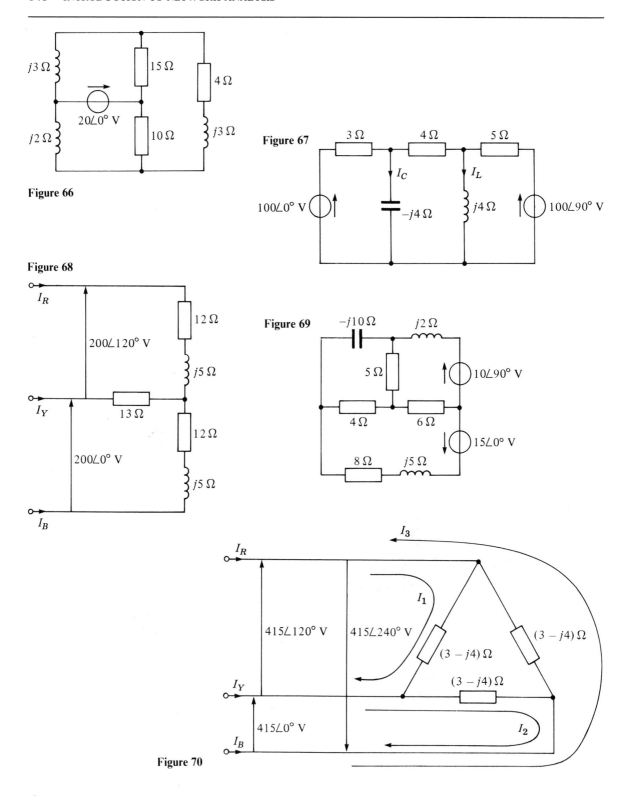

Figure 66

Figure 67

Figure 68

Figure 69

Figure 70

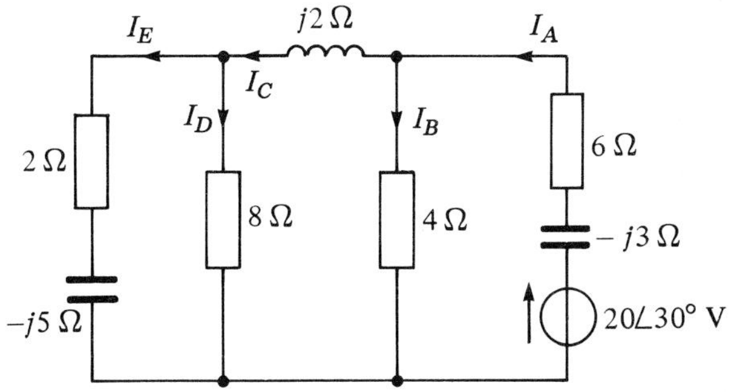

Figure 71

Figure 72

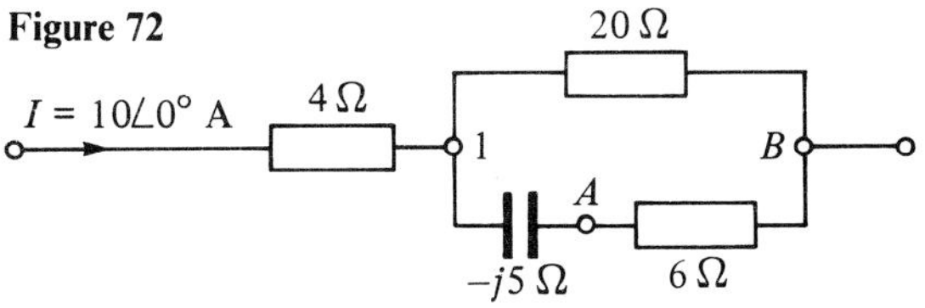

$I = 10\angle 0°$ A

17. For the network shown in Fig. 67, use mesh-current analysis to determine (a) the current in the capacitor, I_C, (b) the current in the inductance I_L, (c) the p.d. across the 4 Ω resistance, and (d) the total active circuit power.
   [(a) 14.5 A   (b) 11.5 A   (c) 71.8 V   (d) 2500 W]

18. Determine the value of the currents I_R, I_Y and I_B in the network shown in Fig. 68 by using mesh-current analysis.
$$\left[\begin{array}{c} I_R = 7.84\angle 71.19° \text{ A}; I_Y = 9.04\angle -37.50° \text{ A};\\ I_B = 9.89\angle -168.81° \text{ A} \end{array}\right]$$

19. In the network of Fig. 69, use mesh-current analysis to determine (a) the current in the capacitor, (b) the current in the 5 Ω resistance, (c) the active power output of the $15\angle 0°$ V source, and (d) the magnitude of the p.d. across the $j2$ Ω inductance.
$$\left[\begin{array}{ll} \text{(a) } 1.03 \text{ A} & \text{(b) } 1.48 \text{ A}\\ \text{(c) } 19.19 \text{ W} & \text{(d) } 3.47 \text{ V} \end{array}\right]$$

20. A balanced 3-phase delta-connected load is shown in Fig. 70. Use mesh-current analysis to determine the values of mesh currents I_1, I_2 and I_3 shown and hence find the line currents I_R, I_Y and I_B.
$$\left[\begin{array}{c} I_1 = 83\angle 173.13° \text{ A}, I_2 = 83\angle 53.13° \text{ A}, I_3 = 83\angle -66.87° \text{ A}\\ I_R = 143.8\angle 143.13° \text{ A}, I_Y = 143.8\angle 23.13° \text{ A}, I_B = 143.8\angle -96.87° \end{array}\right]$$

21. Use mesh-current analysis to determine the value of currents I_A to I_E in the circuit shown in Fig. 71.
$$\left[\begin{array}{c} I_A = 2.40\angle 52.52° \text{ A}; I_B = 1.02\angle 46.18° \text{ A};\\ I_C = 1.39\angle 57.17° \text{ A}; I_D = 0.86\angle 166.32° \text{ A};\\ I_E = 0.996\angle 83.74° \text{ A}. \end{array}\right]$$

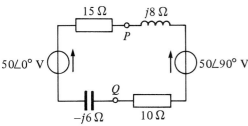

Figure 73

Nodal analysis

22. Repeat problems 1, 2, 5–8, 10, 11, 13–15, 17 and 21 using nodal analysis.
23. Determine for the network shown in Fig. 72 the voltage at node 1 and the voltage V_{AB}.
   $[V_1 = 59.0\angle -28.92° \text{ V}; V_{AB} = 45.3\angle 10.89° \text{ V}.]$
24. Determine the voltage V_{PQ} in the network shown in Fig. 73.
   $[V_{PQ} = 56\angle 51° \text{ V}.]$

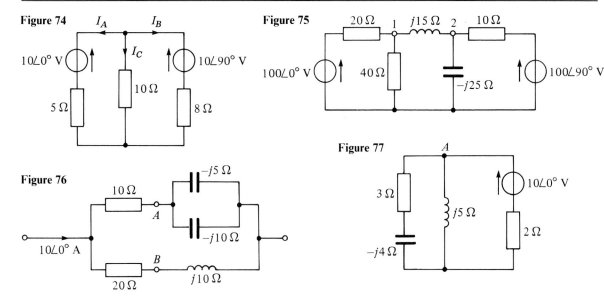

Figure 74

Figure 75

Figure 76

Figure 77

25. Use nodal analysis to determine the currents I_A, I_B and I_C shown in the network of Fig. 74.
$$\left[\begin{array}{c} I_A = 1.21\angle\,150.94^\circ \text{ A}; I_B = 1.06\angle -56.31^\circ \text{ A}; \\ I_C = 0.56\angle\,32.01^\circ \text{ A}. \end{array} \right]$$

26. For the network shown in Fig. 75 determine (a) the voltages at nodes 1 and 2, (b) the current in the 40 Ω resistance, (c) the current in the 20 Ω resistance, and (d) the magnitude of the active power dissipated in the 10 Ω resistance.
$$\left[\begin{array}{l} \text{(a)} \ \ V_1 = 88.10\angle\,33.88^\circ \text{ V}, V_2 = 58.73\angle\,72.34^\circ \text{ V} \\ \text{(b)} \ \ 2.20\angle\,33.88^\circ \text{ A, away from node 1,} \\ \text{(c)} \ \ 2.80\angle\,118.64^\circ \text{ A, away from node 1,} \\ \text{(d)} \ \ 226 \text{ W.} \end{array} \right]$$

27. Determine the voltage V_{AB} in the network of Fig. 76, using nodal analysis.
$$[V_{AB} = 54.3\angle -102.54^\circ \text{ V}]$$

28. Use nodal analysis in the network of Fig. 77 to determine (a) the voltage at node A, (b) the current in each branch, (c) the active power delivered by the $10\angle\,0^\circ$ V source, and (d) the active power dissipated in each of the resistances.
$$\left[\begin{array}{ll} \text{(a)} \ \ 8.05\angle\,3.69^\circ \text{ V} & \text{(b)} \ \ 1.61\angle\,56.82^\circ \text{ A}; \\ 1.61\angle -86.31^\circ \text{ A}; 1.02\angle\,165.26^\circ \text{ A}; \\ \text{(c)} \ \ 9.85 \text{ W} & \text{(d)} \ \ P_3 = 7.78 \text{ W}, P_2 = 2.07 \text{ W}. \end{array} \right]$$

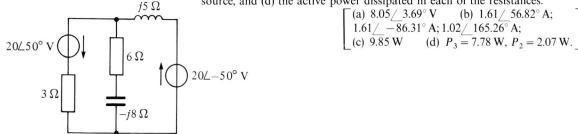

Figure 78

Superposition theorem

29. Repeat problems 1, 5, 8, 9, 14, 17, 25 and 28 using the superposition theorem.
30. Two batteries each of emf 15 V are connected in parallel to supply a load of resistance 2.0 Ω. The internal resistances of the batteries are 0.5 Ω and 0.3 Ω. Determine, using the superposition theorem, the current in the load and the current supplied by each battery.

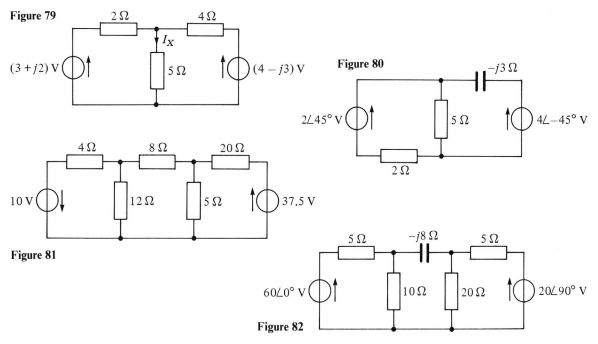

Figure 79

Figure 80

Figure 81

Figure 82

[6.86 A; 2.57 A; 4.29 A]

31. Use the superposition theorem to determine the magnitude of the current flowing in the capacitive branch of the network shown in Fig. 78.

[2.584 A]

32. A.C. sources of $20\underline{/90°}$ V and internal resistance $10\,\Omega$ and $30\underline{/0°}$ V and internal resistance $12\,\Omega$ are connected in parallel across an $8\,\Omega$ load. Use the superposition theorem to determine (a) the current in the $8\,\Omega$ load, and (b) the current in each voltage source.

$$\left[\begin{array}{ll}\text{(a)} \ 1.30\,\text{A} & \text{(b)} \ 20\underline{/90°}\,\text{V source discharges at } 1.58\underline{/120.96°}\,\text{A} \\ & 30\underline{/0°}\,\text{V source discharges at } 1.90\underline{/-16.51°}\,\text{A}\end{array}\right]$$

33. Use the superposition theorem to determine current I_x flowing in the $5\,\Omega$ resistance of the network shown in Fig. 79.

[$0.529\underline{/5.71°}$ A]

34. For the network shown in Fig. 80, determine, using the superposition theorem, (a) the current flowing in the capacitor, (b) the current flowing in the $2\,\Omega$ resistance, (c) the p.d. across the $5\,\Omega$ resistance, and (d) the total active circuit power.

[(a) 1.28 A   (b) 0.74 A   (c) 3.01 V   (d) 2.91 W]

35. (a) Use the superposition theorem to determine the current in the $12\,\Omega$ resistance of the network shown in Fig. 81. Determine also the p.d. across the $8\,\Omega$ resistance and the power dissipated in the $20\,\Omega$ resistance.

(b) If the 37.5 V source in Fig. 81 is reversed in direction, determine the current in the $12\,\Omega$ resistance.

[(a) 0.375 A, 8 V, 57.8 W   (b) 0.625 A]

36. For the network shown in Fig. 82, use the superposition theorem to determine (a) the current in the capacitor, (b) the p.d. across the $10\,\Omega$ resistance, (c) the active power dissipated in the $20\,\Omega$ resistance, and (d) the total active circuit power.

[(a) 3.97 A   (b) 28.7 V   (c) 36.4 W   (d) 371.6 W]

8 Thévenin's and Norton's theorems

1. Introduction

Many of the networks analysed in chapter 7 using Kirchhoff's laws, mesh-current and nodal analysis and the superposition theorem can be analysed more quickly and easily by using Thévenin's or Norton's theorems. Each of these theorems involves replacing what may be a complicated network of sources and linear impedances with a simple equivalent circuit. A set procedure may be followed when using each theorem, the procedures themselves requiring a knowledge of basic circuit theory.

2. Thévenin's theorem

Thévenin's theorem states:

"The current which flows in any branch of a network is the same as that which would flow in the branch if it were connected across a source of electrical energy, the emf of which is equal to the potential difference which would appear across the branch if it were open-circuited, and the internal impedance of which is equal to the impedance which appears across the open-circuited branch terminals when all sources are replaced by their internal impedances."

The theorem applies to any linear active network ("linear" meaning that the measured values of circuit components are independent of the direction and magnitude of the current flowing in them, and "active" meaning that it contains a source, or sources, of emf).

The above statement of Thévenin's theorem simply means that a complicated network with output terminals AB, as shown in Fig. 1(a), can be replaced by a single voltage source E in series with an impedance z, as shown in Fig. 1(b). E is the open-circuit voltage measured at terminals AB and z is the equivalent impedance of the network at the terminals AB when all internal sources of emf are made zero. The polarity of voltage E is chosen so that the current flowing through an impedance connected between A and B will have the same direction as would result if the impedance had been connected between A and B of the original network. Fig. 1(b) is known as the **Thévenin equivalent circuit.**

The following four-step **procedure** can be adopted when determining, by means of Thévenin's theorem, the current flowing in a branch containing impedance Z_L of an active network:

(i) remove the impedance Z_L from that branch;

Figure 1 The Thévenin equivalent circuit

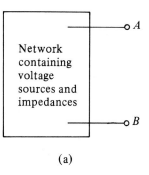

(a)

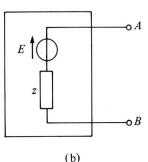

(b)

Thévenin
equivalent circuit

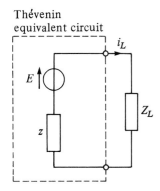

Figure 2

Figure 3

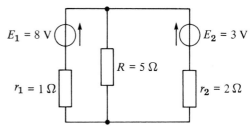

(ii) determine the open-circuit voltage E across the break;
(iii) remove each source of emf and replace it by its internal impedance (if it has zero internal impedance then replace it by a short-circuit), and then determine the internal impedance, z, "looking in" at the break;
(iv) determine the current from the Thévenin equivalent circuit shown in Fig. 2, i.e., **current $i_L = E/(Z_L + z)$.**

A simple d.c. network (Fig. 3) serves to demonstrate how the above procedure is applied to determine the current flowing in the $5\,\Omega$ resistance by using Thévenin's theorem. This is the same network as used in chapter 7 when it was solved using Kirchhoff's laws (see page 117), and by means of the superposition theorem (see page 136). A comparison of methods may be made.

Using the above procedure:
(i) the $5\,\Omega$ resistor is removed, as shown in Fig. 4(a).

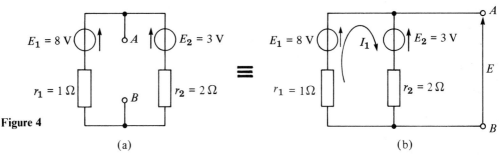

Figure 4

(a)   (b)

(ii) The open-circuit voltage E across the break is now required. The network of Fig. 4(a) is redrawn for convenience as shown in Fig. 4(b), where current,

$$I_1 = \frac{E_1 - E_2}{r_1 + r_2} = \frac{8 - 3}{1 + 2} = \frac{5}{3} \quad \text{or} \quad 1\tfrac{2}{3}\text{A}$$

Hence the open-circuit voltage E is given by

$$E = E_1 - I_1 r_1 \quad \text{i.e.,} \quad E = 8 - (1\tfrac{2}{3})(1) = 6\tfrac{1}{3}\,\text{V}.$$

(Alternatively, $E = E_2 - (-I_1)r_2 = 3 + (1\tfrac{2}{3})(2) = 6\tfrac{1}{3}\,\text{V}$.)
(iii) Removing each source of emf gives the network of Fig. 5. The impedance, z, "looking in" at the break AB is given by

$$z = (1 \times 2)/(1 + 2) = \tfrac{2}{3}\Omega$$

Figure 5

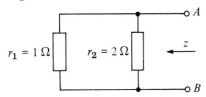

Thévenin
equivalent circuit

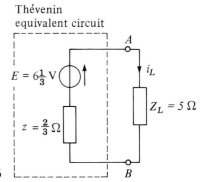

Figure 6

(iv) The Thévenin equivalent circuit is shown in Fig. 6, where current i_L is given by

$$i_L = \frac{E}{Z_L + z} = \frac{6\frac{1}{3}}{5 + \frac{2}{3}} = \mathbf{1.12\ A}$$

To determine the currents flowing in the other two branches of the circuit of Fig. 3, basic circuit theory is used. Thus, from Fig. 7, voltage $V = (1.118)(5) = 5.590\ V$.

Then $V = E_1 - I_A r_1$, i.e., $5.590 = 8 - I_A(1)$, from which current $I_A = 8 - 5.590 = \mathbf{2.41\ A.}$

Similarly, $V = E_2 - I_B r_2$, i.e., $5.590 = 3 - I_B(2)$, from which

$$\text{current } I_B = \frac{3 - 5.590}{2} = \mathbf{-1.29\ A}$$

(i.e., flowing in the direction opposite to that shown in Fig. 7).

The Thévenin theorem procedure used above may be applied to a.c. as well as d.c. networks, as shown below.

An a.c. network is shown in Fig. 8 where it is required to find the current flowing in the $(6 + j8)\,\Omega$ impedance by using Thévenin's theorem. Using the above procedure:

(i) The $(6 + j8)\,\Omega$ impedance is removed, as shown in Fig. 9(a).

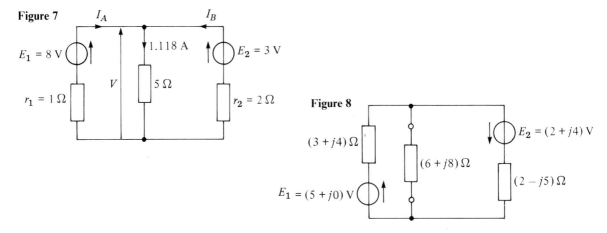

Figure 7

Figure 8

Figure 9

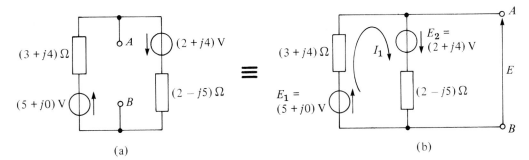

(a)   (b)

(ii) The open-circuit voltage across the break is now required. The network is redrawn for convenience as shown in Fig. 9(b), where

$$\text{current } I_1 = \frac{(5+j0)+(2+j4)}{(3+j4)+(2-j5)} = \frac{(7+j4)}{(5-j)} = 1.581\underline{/\,41.05°}\text{ A.}$$

Hence open-circuit voltage across AB, $E = E_1 - I_1(3+j4)$, i.e.,

$$E = (5+j0) - (1.581\underline{/\,41.05°})(5\underline{/\,53.13°})$$

from which

$$E = 9.657\underline{/\,-54.73°}\text{ V}$$

(iii) From Fig. 10, the impedance z "looking in" at terminals AB is given by

$$z = \frac{(3+j4)(2-j5)}{(3+j4)+(2-j5)} = 5.281\underline{/\,-3.76°}\,\Omega \quad \text{or} \quad (5.270-j0.346)\,\Omega$$

Figure 10

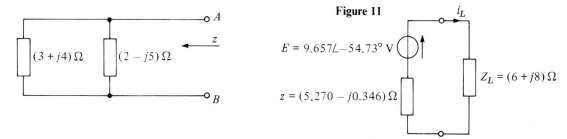

(iv) The Thévenin equivalent circuit is shown in Fig. 11, from which

$$\text{current } i_L = \frac{E}{Z_L+z} = \frac{9.657\underline{/\,-54.73°}}{(6+j8)+(5.270-j0.346)}$$

Thus, current in $(6+j8)\,\Omega$ impedance,

$$i_L = \frac{9.657\underline{/\,-54.73°}}{13.623\underline{/\,34.18°}} = \mathbf{0.71\underline{/\,-88.91°}\text{ A}}$$

The network of Fig. 8 is analysed using Kirchhoff's laws in problem 3, page 120, and by the superposition theorem in problem 4, page 141. The above analysis using Thévenin's theorem is seen to be much quicker.

Worked problems on Thévenin's theorem

Problem 1. For the circuit shown in Fig. 12, use Thévenin's theorem to determine (a) the current flowing in the capacitor, and (b) the p.d. across the 150 kΩ resistor.

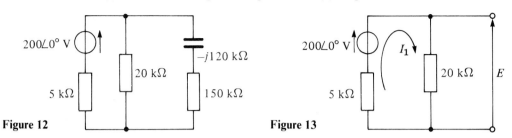

Figure 12   **Figure 13**

(a) (i) Initially the $(150 - j120)\,$kΩ impedance is removed from the circuit as shown in Fig. 13.

Note that, to find the current in the capacitor, only the capacitor needs to have been initially removed from the circuit. However, removing each of the components from the branch through which the current is required will often result in a simpler solution.

(ii) From Fig. 13,

$$\text{current } I_1 = \frac{200\angle 0°}{(5000 + 20000)} = 8\,\text{mA}$$

The open-circuit emf E is equal to the p.d. across the 20 kΩ resistor, i.e., $E = (8 \times 10^{-3})(20 \times 10^3) = 160\,$V.

Figure 14

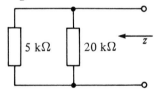

(iii) Removing the $200\angle 0°$ V source gives the network shown in Fig. 14. The impedance, z, "looking in" at the open-circuited terminals is given by

$$z = \frac{5 \times 20}{5 + 20}\text{kΩ} = 4\,\text{kΩ}$$

(iv) The Thévenin equivalent circuit is shown in Fig. 15, where current i_L is given by

$$i_L = \frac{E}{Z_L + z} = \frac{160}{(150 - j120) \times 10^3 + 4 \times 10^3} = \frac{160}{195.23 \times 10^3 \angle -37.93°}$$
$$= 0.82\angle 37.93°\,\text{mA}$$

Figure 15

Thus the current flowing in the capacitor is 0.82 mA.

(b) P.d. across the 150 kΩ resistor, $V_0 = i_L R = (0.82 \times 10^{-3})(150 \times 10^3) = 123\,$V.

Thévenin equivalent circuit

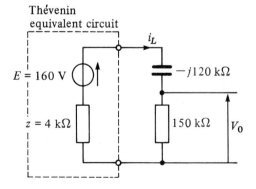

Figure 16

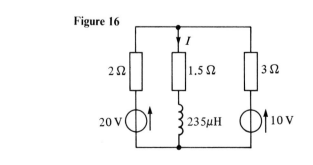

Problem 2. Determine, for the network shown in Fig. 16, the value of current I. Each of the voltage sources has a frequency of 2 kHz.

(i) The impedance through which current I is flowing is initially removed from the network, as shown in Fig. 17.

(ii) From Fig. 17,

$$\text{current, } I_1 = \frac{20 - 10}{2 + 3} = 2 \text{ A}$$

Hence the open circuit emf $E = 20 - I_1(2) = 20 - 2(2) = 16$ V. (Alternatively, $E = 10 + I_1(3) = 10 + (2)(3) = 16$ V.)

(iii) When the sources of emf are removed from the circuit, the impedance, z, "looking in" at the break is given by

$$z = \frac{2 \times 3}{2 + 3} = 1.2 \, \Omega$$

Figure 17

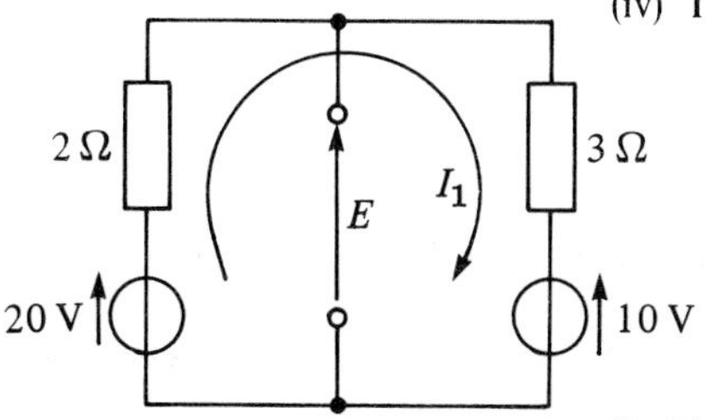

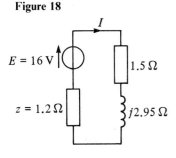

(iv) The Thévenin equivalent circuit is shown in Fig. 18, where inductive reactance,

$$X_L = 2\pi f L = 2\pi (2000)(235 \times 10^{-6}) = 2.95 \, \Omega$$

Hence

$$\text{current } I = \frac{16}{(1.2 + 1.5 + j2.95)} = \frac{16}{4.0 \angle 47.53°} = \mathbf{4.0 \angle -47.53° \text{ A}}$$

$$\text{or } \mathbf{(2.70 - j2.95) \text{ A}}$$

Problem 3. Use Thévenin's theorem to determine the power dissipated in the 48 Ω resistor of the network shown in Fig. 19.

The power dissipated by a current I flowing through a resistor R is given by $I^2 R$, hence initially the current flowing in the 48 Ω resistor is required.

(i) The $(48 + j144) \, \Omega$ impedance is initially removed from the network as shown in Fig. 20.

(ii) From Fig. 20,

$$\text{current, } i = \frac{50 \angle 0°}{(300 - j400)} = 0.1 \angle 53.13° \text{ A}$$

Hence the open-circuit voltage

$$E = i(300) = (0.1 \angle 53.13°)(300) = 30 \angle 53.13° \text{ V}$$

(iii) When the $50 \angle 0°$ V source shown in Fig. 20 is removed, the impedance, z, is given by

Figure 18

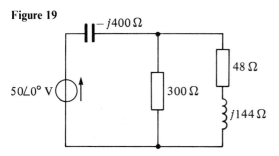

Figure 19

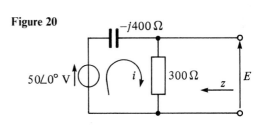

Figure 20

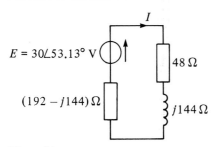

Figure 21

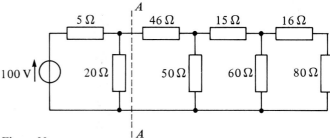

Figure 22

$$z = \frac{(-j400)(300)}{(300 - j400)} = \frac{(400\underline{/-90°})(300)}{500\underline{/-53.13°}} = 240\underline{/-36.87°}\,\Omega \quad \text{or} \quad (192 - j144)\,\Omega$$

(iv) The Thévenin equivalent circuit is shown in Fig. 21 connected to the $(48 + j144)\,\Omega$ load.

$$\text{Current } I = \frac{30\underline{/53.13°}}{(192 - j144) + (48 + j144)} = \frac{30\underline{/53.13°}}{240\underline{/0°}} = 0.125\underline{/53.13°}\,\text{A}$$

Hence the power dissipated in the 48 Ω resistor $= I^2R = (0.125)^2(48) = 0.75$ W.

Problem 4. For the network shown in Fig. 22, use Thevenin's theorem to determine the current flowing in the 80 Ω resistor.

One method of analysing a multi-branch network as shown in Fig. 22 is to use Thévenin's theorem on one part of the network at a time. For example, the part of the circuit to the left of AA may be reduced to a Thévenin equivalent circuit. From Fig. 23,

$$E_1 = \left(\frac{20}{20 + 5}\right)100 = 80\,\text{V, by voltage division,}$$

and

$$z_1 = \frac{20 \times 5}{20 + 5} = 4\,\Omega$$

Thus the network of Fig. 22 reduces to that of Fig. 24. The part of the network shown in Fig. 24 to the left of BB may be reduced to a Thévenin equivalent circuit, where

$$E_2 = \left(\frac{50}{50 + 46 + 4}\right)(80) = 40\,\text{V}$$

and

$$z_2 = \frac{50 \times 50}{50 + 50} = 25\,\Omega$$

Thus the original network reduces to that shown in Fig. 25.
    The part of the network shown in Fig. 25 to the left of CC may be reduced to a Thévenin equivalent circuit, where

$$E_3 = \left(\frac{60}{60 + 25 + 15}\right)(40) = 24\,\text{V} \quad \text{and} \quad z_3 = \frac{(60)(40)}{(60 + 40)} = 24\,\Omega$$

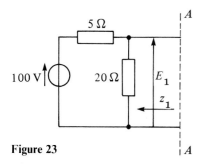

Figure 23

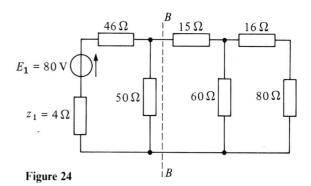

Figure 24

Thus the original network reduces to that of Fig. 26, from which **the current in the 80 Ω resistor** is given by

$$I = \left(\frac{24}{80 + 16 + 24}\right) = 0.20 \text{ A}$$

Problem 5. Determine the Thévenin equivalent circuit with respect to terminals AB of the circuit shown in Fig. 27. Hence determine (a) the magnitude of the current flowing in a $(3.75 + j11)\,\Omega$ impedance connected across terminals AB, and (b) the magnitude of the p.d. across the $(3.75 + j11)\,\Omega$ impedance.

Current I_1 shown in Fig. 27 is given by

$$I_1 = \frac{24\angle 0°}{(4 + j3 - j3)} = \frac{24\angle 0°}{4\angle 0°} = 6\angle 0° \text{ A}$$

Figure 25

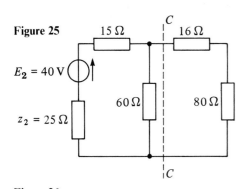

Figure 26

Figure 27

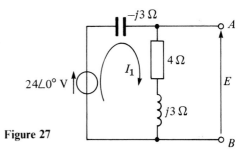

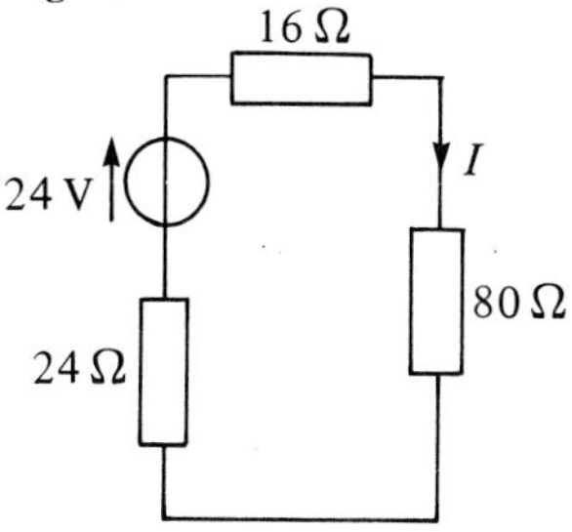

The Thévenin equivalent voltage, i.e., the open-circuit voltage across terminals AB, is given by

$$E = I_1(4 + j3) = (6\angle 0°)(5\angle 36.87°) = 30\angle 36.87° \text{ V}$$

When the $24\angle 0°$ V source is removed, the impedance z "looking in" at AB is given by

$$z = \frac{(4 + j3)(-j3)}{(4 + j3 - j3)} = \frac{9 - j12}{4} = (2.25 - j3.0)\,\Omega$$

Thus the Thévenin equivalent circuit is as shown in Fig. 28.

Thévenin equivalent circuit

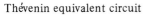

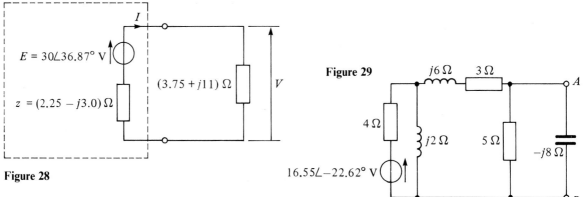

Figure 28

Figure 29

(a) When a $(3.75 + j11)\,\Omega$ impedance is connected across terminals AB, the current I flowing in the impedance is given by

$$I = \frac{30\underline{/\,36.87°}}{(3.75 + j11) + (2.25 - j3.0)} = \frac{30\underline{/\,36.87°}}{10\underline{/\,53.13°}} = 3\underline{/-16.26°}\ \text{A}$$

Hence the current flowing in the $(3.75 + j11)\,\Omega$ impedance is 3 A.

(b) P.d. across the $(3.75 + j11)\,\Omega$ impedance is given by

$$V = (3\underline{/-16.26°})(3.75 + j11) = (3\underline{/-16.26°})(11.62\underline{/\,71.18°})$$
$$= 34.86\underline{/\,54.92°}\ \text{V}$$

Hence the magnitude of the p.d. across the impedance is 34.9 V

Problem 6. Use Thévenin's theorem to determine the current flowing in the capacitor of the network shown in Fig. 29.

(i) The capacitor is removed from branch AB, as shown in Fig. 30.
(ii) The open-circuit voltage, E, shown in Fig. 30, is given by $(I_2)(5)$. I_2 may be determined by current division if I_1 is known. (Alternatively, E may be determined by the method used in problem 4.)

Current $I_1 = V/Z$, where Z is the total circuit impedance and $V = 16.55\underline{/-22.62°}\ \text{V}$.

$$\text{Impedance, } Z = 4 + \frac{(j2)(8 + j6)}{j2 + 8 + j6} = 4 + \frac{-12 + j16}{8 + j8} = 4.596\underline{/\,22.38°}\ \Omega$$

Hence

$$I_1 = \frac{16.55\underline{/-22.62°}}{4.596\underline{/\,22.38°}} = 3.60\underline{/-45°}\ \text{A}$$

and

$$I_2 = \left(\frac{j2}{j2 + 3 + j6 + 5}\right)I_1 = \frac{(2\underline{/\,90°})(3.60\underline{/-45°})}{11.314\underline{/\,45°}} = 0.636\underline{/\,0°}\ \text{A}$$

(An alternative method of finding I_2 is to use Kirchhoff's laws or mesh-current or nodal analysis on Fig. 30.)

Hence $E = (I_2)(5) = (0.636\underline{/\,0°})(5) = 3.18\underline{/\,0°}\ \text{V}$.

Figure 30

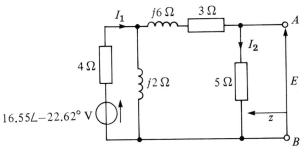

Figure 31   Thévenin equivalent circuit

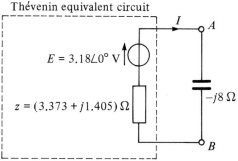

(iii) If the $16.55\underline{/\,-22.62°}$ V source is removed from Fig. 30, the impedance, z, "looking in" at AB is given by

$$z = \frac{5[((4 \times j2)/(4 + j2)) + (3 + j6)]}{5 + [((4 \times j2)/(4 + j2)) + 3 + j6]} = \frac{5(3.8 + j7.6)}{8.8 + j7.6}$$

i.e.,

$$z = 3.654\underline{/\,22.61°}\,\Omega \quad \text{or} \quad (3.373 + j1.405)\,\Omega$$

(iv) The Thévenin equivalent circuit is shown in Fig. 31, where the current flowing in the capacitor, I, is given by

$$I = \frac{3.18\underline{/\,0°}}{(3.373 + j1.405) - j8} = \frac{3.18\underline{/\,0°}}{7.408\underline{/\,-62.91°}} = \mathbf{0.43\underline{/\,62.91°}\ A\ in\ the}$$
$$\mathbf{direction\ from\ A\ to\ B.}$$

Problem 7. For the network shown in Fig. 32, derive the Thévenin equivalent circuit with respect to terminals PQ, and hence determine the power dissipated by a $2\,\Omega$ resistor connected across PQ.

Figure 32

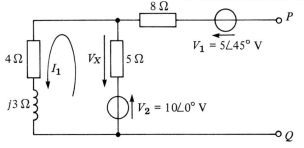

Figure 33

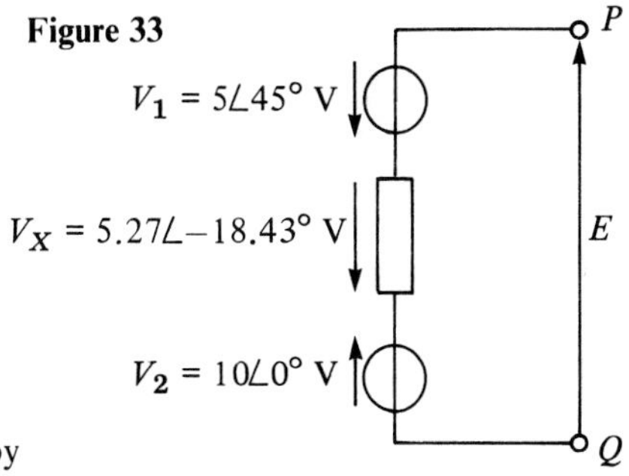

Current I_1 shown in Fig. 32 is given by

$$I_1 = \frac{10\underline{/\,0°}}{(5 + 4 + j3)} = 1.054\underline{/\,-18.43°}\ A.$$

Hence the voltage drop across the $5\,\Omega$ resistor is given by $V_X = (I_1)(5) = 5.27\underline{/\,-18.43°}$ V, and is in the direction shown in Fig. 32, i.e., the direction opposite to that in which I_1 is flowing.

The open-circuit voltage E across PQ is the phasor sum of V_1, V_x and V_2, as shown in Fig. 33. Thus

$$E = 10\underline{/\,0°} - 5\underline{/\,45°} - 5.27\underline{/\,-18.43°}$$
$$= (1.465 - j1.869)\,V \quad \text{or} \quad 2.375\underline{/\,-51.91°}\,V$$

The impedance, z, "looking in" at terminals PQ with the voltage sources removed is given by

$$z = 8 + \frac{5(4 + j3)}{(5 + 4 + j3)} = 8 + 2.635\underline{/\ 18.44°} = (10.50 + j0.833)\,\Omega.$$

The Thévenin equivalent circuit is shown in Fig. 34 with the $2\,\Omega$ resistance connected across terminals PQ.

The current flowing in the $2\,\Omega$ resistance is given by

$$I = \frac{2.375\underline{/\ -51.91°}}{(10.50 + j0.833) + 2} = 0.1896\underline{/\ -55.72°}\text{ A}$$

The power P dissipated in the $2\,\Omega$ resistor is given by

$$P = I^2 R = (0.1896)^2(2) = \textbf{0.0719 W} \equiv \textbf{72 mW},\text{ correct to two significant figures.}$$

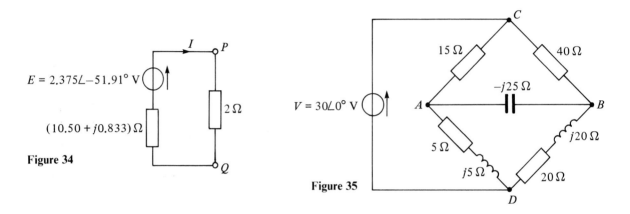

Figure 34

Figure 35

Problem 8. For the a.c. bridge network shown in Fig. 35, determine the current flowing in the capacitor, and its direction, by using Thévenin's theorem. Assume the $30\underline{/\ 0°}$ V source to have negligible internal impedance.

(i) The $-j25\,\Omega$ capacitor is initially removed from the network, as shown in Fig. 36.

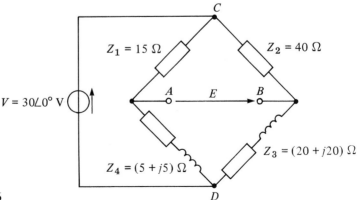

Figure 36

(ii) P.d. between A and C,

$$V_{AC} = \left(\frac{Z_1}{Z_1 + Z_4} \right) V = \left(\frac{15}{15 + 5 + j5} \right) (30 \underline{/\ 0°}) = 21.83 \underline{/\ -14.04°} \text{ V}.$$

P.d. between B and C,

$$V_{BC} = \left(\frac{Z_2}{Z_2 + Z_3} \right) V = \left(\frac{40}{40 + 20 + j20} \right) 30 \underline{/\ 0°} = 18.97 \underline{/\ -18.43°} \text{ V}$$

Assuming that point A is at a higher potential than point B, then the p.d. between A and B is

$$21.83 \underline{/\ -14.04°} - 18.97 \underline{/\ -18.43°}$$
$$= (3.181 + j0.704) \text{ V} \quad \text{or} \quad 3.258 \underline{/\ 12.48°} \text{ V},$$

i.e., the open-circuit voltage across AB is given by $E = 3.258 \underline{/\ 12.48°}$ V. Point C is at a potential of $30 \underline{/\ 0°}$ V. Between C and A is a volt-drop of $21.83 \underline{/\ -14.04°}$ V. Hence the voltage at point A is

$$30 \underline{/\ 0°} - 21.83 \underline{/\ -14.04°} = 10.29 \underline{/\ 30.98°} \text{ V}$$

Between C and B is a voltage drop of $18.97 \underline{/\ -18.43°}$ V. Hence the voltage at point B is $30 \underline{/\ 0°} - 18.97 \underline{/\ -18.43°} = 13.42 \underline{/\ 26.57°}$ V.

Since the magnitude of the voltage at B is higher than at A, current must flow in the direction B to A.

(iii) Replacing the $30 \underline{/\ 0°}$ V source with a short-circuit (i.e., zero internal impedance) gives the network shown in Fig. 37(a). The network is shown redrawn in Fig. 37(b) and simplified in Fig. 37(c). Hence the impedance, z, "looking in" at terminals AB is given by

$$z = \frac{(15)(5 + j5)}{(15 + 5 + j5)} + \frac{(40)(20 + j20)}{(40 + 20 + j20)} = 5.145 \underline{/\ 30.96°} + 17.889 \underline{/\ 26.57°}$$

i.e., $z = (20.41 + j10.65) \Omega$.

Figure 37

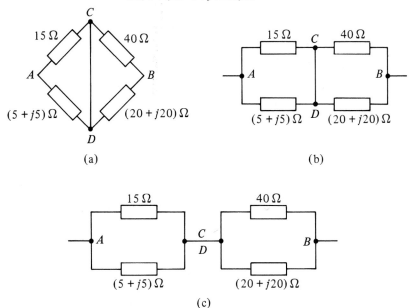

(a)

(b)

(c)

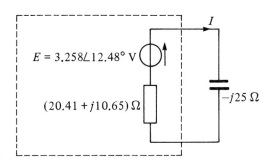

Figure 38

(iv) The Thévenin equivalent circuit is shown in Fig. 38, where current I is given by

$$I = \frac{3.258 \angle 12.48°}{(20.41 + j10.65) - j25} = \frac{3.258 \angle 12.48°}{24.95 \angle -35.11°} = 0.131 \angle 47.59° \text{ A}$$

Thus a current of 131 mA flows in the capacitor in a direction from B to A.

Further problems on Thevenin's theorem may be found in section 5, problems 1 to 13, page 177.

3. Norton's theorem

A source of electrical energy can be represented by a source of emf in series with an impedance. In section 2, the Thévenin constant-voltage source consisted of a constant emf E, which may be alternating or direct, in series with an internal impedance z. However, this is not the only form of representation. A source of electrical energy can also be represented by a constant-current source, which may be alternating or direct, in parallel with an impedance. It is shown in section 4 that the two forms are in fact equivalent.

Norton's theorem states:

"The current that flows in any branch of a network is the same as that which would flow in the branch if it were connected across a source of electrical energy, the short-circuit current of which is equal to the current that would flow in a short-circuit across the branch, and the internal impedance of which is equal to the impedance which appears across the open-circuited branch terminals."

The above statement simply means that any linear active network with output terminals AB, as shown in Fig. 39(a), can be replaced by a current

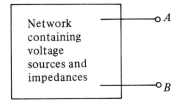

Figure 39 The Norton equivalent circuit   (a)   (b)

Norton equivalent circuit

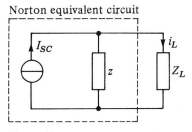

Figure 40

Figure 41

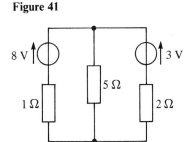

Figure 42

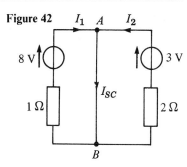

source in parallel with an impedance z as shown in Fig. 39(b). The equivalent current source I_{SC} (note the symbol in Fig. 39(b) as per BS 3939: 1985) is the current through a short-circuit applied to the terminals of the network. The impedance z is the equivalent impedance of the network at the terminals AB when all internal sources of emf are made zero. Figure 39(b) is known as the **Norton equivalent circuit.**

The following four-step procedure may be adopted when determining the current flowing in an impedance Z_L of a branch AB of an active network, using Norton's theorem:

  (i) short-circuit branch AB;
 (ii) determine the short-circuit current, I_{SC};
(iii) remove each source of emf and replace it by its internal impedance (or, if a current source exists, replace with an open circuit), then determine the impedance, z, "looking in" at a break made between A and B;
 (iv) determine the value of the current i_L flowing in impedance Z_L from the Norton equivalent network shown in Fig. 40, i.e.,

$$i_L = \left(\frac{z}{Z_L + z}\right) I_{SC}$$

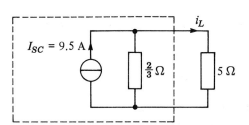

Figure 43

A simple d.c. network (Fig. 41) serves to demonstrate how the above procedure is applied to determine the current flowing in the $5\,\Omega$ resistance by using Norton's theorem:

  (i) The $5\,\Omega$ branch is short-circuited, as shown in Fig. 42.
 (ii) From Fig. 42, $I_{SC} = I_1 + I_2 = \frac{8}{1} + \frac{3}{2} = 9.5\,\text{A}$.
(iii) If each source of emf is removed the impedance "looking in" at a break made between A and B is given by $z = (1 \times 2)/(1 + 2) = \frac{2}{3}\,\Omega$.
 (iv) From the Norton equivalent network shown in Fig. 43, the current in the $5\,\Omega$ resistance is given by $i_L = (\frac{2}{3}/(5 + \frac{2}{3}))\,9.5 = \textbf{1.12 A,}$ as obtained previously using Kirchhoff's laws, the superposition theorem and by Thévenin's theorem.

As with Thévenin's theorem, Norton's theorem may be used with a.c. as well as d.c. networks, as shown below.

Figure 44

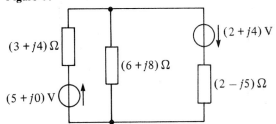

Figure 45

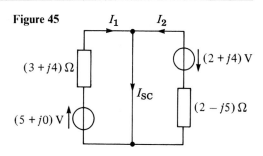

An a.c. network is shown in Fig. 44 where it is required to find the current flowing in the $(6 + j8)\,\Omega$ impedance by using Norton's theorem. Using the above procedure:

(i) Initially the $(6 + j8)\,\Omega$ impedance is short-circuited, as shown in Fig. 45.

(ii) From Fig. 45,

$$I_{SC} = I_1 + I_2 = \frac{(5 + j0)}{(3 + j4)} + \frac{(-(2 + j4))}{(2 - j5)}$$

$$= 1\underline{/-53.13°} - \frac{4.472\underline{/\,63.43°}}{5.385\underline{/-68.20°}}$$

$$= (1.152 - j1.421)\,\text{A} \quad \text{or} \quad 1.829\underline{/-50.97°}\,\text{A}$$

(iii) If each source of emf is removed, the impedance, z, "looking in" at a break made between A and B is given by

$$z = \frac{(3 + j4)(2 - j5)}{(3 + j4) + (2 - j5)} = 5.28\underline{/-3.76°}\,\Omega \quad \text{or} \quad (5.269 - j0.346)\,\Omega$$

(iv) From the Norton equivalent network shown in Fig. 46, the current is given by

$$i_L = \left(\frac{z}{Z_L + z}\right)I_{SC} = \left(\frac{5.28\underline{/-3.76°}}{(6 + j8) + (5.269 - j0.346)}\right)1.829\underline{/-50.97°}$$

i.e., **current in $(6 + j8)\,\Omega$ impedance, $i_L = 0.71\underline{/-88.91°}\,\text{A}$.**

Figure 46

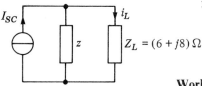

Worked problems on Norton's theorem

Problem 1. Use Norton's theorem to determine the value of current I in the circuit shown in Fig. 47.

(i) The branch containing $2.8\,\Omega$ resistor is short-circuited, as shown in Fig. 48.

(ii) The $3\,\Omega$ resistor in parallel with a short-circuit is the same as $3\,\Omega$ in parallel with 0 giving an equivalent impedance of $(3 \times 0)/(3 + 0) = 0$. Hence the network reduces to that shown in Fig. 49, where $I_{SC} = 5/2 = 2.5\,\text{A}$.

(iii) If the 5 V source is removed from the network the input impedance, z, "looking-in" at a break made in AB of Fig. 48 gives $z = (2 \times 3)/(2 + 3) = 1.2\,\Omega$ (see Fig. 50).

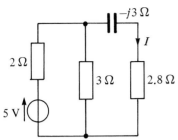

−j3 Ω

I

2 Ω

3 Ω   2.8 Ω

5 V

Figure 47

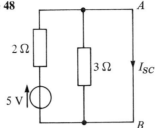

Figure 48   A

2 Ω

3 Ω I_{SC}

5 V

B

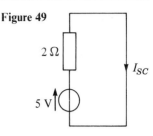

Figure 49

2 Ω

I_{SC}

5 V

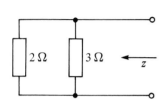

2 Ω   3 Ω

z

Figure 50

(iv) The Norton equivalent network is shown in Fig. 51, where current I is given by

$$I = \left(\frac{1.2}{1.2 + (2.8 - j3)} \right)(2.5) = \frac{3}{4 - j3} = \mathbf{0.60 \underline{/\ 36.87°}\ A}$$

Problem 2. For the circuit shown in Fig. 52 determine the current flowing in the inductive branch by using Norton's theorem.

(i) The inductive branch is initially short-circuited, as shown in Fig. 53.
(ii) From Fig. 53,

$$I_{SC} = I_1 + I_2 = \tfrac{20}{2} + \tfrac{10}{3} = 13.\dot{3}\ A$$

(iii) If the voltage sources are removed, the impedance, z, "looking in" at a break made in AB is given by $z = (2 \times 3)/(2 + 3) = 1.2\ \Omega$.
(iv) The Norton equivalent network is shown in Fig. 54, where current I is given by

$$I = \left(\frac{1.2}{1.2 + 1.5 + j2.95} \right)(13.\dot{3}) = \frac{16}{2.7 + j2.95}$$
$$= \mathbf{4.0 \underline{/\ -47.53°}\ A} \quad \text{or} \quad \mathbf{(2.7 - j2.95)\ A}$$

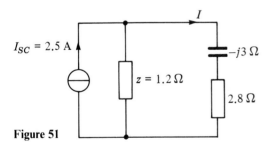

$I_{SC} = 2.5$ A

I

−j3 Ω

z = 1.2 Ω

2.8 Ω

Figure 51

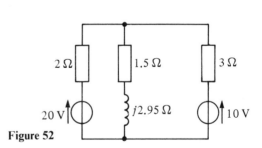

2 Ω   1.5 Ω   3 Ω

20 V   j2.95 Ω   10 V

Figure 52

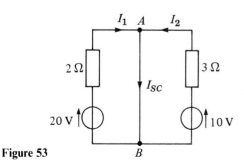

I_1 A I_2

2 Ω   3 Ω

I_{SC}

20 V   10 V

Figure 53   B

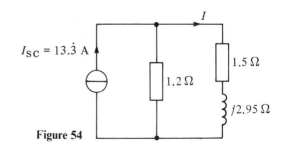

I

$I_{SC} = 13.\dot{3}$ A

1.5 Ω

1.2 Ω

j2.95 Ω

Figure 54

Figure 55

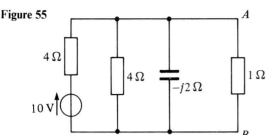

Figure 56

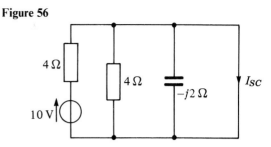

Problem 3. Use Norton's theorem to determine the magnitude of the p.d. across the $1\,\Omega$ resistance of the network shown in Fig. 55.

(i) The branch containing the $1\,\Omega$ resistance is initially short-circuited, as shown in Fig. 56.
(ii) $4\,\Omega$ in parallel with $-j2\,\Omega$ in parallel with $0\,\Omega$ (i.e., the short-circuit) is equivalent to 0, giving the equivalent circuit of Fig. 57. Hence $I_{SC} = 10/4 = 2.5\,\text{A}$.
(iii) The 10 V source is removed from the network of Fig. 55, as shown in Fig. 58, and the impedance z, "looking in" at a break made in AB is given by

$$\frac{1}{z} = \frac{1}{4} + \frac{1}{4} + \frac{1}{-j2} = \frac{-j-j+2}{-j4} = \frac{2-j2}{-j4}$$

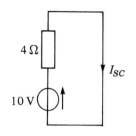

Figure 57

from which

$$z = \frac{-j4}{2-j2} = \frac{8-j8}{8} = (1-j1)\,\Omega$$

(iv) The Norton equivalent network is shown in Fig. 59, from which current I is given by

$$I = \left(\frac{1-j1}{(1-j1)+1}\right)(2.5) = 1.58\underline{/-18.43°}\,\text{A}$$

Hence the magnitude of the p.d. across the $1\,\Omega$ resistor is given by $IR = (1.58)(1) = \textbf{1.58 V}$.

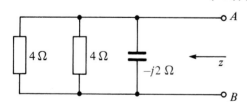

Figure 58

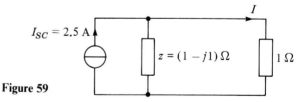

Figure 59

Problem 4. For the network shown in Fig. 60, obtain the Norton equivalent network at terminals AB. Hence determine the power dissipated in a $5\,\Omega$ resistor connected between A and B.

(i) Terminals AB are initially short-circuited, as shown in Fig. 61.
(ii) The circuit impedance Z presented to the $20\underline{/\,0°}\,\text{V}$ source is given by

$$Z = 2 + \frac{(4+j3)(-j3)}{(4+j3)+(-j3)} = 2 + \frac{9-j12}{4}$$

$$= (4.25-j3)\,\Omega \quad \text{or} \quad 5.202\underline{/-35.22°}\,\Omega$$

Figure 60

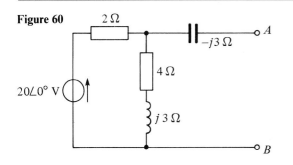

Figure 61

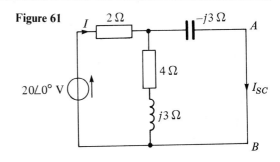

Thus current I in Fig. 61 is given by

$$I = \frac{20\angle\,0°}{5.202\angle\,-35.22°} = 3.845\angle\,35.22°\,\text{A}$$

Hence

$$I_{SC} = \left(\frac{(4+j3)}{(4+j3)-j3}\right)(3.845\angle\,35.22°) = 4.806\angle\,72.09°\,\text{A}$$

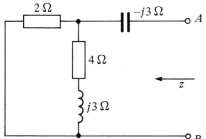

Figure 62

Norton equivalent circuit

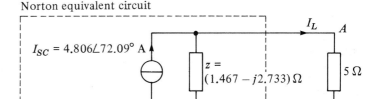

Figure 63

(iii) Removing the $20\angle\,0°$ V source of Fig. 60 gives the network of Fig. 62. Impedance, z, "looking in" at terminals AB is given by

$$z = -j3 + \frac{2(4+j3)}{2+4+j3} = -j3 + 1.491\angle\,10.3°$$

$$= (1.467 - j2.733)\,\Omega \quad \text{or} \quad 3.102\angle\,-61.77°\,\Omega$$

(iv) The Norton equivalent network is shown in Fig. 63.

$$\text{Current } I_L = \left(\frac{3.102\angle\,-61.77°}{1.467 - j2.733 + 5}\right)(4.806\angle\,72.09°) = 2.123\angle\,33.23°\,\text{A}$$

Hence the power dissipated in the $5\,\Omega$ resistor is $I_L^2 R = (2.123)^2(5) = \mathbf{22.5\,W}$.

Problem 5. Derive the Norton equivalent network with respect to terminals PQ for the network shown in Fig. 64 and hence determine the magnitude of the current flowing in a $2\,\Omega$ resistor connected across PQ.

This is the same problem as problem 7 on page 161 which was solved by Thévenin's theorem. A comparison of methods may thus be made.

(i) Terminals PQ are initially short-circuited, as shown in Fig. 65.

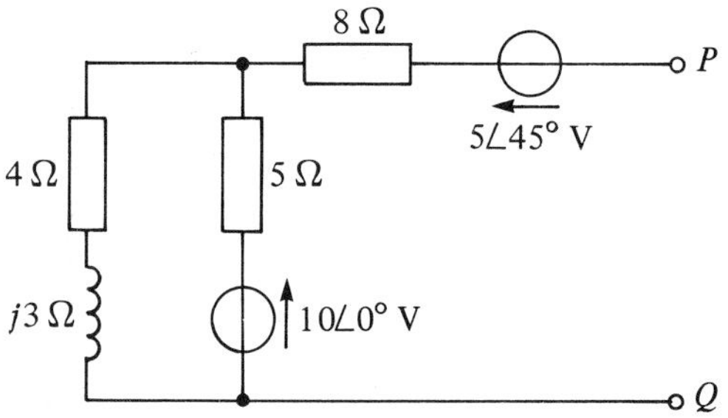

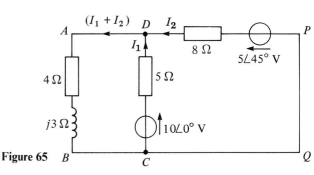

Figure 64

Figure 65

(ii) Currents I_1 and I_2 are shown labelled. Kirchhoff's laws are used.
For loop $ABCD$, and moving anticlockwise, $10\underline{/0°} = 5I_1 + (4+j3)(I_1+I_2)$, i.e.,

$$(9+j3)I_1 + (4+j3)I_2 - 10 = 0 \tag{1}$$

For loop $DPQC$, and moving clockwise, $10\underline{/0°} - 5\underline{/45°} = 5I_1 - 8I_2$, i.e.,

$$5I_1 - 8I_2 + (5\underline{/45°} - 10) = 0 \tag{2}$$

Solving equations (1) and (2) by using determinants gives

$$\frac{I_1}{\begin{vmatrix} (4+j3) & -10 \\ -8 & (5\underline{/45°}-10) \end{vmatrix}} = \frac{-I_2}{\begin{vmatrix} (9+j3) & -10 \\ 5 & (5\underline{/45°}-10) \end{vmatrix}} = \frac{1}{\begin{vmatrix} 9+j3 & (4+j3) \\ 5 & -8 \end{vmatrix}}$$

from which

$$I_2 = \frac{-\begin{vmatrix} (9+j3) & -10 \\ 5 & (5\underline{/45°}-10) \end{vmatrix}}{\begin{vmatrix} (9+j3) & (4+j3) \\ 5 & -8 \end{vmatrix}} = \frac{-[(9+j3)(5\underline{/45°}-10)+50]}{[-72-j24-20-j15]}$$

$$= \frac{-[22.52\underline{/146.52°}]}{[99.925\underline{/-157.03°}]} = -0.225\underline{/303.55°}$$

$$\text{or } -0.225\underline{/-56.45°}$$

Hence the short-circuit current $I_{SC} = 0.225\underline{/-56.45°}$ A flowing from P to Q.
(iii) The impedance, z, "looking in" at a break made between P and Q is given by $z = (10.50 + j0.833)\,\Omega$ (see problem 7, page 161).

Figure 66

$$I_{SC} = 0.225\underline{/-56.49°} \text{ A}$$

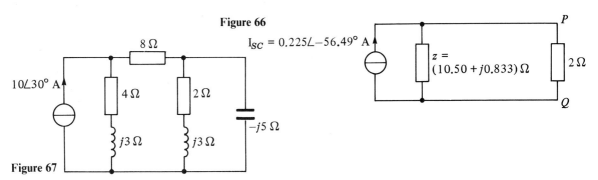

Figure 67

Figure 68

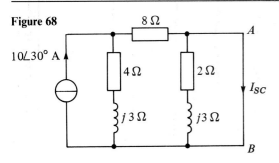

Figure 69

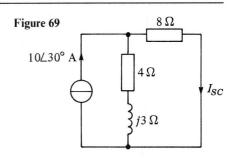

(iv) The Norton equivalent circuit is shown in Fig. 66, where current I is given by

$$I = \left(\frac{10.50 + j0.833}{10.50 + j0.833 + 2}\right)(0.225\angle -56.45°) = 0.19\angle -55.72° \text{ A}$$

Hence the magnitude of the current flowing in the $2\,\Omega$ resistor is 0.19 A.

Problem 6. Use Norton's theorem to determine the magnitude of the current flowing in the capacitor shown in the circuit of Fig. 67.

(i) The branch containing the capacitor is initially short-circuited, as shown in Fig. 68.

(ii) The equivalent network of Fig. 68 is shown in Fig. 69. Thus the short-circuit current is given by

$$I_{SC} = \left(\frac{4 + j3}{4 + j3 + 8}\right)(10\angle 30°) = 4.042\angle 52.83° \text{ A}$$

(iii) When determining the impedance, z, at a break made between A and B of Fig. 68, the $10\angle 30°$ A current source is removed and replaced by an open-circuit, as shown in Fig. 70.

$$\text{Impedance } z = \frac{(12 + j3)(2 + j3)}{(12 + j3) + (2 + j3)} = 2.928\angle 47.15° \,\Omega \quad \text{or} \quad (1.991 + j2.147)\,\Omega$$

(iv) The Norton equivalent network is shown in Fig. 71, where current I is given by

$$I = \left(\frac{2.928 \angle 47.15°}{1.991 + j2.147 - j5}\right)(4.042\angle 52.83°) = 3.40\angle 155.07° \text{A}$$

Hence the magnitude of the current flowing in the capacitor is 3.40 A.

Further problems on Norton's theorem may be found in section 5, problems 14 to 18, page 181.

Figure 70

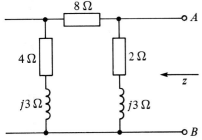

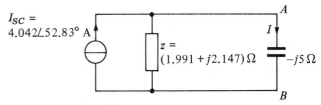

Figure 71

4. Thévenin and Norton equivalent networks

It is seen in sections 2 and 3 that when Thévenin's and Norton's theorems are applied to the same circuit, identical results are obtained. Thus the Thévenin and Norton networks shown in Fig. 72 are equivalent to each other. The impedance "looking in" at terminals AB is the same in each of the networks; i.e., z.

If terminals AB in Fig. 72(a) are short-circuited, the short-circuit current is given by E/z. If terminals AB in Fig. 72(b) are short-circuited, the short-circuit current is I_{SC}. Thus $\mathbf{I_{SC} = E/z}$.

Figure 72 Equivalent Thévenin and Norton circuits

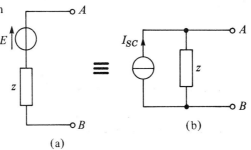

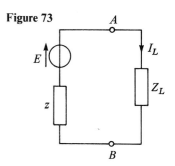

Figure 73

Figure 73 shows a source of emf E in series with an impedance z feeding a load impedance Z_L. From Fig. 73,

$$I_L = \frac{E}{z + Z_L} = \frac{E/z}{(z + Z_L)/z} = \left(\frac{z}{z + Z_L}\right)\frac{E}{z}$$

i.e.,

$$I_L = \left(\frac{z}{z + Z_L}\right)I_{SC}, \text{ from above.}$$

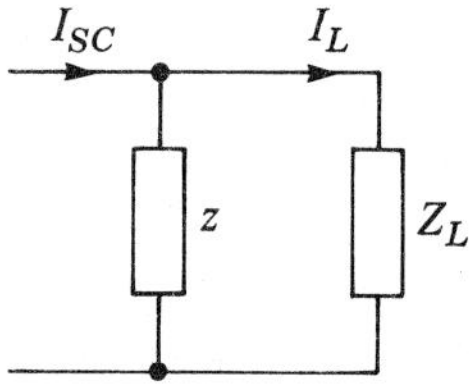

From Fig. 74 it can be seen that, when viewed from the load, the source appears as a source of current I_{SC} which is divided between z and Z_L connected in parallel.

Thus it is shown that the two representations shown in Fig. 72 are equivalent.

Figure 74

Figure 75

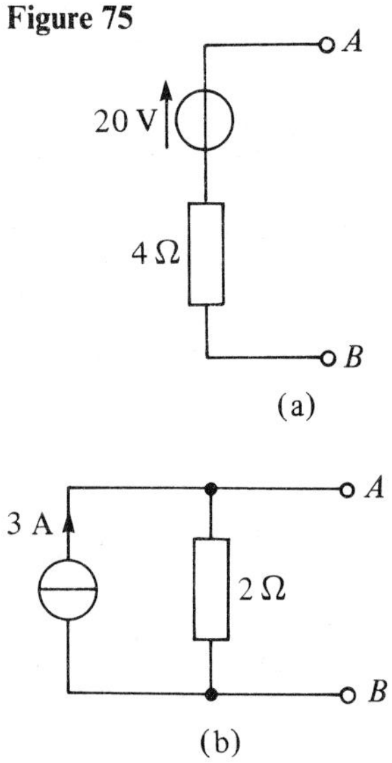

Worked problems on Thévenin and Norton equivalent networks

Problem 1. (a) Convert the circuit shown in Fig. 75(a) to an equivalent Norton network. (b) Convert the network shown in Fig. 75(b) to an equivalent Thévenin circuit.

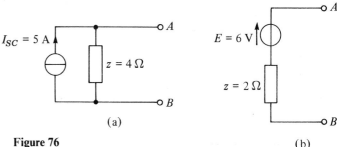

Figure 76

(a) If the terminals AB of Fig. 75(a) are short-circuited, the short-circuit current, $I_{SC} = 20/4 = 5$ A. The impedance "looking in" at terminals AB is $4\,\Omega$. Hence the equivalent Norton network is as shown in Fig. 76(a).

(b) The open-circuit voltage E across terminals AB in Fig. 75(b) is given by $E = (I_{SC})(z) = (3)(2) = 6$ V. The impedance "looking in" at terminals AB is $2\,\Omega$. Hence the equivalent Thévenin circuit is as shown in Fig. 76(b).

Problem 2. (a) Convert the circuit to the left of terminals AB in Fig. 77 to an equivalent Thévenin circuit by initially converting to a Norton equivalent circuit. (b) Determine the magnitude of the current flowing in the $(1.8 + j4)\,\Omega$ impedance connected between terminals A and B of Fig. 77.

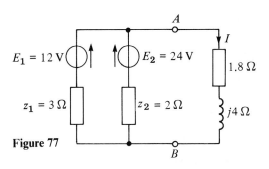

Figure 77

Figure 78

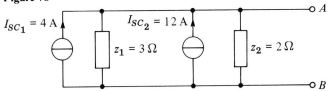

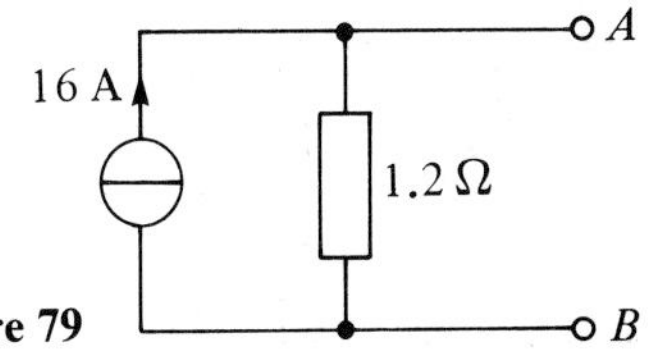

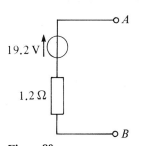

Figure 80

Figure 79

(a) For the branch containing the 12 V source, conversion to a Norton equivalent network gives $I_{SC_1} = 12/3 = 4$ A and $z_1 = 3\,\Omega$. For the branch containing the 24 V source, conversion to a Norton equivalent circuit gives $I_{SC_2} = 24/2 = 12$ A and $z_2 = 2\,\Omega$.

Thus Fig. 78 shows a network equivalent to Fig. 77. From Fig. 78, the total short-circuit current is $4 + 12 = 16$ A, and the total impedance is given by $(3 \times 2)/(3 + 2) = 1.2\,\Omega$. Thus Fig. 78 simplifies to Fig. 79.

The open-circuit voltage across AB of Fig. 79, $E = (16)(1.2) = 19.2$ V, and the impedance "looking in" at AB, $z = 1.2\,\Omega$. Hence the Thévenin equivalent circuit is as shown in Fig. 80.

(b) When the $(1.8 + j4)\,\Omega$ impedance is connected to terminals AB of Fig. 80, the current I flowing is given by

$$I = \frac{19.2}{(1.2 + 1.8 + j4)} = 3.84\angle -53.13°\,\text{A}$$

Hence the current flowing in the $(1.8 + j4)\,\Omega$ impedance is 3.84 A.

Problem 3. Determine, by successive conversions between Thévenin's and Norton's equivalent networks, a Thévenin equivalent circuit for terminals AB of Fig. 81. Hence determine the magnitude of the current flowing in the capacitive branch connected to terminals AB.

Figure 81

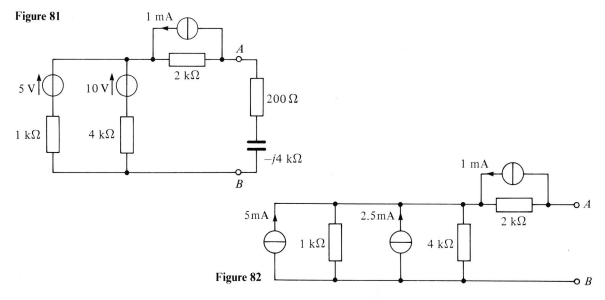

Figure 82

For the branch containing the 5 V source, converting to a Norton equivalent network gives $I_{SC} = 5/1000 = 5 \text{ mA}$ and $z = 1 \text{ k}\Omega$. For the branch containing the 10 V source, converting to a Norton equivalent network gives $I_{SC} = 10/4000 = 2.5 \text{ mA}$ and $z = 4 \text{ k}\Omega$. Thus the circuit of Fig. 81 converts to that of Fig. 82.

The above two Norton equivalent networks shown in Fig. 82 may be combined, since the total short-circuit current is $(5 + 2.5) = 7.5 \text{ mA}$ and the total impedance z is given by $(1 \times 4)/(1 + 4) = 0.8 \text{ k}\Omega$. This results in the network of Fig. 83.

Both of the Norton equivalent networks shown in Fig. 83 may be converted to Thévenin equivalent circuits. Open-circuit voltage across CD is

$$(7.5 \times 10^{-3})(0.8 \times 10^3) = 6 \text{ V}$$

and the impedance "looking in" at CD is $0.8 \text{ k}\Omega$. Open-circuit voltage across EF is $(1 \times 10^{-3})(2 \times 10^3) = 2 \text{ V}$ and the impedance "looking in" at EF is $2 \text{ k}\Omega$. Thus Fig. 83 converts to Fig. 84.

Combining the two Thévenin circuits gives emf $E = 6 - 2 = 4 \text{ V}$, and impedance $z = (0.8 + 2) = 2.8 \text{ k}\Omega$. Thus the Thévenin equivalent circuit for terminals AB of Fig. 81 is as shown in Fig. 85.

If an impedance $(200 - j4000) \, \Omega$ is connected across terminals AB, then the current I flowing is given by

$$I = \frac{4}{2800 + (200 - j4000)} = \frac{4}{5000 \underline{/-53.13^\circ}} = 0.80 \underline{/53.13^\circ} \text{ mA}$$

i.e., **the current in the capacitive branch in 0.80 mA.**

Problem 4. (a) Determine an equivalent Thévenin circuit for terminals AB of the network shown in Fig. 86. (b) Calculate the power dissipated in a $(600 - j800) \, \Omega$ impedance connected between A and B of Fig. 86.

(a) Converting the Thévenin circuit to a Norton network gives

$$I_{SC} = \frac{5}{j1000} = -j5 \text{ mA} \quad \text{or} \quad 5 \underline{/-90^\circ} \text{ mA} \quad \text{and} \quad z = j1 \text{ k}\Omega$$

Figure 83

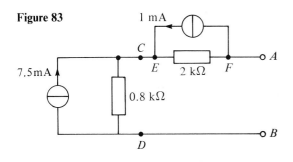

Figure 84

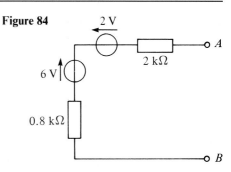

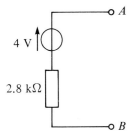

Figure 85

Thus Fig. 86 converts to that shown in Fig. 87. The two Norton equivalent networks may be combined, giving

$$I_{sc} = 4 + 5\underline{/-90°} = (4 - j5)\,\text{mA} \quad \text{or} \quad 6.403\underline{/-51.34°}\,\text{mA}$$

and

$$z = \frac{(2)(j1)}{(2 + j1)} = (0.4 + j0.8)\,\text{k}\Omega \quad \text{or} \quad 0.894\underline{/63.43°}\,\text{k}\Omega$$

This results in the equivalent network shown in Fig. 88. Converting to an equivalent Thévenin circuit gives open circuit emf across AB,

$$E = (6.403 \times 10^{-3}\underline{/-51.34°})(0.894 \times 10^{3}\underline{/63.43°}) = 5.724\underline{/12.09°}\,\text{V}$$

and

$$\text{impedance } z = 0.894\underline{/63.43°}\,\text{k}\Omega \quad \text{or} \quad (400 + j800)\,\Omega$$

Thus the Thévenin equivalent circuit is as shown in Fig. 89.

(b) When a $(600 - j800)\,\Omega$ impedance is connected across AB, the current I flowing is given by

$$I = \frac{5.724\underline{/12.09°}}{(400 + j800) + (600 - j800)} = 5.724\underline{/12.09°}\,\text{mA}$$

Hence the power P dissipated in the $(600 - j800)\,\Omega$ impedance is given by
$P = I^2 R = (5.724 \times 10^{-3})^2\,(600) = \mathbf{19.7\,mW.}$

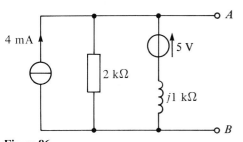

Figure 86

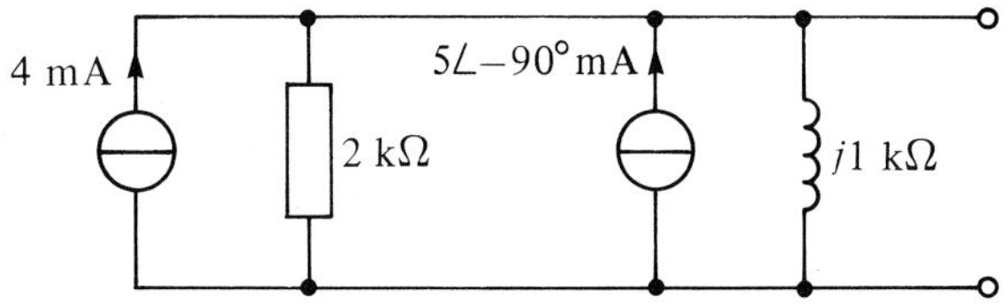

Figure 87

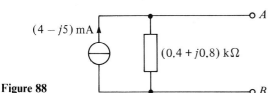

Figure 88

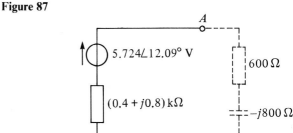

Figure 89

Problem 5. For the network shown in Fig. 90 obtain (a) the Thévenin equivalent circuit, and (b) the Norton equivalent network. (c) Show that if a 1Ω resistance is connected across terminals AB, the same current will flow in each case.

(a) From Fig. 90,

$$\text{current } i = \frac{5 \angle 0°}{2 + 4 + j3} = (\tfrac{2}{3} - j\tfrac{1}{3}) \text{A} \quad \text{or} \quad 0.745 \angle -26.57° \text{A}$$

The open-circuit voltage across AB,

$$E = i(4 + j3) = (0.745 \angle -26.57°)(5 \angle 36.87°) = 3.725 \angle 10.30° \text{ V}$$

(Alternatively,

$$E = 5 \angle 0° - 2i = 5 - (\tfrac{4}{3} - j\tfrac{2}{3}) = 3.727 \angle 10.3° \text{ V})$$

The impedance "looking in" at terminals AB with the $5 \angle 0°$ V source removed is given by

$$z = \frac{2(4 + j3)}{(2 + 4 + j3)} + j1.5 = 1.491 \angle 10.3° + j1.5$$

$$= (1.467 + j1.767) \Omega$$

The equivalent Thévenin circuit is shown in Fig. 91.

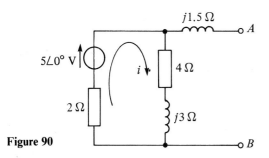

Figure 90

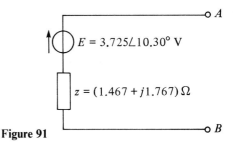

Figure 91

(b) The terminals AB are short-circuited, as shown in Fig. 92. Short-circuit current,

$$I_{sc} = \left(\frac{4 + j3}{4 + j3 + j1.5} \right) I$$

Current $I = 5 \angle 0° / Z_T$, where Z_T is the total circuit impedance.

$$Z_T = 2 + \frac{(4 + j3)(j1.5)}{(4 + j3 + j1.5)} = 2 + \frac{7.5 \angle 126.87°}{6.021 \angle 48.37°}$$

$$= (2.248 + j1.221) \Omega \quad \text{or} \quad 2.558 \angle 28.51° \Omega$$

Hence

$$\text{current, } I = \frac{5 \angle 0°}{2.558 \angle 28.51°} = 1.955 \angle -28.51° \text{A}$$

Thus

$$I_{sc} = \left(\frac{4 + j3}{4 + j4.5} \right)(1.955 \angle -28.51°) = 1.623 \angle -40.01° \text{A}$$

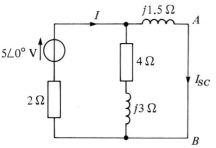

Figure 92

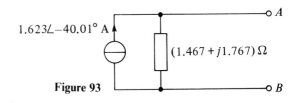

Figure 93

The impedance "looking in" at the open-circuited terminals AB with the $5\underline{/\ 0°}$ V source removed is the same as in part (a), above, i.e., $z = (1.467 + j1.767)\,\Omega$. Thus the equivalent Norton network is as shown in Fig. 93.

(c) For the Thévenin equivalent circuit of Fig. 91, the current flowing in a $1\,\Omega$ resistor connected between A and B is given by

$$I = \frac{3.725\underline{/\ 10.30°}}{(1.467 + j1.767) + 1} = 1.23\underline{/\ -25.31°}\ \text{A}$$

For the Norton equivalent network of Fig. 93, the current flowing in a $1\,\Omega$ resistor connected between A and B is given by

$$I = \left(\frac{1.467 + j1.767}{(1.467 + j1.767) + 1}\right)(1.623\underline{/\ -40.01°}) = 1.23\underline{/\ -25.32°}\ \text{A}$$

Further problems on Thévenin and Norton equivalent networks may be found in the following section (5), problems 19 to 25, page 181.

5. Further problems

Thévenin's theorem

1. Use Thévenin's theorem to determine the current flowing in the $10\,\Omega$ resistor of the d.c. network shown in Fig. 94.

[0.85 A]

2. Determine, using Thévenin's theorem, the values of currents I_1, I_2 and I_3 of the network shown in Fig. 95.

[$I_1 = 2.8$ A, $I_2 = 4.8$ A, $I_3 = 7.6$ A]

3. Determine the Thévenin equivalent circuit with respect to terminals AB of the network shown in Fig. 96. Hence determine the magnitude of the current flowing in a $(4 - j7)\,\Omega$ impedance connected across terminals AB and the power delivered to this impedance.

[$E = 15.37\underline{/\ -38.66°}, z = (3.20 + j4.00)\,\Omega; 1.97$ A; 15.5 W]

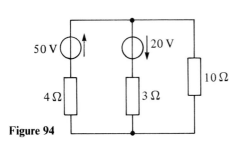

Figure 94

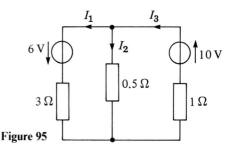

Figure 95

Figure 96

Figure 97

Figure 98

Figure 99

Figure 100

Figure 101

Figure 102

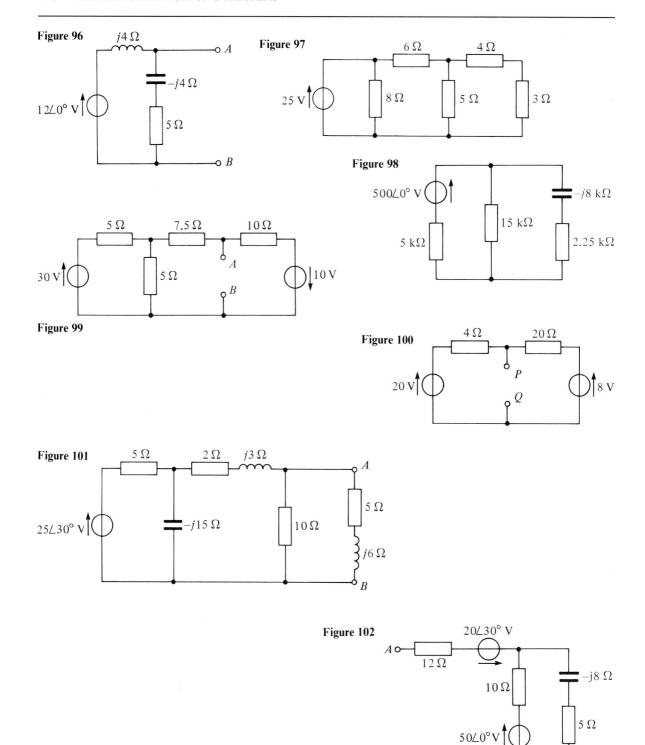

4. For the network shown in Fig. 97 use Thévenin's theorem to determine the current flowing in the 3 Ω resistance.

[1.17 A]

5. Use Thévenin's theorem to determine the magnitude of the current flowing in the capacitive branch of the network shown in Fig. 98. Find also the p.d. across the capacitor.

[37.5 mA; 300 V]

6. Derive for the network shown in Fig. 99 the Thévenin equivalent circuit at terminals AB, and hence determine the current flowing in a 20 Ω resistance connected between A and B.

[$E = 2.5$ V, $z = 5$ Ω; 0.1 A]

7. In the d.c. network shown in Fig. 100, resistances of 3 Ω, 8 Ω and 10 Ω are connected in turn to terminals PQ. Derive for the network the Thévenin equivalent circuit and hence determine the power delivered to each of the three resistors.

[24.2 W; 20.2 W; 18.2 W]

8. Determine for the network shown in Fig. 101 the Thévenin equivalent circuit with respect to terminals AB, and hence determine the current flowing in the $(5 + j6)$ Ω impedance connected between A and B.

[$E = 14.3 \underline{/\,6.38°}$, $z = (3.99 + j0.55)$ Ω; 1.29 A]

9. For the network shown in Fig. 102, derive the Thévenin equivalent circuit with respect to terminals AB, and hence determine the magnitude of the current flowing in a $(2 + j13)$ Ω impedance connected between A and B.

[1.157 A]

10. Use Thévenin's theorem to determine the power dissipated in the 4 Ω resistance of the network shown in Fig. 103.

[0.24 W]

11. Find the Thévenin equivalent circuit at terminals PQ for the network shown in Fig. 104 and hence determine the magnitude of the current flowing through a 3 Ω resistance connected across PQ.

[$E = 9.52 \underline{/\,6.46°}$ V, $z = 4.39 \underline{/\,0°}$ Ω; 1.29 A]

12. For the bridge network shown in Fig. 105 use Thévenin's theorem to determine the current flowing in the $(4 + j3)$ Ω impedance and its direction. Assume that the $20 \underline{/\,0°}$ V source has negligible internal impedance.

[0.12 A from Q to P]

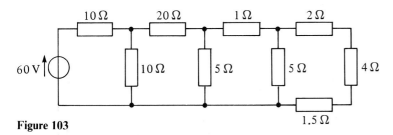

Figure 103

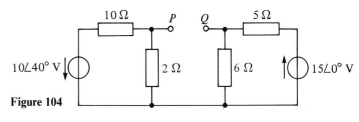

Figure 104

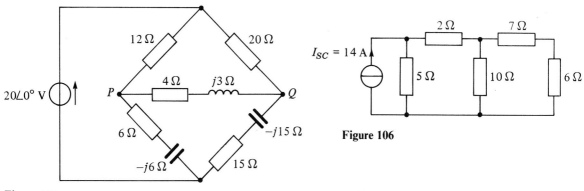

Figure 106

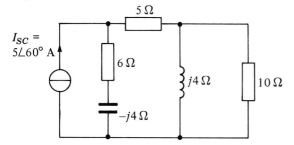

Figure 105

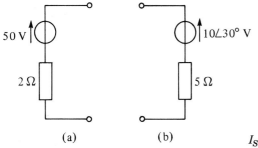

Figure 107

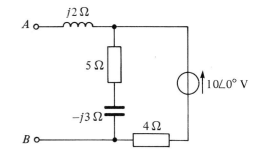

Figure 108

Figure 109

Figure 111

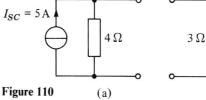

Figure 110   (a)   (b)

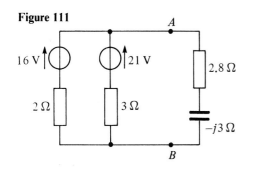

13. Repeat problems 1 to 11, 13 to 15 and 23 to 28 of section 7, chapter 7, pp. 145–150, using Thévenin's theorem and compare the method of solution with that used for Kirchhoff's laws, mesh-current and nodal analysis and the super-position theorem.

Norton's theorem

14. Repeat problems 1 to 12 above using Norton's theorem instead of Thévenin's theorem.

15. Determine the current flowing in the $10\,\Omega$ resistance of the network shown in Fig. 106 by using Norton's theorem.

[3.13 A]

16. For the network shown in Fig. 107, use Norton's theorem to determine the current flowing in the $10\,\Omega$ resistance.

[1.09 A]

17. Determine for the network shown in Fig. 108 the Norton equivalent network at terminals AB. Hence determine the current flowing in a $(2+j4)\,\Omega$ impedance connected between A and B.

$[I_{SC} = 2.17\underline{/\,-43.96°}\,\text{A}, z = (2.40 + j1.47)\,\Omega; 0.87\,\text{A}]$

18. Repeat problems 1 to 11, 13 to 15 and 23 to 28 of section 7, chapter 7, pp. 145–150, using Norton's theorem.

Thévenin and Norton equivalent networks

19. Convert the circuits shown in Fig. 109 to Norton equivalent networks.

$$\left[\begin{array}{l} \text{(a) } I_{SC} = 25\,\text{A}, z = 2\,\Omega \\ \text{(b) } I_{SC} = 2\underline{/\,30°}, z = 5\,\Omega \end{array}\right]$$

20. Convert the network shown in Fig. 110 to Thévenin equivalent circuits.

[(a) $E = 20\,\text{V}, z = 4\,\Omega$;   (b) $E = 12\underline{/\,50°}\,\text{V}, z = 3\,\Omega$]

21. (a) Convert the network to the left of terminals AB in Fig. 111 to an equivalent Thévenin circuit by initially converting to a Norton equivalent network.
    (b) Determine the current flowing in the $(2.8 - j3)\,\Omega$ impedance connected between A and B in Fig. 111.

[(a) $E = 18\,\text{V}, z = 1.2\,\Omega$ (b) 3.6 A]

22. Determine, by successive conversions between Thévenin and Norton equivalent networks, a Thévenin equivalent circuit for terminals AB of Fig. 112. Hence determine the current flowing in a $6\,\Omega$ resistor connected between A and B.

$[E = 9\frac{1}{3}\text{V}, z = 1\,\Omega; 1\frac{1}{3}\,\text{A}]$

23. Derive an equivalent Thévenin circuit for terminals AB of the network shown in Fig. 113. Hence determine the p.d. across AB when a $(3 + j4)\,\text{k}\Omega$ impedance is connected between these terminals.

$[E = 4.81\underline{/\,-41.63°}\,\text{V}, z = (800 + j400)\,\Omega; 4.14\,\text{V}]$

Figure 112

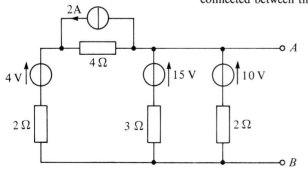

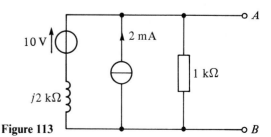

Figure 113

Figure 114   **Figure 115**

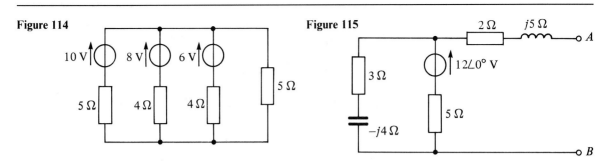

24. For the network shown in Fig. 114, convert each branch containing a voltage source to its Norton equivalent and hence determine the current flowing in the 5 Ω resistance.

[1.22 A]

25. For the network shown in Fig. 115, derive (a) the Thévenin equivalent circuit, and (b) the Norton equivalent network. (c) A 6 Ω resistance is connected between A and B. Determine the current flowing in the 6 Ω resistance by using both the Thévenin and Norton equivalent circuits.

$$\left[\begin{array}{l} \text{(a) } E = 6.71\underline{/-26.56°}\text{ V}, z = (4.50 + j3.75)\,\Omega \\ \text{(b) } I_{SC} = 1.15\underline{/-66.37°}, z = (4.50 + j3.75)\,\Omega \\ \text{(c) } 0.60\text{ A} \end{array}\right]$$

9 Delta–star and star–delta transformations, and the maximum power transfer theorems

1. Introduction

By using Kirchhoff's laws, mesh-current analysis, nodal analysis or the superposition theorem, currents and voltages in many network can be determined as shown in chapter 7. Thévenin's and Norton's theorems, introduced in chapter 8, provide an alternative method of solving networks and often with considerably reduced numerical calculations. Also, these latter theorems are especially useful when only the current in a particular branch of a complicated network is required. Delta–star and star–delta transformations may be applied in certain types of circuit to simplify them before application of circuit theorems.

2. Delta and star connections

The network shown in Fig. 1(a) consisting of three impedances Z_A, Z_B and Z_C is said to be **π-connected.** This network can be redrawn as shown in Fig. 1(b), where the arrangement is referred to as **delta-connected** or **mesh-connected.**

Figure 1 (a) π-connected network
(b) Delta-connected network

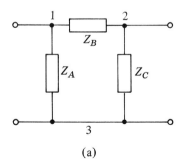

(a)

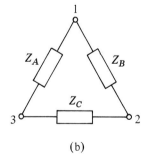

(b)

The network shown in Fig. 2(a), consisting of three impedances, Z_1, Z_2 and Z_3, is said to be **T-connected**. This network can be redrawn as shown in Fig. 2(b), where the arrangement is referred to as **star-connected.**

3. Delta–star transformation

It is possible to replace the delta connection shown in Fig. 3(a) by an equivalent star connection as shown in Fig. 3(b) such that the impedance measured between any pair of terminals (1–2, 2–3 or 3–1) is the same in star as in delta. The equivalent star network will consume the same power and operate at the same power factor as the original delta network. A delta–star transformation may alternatively be termed "π to T transformation".

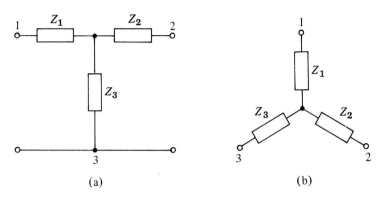

Figure 2 (a) T-connected
network
(b) Star-connected
network

(a)   (b)

Figure 3

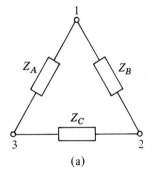

(a)

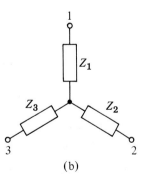

(b)

Considering terminals 1 and 2 of Fig. 3(a), the equivalent impedance is given by the impedance Z_B in parallel with the series combination of Z_A and Z_C, i.e.,

$$\frac{Z_B(Z_A + Z_C)}{Z_B + Z_A + Z_C}.$$

In Fig. 3(b), the equivalent impedance between terminals 1 and 2 is Z_1 and Z_2 in series, i.e., $Z_1 + Z_2$. Thus,

$$\qquad\qquad\quad Delta \qquad\qquad Star$$
$$Z_{12} = \frac{Z_B(Z_A + Z_C)}{Z_B + Z_A + Z_C} = Z_1 + Z_2 \qquad (1)$$

By similar reasoning,

$$Z_{23} = \frac{Z_C(Z_A + Z_B)}{Z_C + Z_A + Z_B} = Z_2 + Z_3 \qquad (2)$$

and

$$Z_{31} = \frac{Z_A(Z_B + Z_C)}{Z_A + Z_B + Z_C} = Z_3 + Z_1 \qquad (3)$$

Hence we have three simultaneous equations to be solved for Z_1, Z_2 and Z_3. Equation (1) − equation (2) gives

$$\frac{Z_A Z_B - Z_A Z_C}{Z_A + Z_B + Z_C} = Z_1 - Z_3 \qquad (4)$$

Equation (3) + equation (4) gives

$$\frac{2Z_A Z_B}{Z_A + Z_B + Z_C} = 2Z_1$$

from which

$$Z_1 = \frac{Z_A Z_B}{Z_A + Z_B + Z_C}$$

Similarly, equation (2) − equation (3) gives

$$\frac{Z_B Z_C - Z_A Z_B}{Z_A + Z_B + Z_C} = Z_2 - Z_1 \tag{5}$$

Equation (1) + equation (5) gives

$$\frac{2 Z_B Z_C}{Z_A + Z_B + Z_C} = 2 Z_2$$

from which

$$Z_2 = \frac{Z_B Z_C}{Z_A + Z_B + Z_C}$$

Finally, equation (3) − equation (1) gives

$$\frac{Z_A Z_C - Z_B Z_C}{Z_A + Z_B + Z_C} = Z_3 - Z_2 \tag{6}$$

Equation (2) + equation (6) gives

$$\frac{2 Z_A Z_C}{Z_A + Z_B + Z_C} = 2 Z_3$$

from which

$$Z_3 = \frac{Z_A Z_C}{Z_A + Z_B + Z_C}$$

Summarising, the star section shown in Fig. 3(b) is equivalent to the delta section shown in Fig. 3(a) when

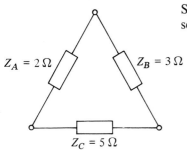

$$Z_1 = \frac{Z_A Z_B}{Z_A + Z_B + Z_C} \tag{7}$$

$$Z_2 = \frac{Z_B Z_C}{Z_A + Z_B + Z_C} \tag{8}$$

and

$$Z_3 = \frac{Z_A Z_C}{Z_A + Z_B + Z_C} \tag{9}$$

It is noted that impedance Z_1 is given by the product of the two impedances in delta joined to terminal 1 (i.e., Z_A and Z_B), divided by the sum of the three impedances; impedance Z_2 is given by the product of the two impedances in delta joined to terminal 2 (i.e., Z_B and Z_C), divided by the sum of the three impedances; and impedance Z_3 is given by the product of the two impedances in delta joined to terminal 3 (i.e., Z_A and Z_C), divided by the sum of the three impedances.

Thus, for example, the star equivalent of the resistive delta network shown in Fig. 4 is given by

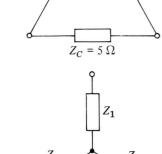

Figure 4

$$Z_1 = \frac{(2)(3)}{2+3+5} = 0.6\,\Omega, \qquad Z_2 = \frac{(3)(5)}{2+3+5} = 1.5\,\Omega$$

and

$$Z_3 = \frac{(2)(5)}{2+3+5} = 1.0\,\Omega$$

4. Star–delta transformation

It is possible to replace the star section shown in Fig. 3(b) by an equivalent delta section as shown in Fig. 3(a). Such a transformation is also known as "T to π transformation".

From equations (7), (8) and (9),

$$Z_1Z_2+Z_2Z_3+Z_3Z_1 = \frac{Z_AZ_B^2Z_C + Z_AZ_BZ_C^2 + Z_A^2Z_BZ_C}{(Z_A+Z_B+Z_C)^2}$$

$$= \frac{Z_AZ_BZ_C(Z_B+Z_C+Z_A)}{(Z_A+Z_B+Z_C)^2}$$

$$= \frac{Z_AZ_BZ_C}{Z_A+Z_B+Z_C} \tag{10}$$

$$= Z_A\left(\frac{Z_BZ_C}{Z_A+Z_B+Z_C}\right) = Z_A(Z_2) \quad \text{from equation (8)}$$

Hence

$$Z_A = \frac{Z_1Z_2+Z_2Z_3+Z_3Z_1}{Z_2}$$

From equation (10),

$$Z_1Z_2+Z_2Z_3+Z_3Z_1 = Z_B\left(\frac{Z_AZ_C}{Z_A+Z_B+Z_C}\right) = Z_B(Z_3) \quad \text{from equation (9)}$$

Hence

$$Z_B = \frac{Z_1Z_2+Z_2Z_3+Z_3Z_1}{Z_3}$$

Also from equation (10),

$$Z_1Z_2+Z_2Z_3+Z_3Z_1 = Z_C\left(\frac{Z_AZ_B}{Z_A+Z_B+Z_C}\right) = Z_C(Z_1) \quad \text{from equation (7)}$$

Hence

$$Z_C = \frac{Z_1Z_2+Z_2Z_3+Z_3Z_1}{Z_1}$$

Summarising, the delta section shown in Fig. 3(a) is equivalent to the star section shown in Fig. 3(b) when

$$Z_A = \frac{Z_1Z_2+Z_2Z_3+Z_3Z_1}{Z_2} \tag{11}$$

$$Z_B = \frac{Z_1Z_2+Z_2Z_3+Z_3Z_1}{Z_3} \tag{12}$$

and

$$Z_C = \frac{Z_1Z_2+Z_2Z_3+Z_3Z_1}{Z_1} \tag{13}$$

It is noted that the numerator in each expression is the sum of the products of the star impedances taken in pairs. The denominator of the expression for Z_A, which is connected between terminals 1 and 3 of Fig. 3(a), is Z_2, which is connected to terminal 2 of Fig. 3(b). Similarly, the denominator of the

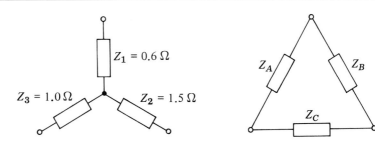

Figure 5

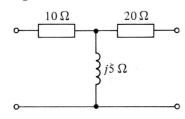

Figure 6

expression for Z_B, which is connected between terminals 1 and 2 of Fig. 3(a), is Z_3, which is connected to terminal 3 of Fig. 3(b). Also the denominator of the expression for Z_C, which is connected between terminals 2 and 3 of Fig. 3(a), is Z_1, which is connected to terminal 1 of Fig. 3(b).

Thus, for example, the delta equivalent of the resistive star circuit shown in Fig. 5 is given by

$$Z_A = \frac{(0.6)(1.5) + (1.5)(1.0) + (1.0)(0.6)}{1.5} = \frac{3.0}{1.5} = 2\,\Omega$$

$$Z_B = \frac{3.0}{1.0} = 3\,\Omega \quad \text{and} \quad Z_C = \frac{3.0}{0.6} = 5\,\Omega.$$

Worked problems on delta–star and star–delta transformations

Problem 1. Determine the delta-connected equivalent network for the star-connected impedances shown in Fig. 6.

Figure 7(a) shows the network of Fig. 6 redrawn and Fig. 7(b) shows the equivalent delta connection containing impedances Z_A, Z_B and Z_C. From equation (11),

$$Z_A = \frac{Z_1 Z_2 + Z_2 Z_3 + Z_3 Z_1}{Z_2} = \frac{(10)(20) + (20)(j5) + (j5)(10)}{20}$$
$$= \frac{200 + j150}{20} = (10 + j7.5)\,\Omega$$

From equation (12),

$$Z_B = \frac{(200 + j150)}{Z_3} = \frac{200 + j150}{j5} = \frac{-j5(200 + j150)}{25} = (30 - j40)\,\Omega$$

From equation (13),

$$Z_C = \frac{(200 + j150)}{Z_1} = \frac{(200 + j150)}{10} = (20 + j15)\,\Omega$$

Problem 2. Replace the delta-connected network shown in Fig. 8 by an equivalent star connection.

Let the equivalent star network be as shown in Fig. 9. Then, from equation (7),

$$Z_1 = \frac{Z_A Z_B}{Z_A + Z_B + Z_C} = \frac{(20)(10 + j10)}{20 + 10 + j10 - j20}$$

Figure 7

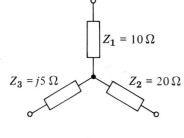

(a)

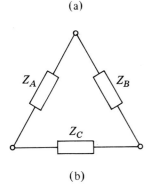

(b)

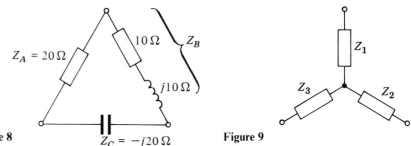

Figure 8   **Figure 9**

$$= \frac{(20)(10+j10)}{(30-j10)} = \frac{(20)(14.14\underline{/\,45°})}{31.62\underline{/-18.43°}}$$
$$= 8.94\underline{/\,63.43°}\,\Omega \quad \text{or} \quad (4+j8)\,\Omega$$

From equation (8),

$$Z_2 = \frac{Z_B Z_C}{Z_A + Z_B + Z_C} = \frac{(10+j10)(-j20)}{31.62\underline{/-18.43°}}$$
$$= \frac{(14.14\underline{/\,45°})(20\underline{/-90°})}{31.62\underline{/-18.43°}} = 8.94\underline{/-26.57°}\,\Omega \quad \text{or} \quad (8-j4)\,\Omega.$$

From equation (9),

$$Z_3 = \frac{Z_A Z_C}{Z_A + Z_B + Z_C} = \frac{(20)(-j20)}{31.62\underline{/-18.43°}}$$
$$= \frac{400\underline{/-90°}}{31.62\underline{/-18.43°}} = 12.65\underline{/-71.57°}\,\Omega \quad \text{or} \quad (4-j12)\,\Omega$$

Problem 3. Three impedances, $Z_1 = 100\underline{/\,0°}\,\Omega$, $Z_2 = 63.25\underline{/\,18.43°}\,\Omega$ and $Z_3 = 100\underline{/-90°}\,\Omega$ are connected in star. Convert the star to an equivalent delta connection.

Figure 10

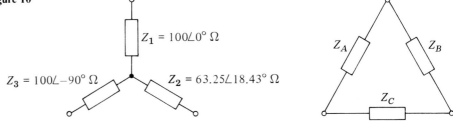

The star-connected network and the equivalent delta network comprising impedances Z_A, Z_B and Z_C are shown in Fig. 10. From equation (11),

$$Z_A = \frac{Z_1 Z_2 + Z_2 Z_3 + Z_3 Z_1}{Z_2}$$
$$= \frac{(100\underline{/\,0°})(63.25\underline{/\,18.43°}) + (63.25\underline{/\,18.43°})(100\underline{/-90°}) + (100\underline{/-90°})(100\underline{/\,0°})}{63.25\underline{/\,18.43°}}$$

$$= \frac{6325\underline{/\ 18.43^\circ} + 6325\underline{/\ -71.57^\circ} + 10000\underline{/\ -90^\circ}}{63.25\underline{/\ 18.43^\circ}}$$

$$= \frac{6000 + j2000 + 2000 - j6000 - j10000}{63.25\underline{/\ 18.43^\circ}} = \frac{8000 - j14000}{63.25\underline{/\ 18.43^\circ}}$$

$$= \frac{16124.5\underline{/\ -60.26^\circ}}{63.25\underline{/\ 18.43^\circ}} = \mathbf{254.9\underline{/\ -78.69^\circ}\ \Omega} \quad \text{or} \quad \mathbf{(50 - j250)\ \Omega}$$

From equation (12),

$$Z_B = \frac{Z_1Z_2 + Z_2Z_3 + Z_3Z_1}{Z_3} = \frac{16124.5\underline{/\ -60.26^\circ}}{100\underline{/\ -90^\circ}}$$

$$= \mathbf{161.3\underline{/\ 29.74^\circ}\ \Omega} \quad \text{or} \quad \mathbf{(140 + j80)\ \Omega}$$

From equation (13),

$$Z_C = \frac{Z_1Z_2 + Z_2Z_3 + Z_3Z_1}{Z_1} = \frac{16124.5\underline{/\ -60.26^\circ}}{100\underline{/\ 0^\circ}}$$

$$= \mathbf{161.3\underline{/\ -60.26^\circ}\ \Omega} \quad \text{or} \quad \mathbf{(80 - j140)\ \Omega}$$

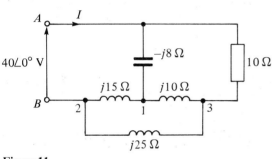

Figure 11

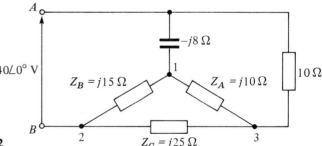

Figure 12

Problem 4. For the network shown in Fig. 11, determine (a) the equivalent circuit impedance across terminals AB, (b) supply current I and (c) the power dissipated in the $10\,\Omega$ resistor.

(a) The network of Fig. 11 is redrawn, as in Fig. 12, showing more clearly the part of the network 1, 2, 3 forming a delta connection. This may be transformed into a star connection as shown in Fig. 13. From equation (7),

$$Z_1 = \frac{Z_AZ_B}{Z_A + Z_B + Z_C} = \frac{(j10)(j15)}{j10 + j15 + j25} = \frac{(j10)(j15)}{(j50)} = j3\,\Omega$$

From equation (8),

$$Z_2 = \frac{Z_BZ_C}{Z_A + Z_B + Z_C} = \frac{(j15)(j25)}{(j50)} = j7.5\,\Omega$$

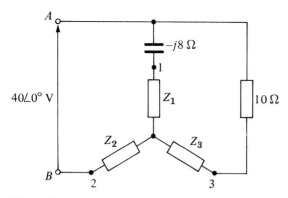

Figure 13

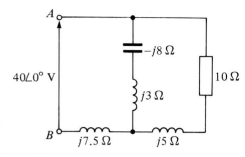

Figure 14

From equation (9),

$$Z_3 = \frac{Z_A Z_C}{Z_A + Z_B + Z_C} = \frac{(j10)(j25)}{(j50)} = j5\,\Omega$$

The equivalent network is shown in Fig. 14 and is further simplified in Fig. 15. $(10 + j5)\,\Omega$ in parallel with $-j5\,\Omega$ gives an equivalent impedance of

$$\frac{(10 + j5)(-j5)}{(10 + j5 - j5)} = (2.5 - j5)\,\Omega.$$

Hence the total circuit equivalent impedance across terminals AB is given by $Z_{AB} = (2.5 - j5) + j7.5 = \mathbf{(2.5 + j2.5)\,\Omega}$ or $\mathbf{3.54\angle\,45°\,\Omega}$

(b) Supply current $I = \dfrac{V}{Z_{AB}} = \dfrac{40\angle\,0°}{3.54\angle\,45°} = \mathbf{11.3\angle -45°\,A}$

(c) Power P dissipated in the $10\,\Omega$ resistance of Fig. 11 is given by $(I_1)^2(10)$, where I_1 (see Fig. 15) is given by

$$I_1 = \left(\frac{-j5}{10 + j5 - j5}\right)11.3\angle -45° = 5.65\angle -135°\,A$$

Hence power $P = (5.65)^2(10) = \mathbf{319\,W}.$

Problem 5. Determine, for the bridge network shown in Fig. 16, (a) the value of the single equivalent resistance that replaces the network between terminals A and B, (b) the current supplied by the 52 V source, and (c) the current flowing in the $8\,\Omega$ resistance.

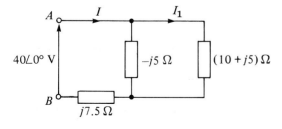

Figure 15

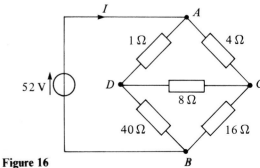

Figure 16

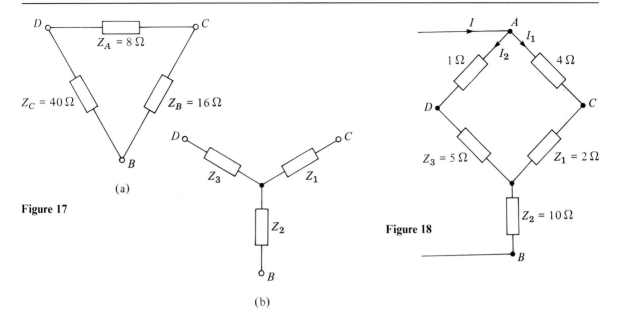

Figure 17

(a)

(b)

Figure 18

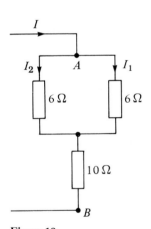

Figure 19

(a) In Fig. 16, no resistances are directly in parallel or directly in series with each other. However, ACD and BCD are both delta connections and either may be converted into an equivalent star connection. The delta network BCD is redrawn in Fig. 17(a) and is transformed into an equivalent star connection as shown in Fig. 17(b), where

$$Z_1 = \frac{(8)(16)}{8 + 16 + 40} = 2\,\Omega \text{ (from equation (7))}$$

$$Z_2 = \frac{(16)(40)}{8 + 16 + 40} = 10\,\Omega \text{ (from equation (8))}$$

$$Z_3 = \frac{(8)(40)}{8 + 16 + 40} = 5\,\Omega \text{ (from equation (9))}$$

The network of Fig. 16 may thus be redrawn as shown in Fig. 18. The $4\,\Omega$ and $2\,\Omega$ resistances are in series with each other, as are the $1\,\Omega$ and $5\,\Omega$ resistors. Hence the equivalent network is as shown in Fig. 19.
The total equivalent resistance across terminals A and B is given by

$$R_{AB} = \frac{(6)(6)}{(6 + 6)} + 10 = \mathbf{13\,\Omega}$$

(b) Current supplied by the 52 V source, i.e., current I in Fig. 19, is given by

$$I = \frac{V}{Z_{AB}} = \frac{52}{13} = \mathbf{4\,A}$$

(c) From Fig. 19, current $I_1 = (6/(6 + 6))\,I = 2\,A$, and current $I_2 = 2\,A$ also. From Fig. 18, p.d. across AC, $V_{AC} = (I_1)(4) = 8\,V$ and p.d. across AD, $V_{AD} = (I_2)(1) = 2\,V$. Hence p.d. between C and D (i.e., p.d. across the $8\,\Omega$ resistance of Fig. 16) is given by $(8 - 2) = 6\,V$.
Thus the current in the $8\,\Omega$ resistance is given by $V_{CD}/8 = 6/8 = \mathbf{0.75\,A}$.

Figure 20

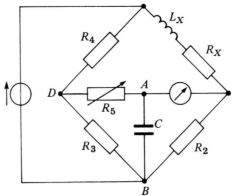

Problem 6. Figure 20 shows an Anderson bridge used to measure, with high accuracy, inductance L_X and series resistance R_X.

(a) Transform the delta ABD into its equivalent star connection and hence determine the balance equations for R_X and L_X.

(b) If $R_2 = R_3 = 1\,\text{k}\Omega$, $R_4 = 500\,\Omega$, $R_5 = 200\,\Omega$ and $C = 2\,\mu\text{F}$, determine the values of R_X and L_X at balance.

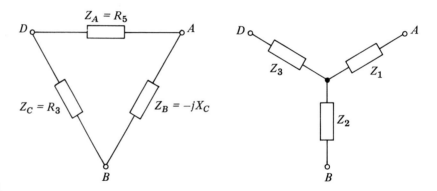

Figure 21

(a) The delta ABD is redrawn separately in Fig. 21, together with its equivalent star connection comprising impedances Z_1, Z_2 and Z_3. From equation (7),

$$Z_1 = \frac{(R_5)(-jX_c)}{R_5 - jX_c + R_3} = \frac{-jR_5X_c}{(R_3 + R_5) - jX_c}$$

From equation (8),

$$Z_2 = \frac{(-jX_c)(R_3)}{R_5 - jX_c + R_3} = \frac{-jR_3X_c}{(R_3 + R_5) - jX_c}$$

From equation (9),

$$Z_3 = \frac{R_5R_3}{(R_3 + R_5) - jX_c}$$

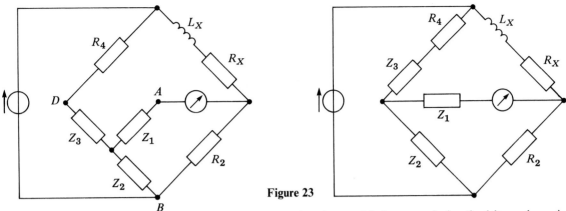

Figure 22

Figure 23

The network of Fig. 20 is redrawn with the star replacing the delta as shown in Fig. 22, and further simplified in Fig. 23. (Note that impedance Z_1 does not affect the balance of the bridge since it is in series with the detector.) At balance,

$$(R_X + jX_{L_X})(Z_2) = (R_2)(R_4 + Z_3)$$

from which

$$(R_X + jX_{L_X}) = \frac{R_2}{Z_2}(R_4 + Z_3) = \frac{R_2 R_4}{Z_2} + \frac{R_2 Z_3}{Z_2}$$

$$= \frac{R_2 R_4}{-jR_3 X_C/((R_3 + R_5) - jX_C)} + \frac{R_2(R_5 R_3/((R_3 + R_5) - jX_C))}{-jR_3 X_C/((R_3 + R_5) - jX_C)}$$

$$= \frac{R_2 R_4((R_3 + R_5) - jX_C)}{-jR_3 X_C} + \frac{R_2 R_5 R_3}{-jR_3 X_C}$$

$$= \frac{jR_2 R_4((R_3 + R_5) - jX_C)}{R_3 X_C} + \frac{jR_2 R_5}{X_C}$$

i.e.,

$$R_X + jX_{L_X} = \frac{jR_2 R_4(R_3 + R_5)}{R_3 X_C} + \frac{R_2 R_4 X_C}{R_3 X_C} + \frac{jR_2 R_5}{X_C}$$

Equating the real parts gives

$$R_X = \frac{R_2 R_4}{R_3}$$

Equating the imaginary parts gives

$$X_{L_X} = \frac{R_2 R_4(R_3 + R_5)}{R_3 X_C} + \frac{R_2 R_5}{X_C}$$

i.e.,

$$\omega L_X = \frac{R_2 R_4 R_3}{R_3(1/\omega C)} + \frac{R_2 R_4 R_5}{R_3(1/\omega C)} + \frac{R_2 R_5}{1/(\omega C)}$$

$$= \omega C R_2 R_4 + \frac{\omega C R_2 R_4 R_5}{R_3} + \omega C R_2 R_5$$

Hence

$$L_X = R_2 C\left(R_4 + \frac{R_4 R_5}{R_3} + R_5\right)$$

(b) When $R_2 = R_3 = 1\,\text{k}\Omega$, $R_4 = 500\,\Omega$, $R_5 = 200\,\Omega$ and $C = 2\mu\text{F}$; then, at balance,

$$R_X = \frac{R_2 R_4}{R_3} = \frac{(1000)(500)}{(1000)} = \mathbf{500\,\Omega}$$

and

$$L_X = R_2 C\left(R_4 + \frac{R_4 R_5}{R_3} + R_5\right)$$

$$= (1000)(2 \times 10^{-6})\left[500 + \frac{(500)(200)}{(1000)} + 200\right]$$

i.e.,

$$L_X = \mathbf{1.60\,H}$$

Figure 24

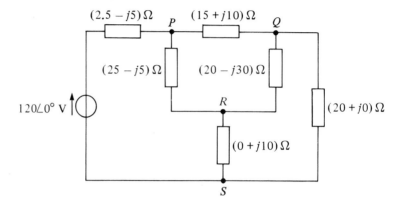

Figure 25

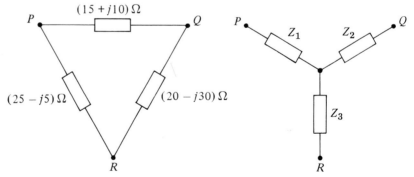

Problem 7. For the network shown in Fig. 24, determine (a) the current flowing in the $(0 + j10)\,\Omega$ impedance, and (b) the power dissipated in the $(20 + j0)\,\Omega$ impedance.

(a) The network may initially be simplified by transforming the delta PQR to its equivalent star connection as represented by impedances Z_1, Z_2 and Z_3 in Fig. 25. From equation (7),

$$Z_1 = \frac{(15 + j10)(25 - j5)}{(15 + j10) + (25 - j5) + (20 - j30)} = \frac{(15 + j10)(25 - j5)}{(60 - j25)}$$

$$= \frac{(18.03\underline{/\,33.69°})(25.50\underline{/\,-11.31°})}{65\underline{/\,-22.62°}} = 7.07\underline{/\,45°}\,\Omega \quad \text{or} \quad (5 + j5)\,\Omega$$

From equation (8),

$$Z_2 = \frac{(15 + j10)(20 - j30)}{(65\angle -22.62°)} = \frac{(18.03\angle 33.69°)(36.06\angle -56.31°)}{65\angle -22.62°}$$

$$= 10.0\angle 0° \,\Omega \quad \text{or} \quad (10 + j0)\,\Omega$$

From equation (9),

$$Z_3 = \frac{(25 - j5)(20 - j30)}{65\angle -22.62°} = \frac{(25.50\angle -11.31°)(36.06\angle -56.31°)}{65\angle -22.62°}$$

$$= 14.15\angle -45° \,\Omega \quad \text{or} \quad (10 - j10)\,\Omega$$

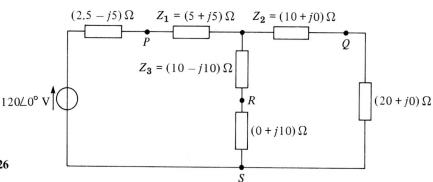

Figure 26

The network is shown redrawn in Fig. 26 and further simplified in Fig. 27, from which

$$\text{current } I_1 = \frac{120\angle 0°}{7.5 + ((10)(30)/(10 + 30))} = \frac{120\angle 0°}{15} = 8 \text{ A}$$

$$\text{current } I_2 = \left(\frac{10}{10 + 30}\right)(8) = 2 \text{ A}$$

and

$$\text{current } I_3 = \left(\frac{30}{10 + 30}\right)(8) = 6 \text{ A}$$

The current flowing in the $(0 + j10)\,\Omega$ impedance of Fig. 24 is the current I_3 shown in Fig. 27, i.e., **6 A.**

Figure 27

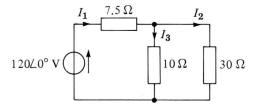

(b) The power P dissipated in the $(20 + j0)\,\Omega$ impedance of Fig. 24 is given by
$P = I_2^2(20) = (2)^2(20) = \textbf{80 W.}$

Further problems on delta–star and star–delta transformations may be found in section 7, problems 1 to 10, page 204.

5. Maximum power transfer theorems

A network that contains linear impedances and one or more voltage or current sources can be reduced to a Thévenin equivalent circuit as shown in chapter 8. When a load is connected to the terminals of this equivalent circuit, power is transferred from the source to the load.

A Thévenin equivalent circuit is shown in Fig. 28 with source internal impedance, $z = (r + jx)\,\Omega$ and complex load $Z = (R + jX)\,\Omega$.

The maximum power transferred from the source to the load depends on the following four conditions.

Condition 1. Let the load consist of a pure variable resistance R (i.e., let $X = 0$). Then current I in the load is given by

$$I = \frac{E}{(r + R) + jx}$$

and

$$\text{magnitude of current, } |I| = \frac{E}{\sqrt{[(r + R)^2 + x^2]}}$$

The active power P delivered to load R is given by

$$P = |I|^2 R = \frac{E^2 R}{(r + R)^2 + x^2}$$

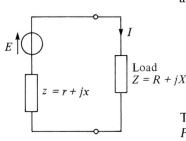

Figure 28

To determine the value of R for maximum power transferred to the load, P is differentiated with respect to R and then equated to zero (this being the normal procedure for finding maximum or minimum values using calculus). Using the quotient rule of differentiation,

$$\frac{dP}{dR} = E^2 \left\{ \frac{[(r + R)^2 + x^2](1) - (R)(2)(r + R)}{[(r + R)^2 + x^2]^2} \right\} = 0 \text{ for a maximum (or minimum) value.}$$

For dP/dR to be zero, the numerator of the fraction must be zero. Hence

$$(r + R)^2 + x^2 - 2R(r + R) = 0$$

i.e.,

$$r^2 + 2rR + R^2 + x^2 - 2Rr - 2R^2 = 0$$

from which

$$r^2 + x^2 = R^2 \tag{14}$$

or

$$R = \sqrt{(r^2 + x^2)} = |\mathbf{z}|$$

Thus, with a variable purely resistive load, the maximum power is delivered to the load if the load resistance R is made equal to the magnitude of the source impedance.

Condition 2. Let both the load and the source impedance be purely resistive (i.e., let $x = X = 0$). From equation (14) it may be seen that the maximum power is transferred when $\mathbf{R} = \mathbf{r}$.

Condition 3. Let the load Z have both variable resistance R and variable reactance X. From Fig. 28,

$$\text{current } I = \frac{E}{(r+R)+j(x+X)} \quad \text{and} \quad |I| = \frac{E}{\sqrt{[(r+R)^2+(x+X)^2]}}$$

The active power P delivered to the load is given by $P = |I|^2 R$ (since power can only be dissipated in a resistance) i.e.,

$$P = \frac{E^2 R}{(r+R)^2+(x+X)^2}$$

If X is adjusted such that $X = -x$ then the value of power is a maximum. If $X = -x$ then

$$P = \frac{E^2 R}{(r+R)^2}$$

$$\frac{dP}{dR} = E^2 \left\{ \frac{(r+R)^2(1)-(R)(2)(r+R)}{(r+R)^4} \right\} = 0 \text{ for a maximum value}$$

Hence

$$(r+R)^2 - 2R(r+R) = 0$$
$$r^2 + 2rR + R^2 - 2Rr - 2R^2 = 0$$

from which $r^2 - R^2 = 0$ and $R = r$.

Thus with the load impedance Z consisting of variable resistance R and variable reactance X, maximum power is delivered to the load when $X = -x$ **and** $R = r$, i.e., when $R + jX = r - jx$. Hence maximum power is delivered to the load when the load impedance is the complex conjugate of the source impedance.

Condition 4. Let the load impedance Z have variable resistance R and fixed reactance X. From Fig. 28, the magnitude of current,

$$|I| = \frac{E}{\sqrt{[(r+R)^2+(x+X)^2]}}$$

and the power dissipated in the load,

$$P = \frac{E^2 R}{(r+R)^2+(x+X)^2}$$

$$\frac{dP}{dR} = E^2 \left\{ \frac{[(r+R)^2+(x+X)^2](1)-(R)(2)(r+R)}{[(r+R)^2+(x+X)^2]^2} \right\}$$
$$= 0 \quad \text{for a maximum value}$$

Hence

$$(r+R)^2 + (x+X)^2 - 2R(r+R) = 0$$
$$r^2 + 2rR + R^2 + (x+X)^2 - 2Rr - 2R^2 = 0$$

from which $R^2 = r^2 + (x+X)^2$ and $R = \sqrt{[r^2+(x+X)^2]}$.

Summary

With reference to Fig. 28:

1. When the load is purely resistive (i.e., $X = 0$) and adjustable, maximum power transfer is achieved when $R = |z| = \sqrt{(r^2 + x^2)}$.
2. When both the load and the source impedance are purely resistive (i.e., $X = x = 0$), maximum power transfer is achieved when $R = r$.
3. When the load resistance R and reactance X are both independently adjustable, maximum power transfer is achieved when $X = -x$ and $R = r$.
4. When the load resistance R is adjustable with reactance X fixed, maximum power transfer is achieved when $R = \sqrt{[r^2 + (x + X)^2]}$.

The maximum power transfer theorems are primarily important where a small source of power is involved—such as, for example, the output from a telephone system (see section 6).

Worked problems on the maximum power transfer theorems

Problem 1. For the circuit shown in Fig. 29 the load impedance Z is a pure resistance. Determine (a) the value of R for maximum power to be transferred from the source to the load, and (b) the value of the maximum power delivered to R.

(a) From condition 1, maximum power transfer occurs when $R = |z|$, i.e., when
$R = |15 + j20| = \sqrt{(15^2 + 20^2)} = 25\,\Omega$.

(b) Current I flowing in the load is given by $I = E/Z_T$, where the total circuit impedance

$$Z_T = z + R = 15 + j20 + 25$$
$$= (40 + j20)\,\Omega \quad \text{or} \quad 44.72\underline{/\,26.57°}\,\Omega$$

Hence

$$I = \frac{120\underline{/\,0°}}{44.72\underline{/\,26.57°}} = 2.683\underline{/\,-26.57°}\,A$$

Thus maximum power delivered, $P = I^2 R = (2.683)^2(25) = \mathbf{180\,W}$.

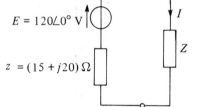

$E = 120\underline{/0°}\,V$

$z = (15 + j20)\,\Omega$

I

Z

Figure 29

Problem 2. If the load impedance Z in Fig. 29 of problem 1 consists of variable resistance R and variable reactance X, determine (a) the value of Z that results in maximum power transfer, and (b) the value of the maximum power.

(a) From condition 3, maximum power transfer occurs when $X = -x$ and $R = r$. Thus if $z = r + jx = (15 + j20)\,\Omega$ then $Z = (15 - j20)\,\Omega$ or $25\underline{/\,-53.13°}\,\Omega$.

(b) Total circuit impedance at maximum power transfer condition, $Z_T = z + Z$, i.e.,

$$Z_T = (15 + j20) + (15 - j20) = 30\,\Omega$$

Hence current in load,

$$I = \frac{E}{Z_T} = \frac{120\underline{/\,0°}}{30} = 4\underline{/\,0°}\,A$$

and maximum power in the load, $P = I^2 R = (4)^2(15) = \mathbf{240\,W}$.

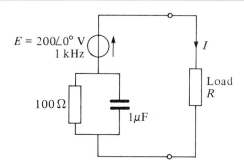

Figure 30

Problem 3. For the network shown in Fig. 30, determine (a) the value of the load resistance R required for maximum power transfer, and (b) the value of the maximum power transferred.

(a) This problem is an example of condition 1, where maximum power transfer is achieved when $R = |z|$. Source impedance z is composed of a $100\,\Omega$ resistance in parallel with a $1\,\mu F$ capacitor.

$$\text{Capacitive reactance, } X_C = \frac{1}{2\pi fC} = \frac{1}{2\pi(1000)(1 \times 10^{-6})} = 159.15\,\Omega$$

Hence

$$\text{source impedance, } z = \frac{(100)(-j159.15)}{(100 - j159.15)} = \frac{15915\angle -90°}{187.96\angle -57.86°}$$

$$= 84.67\angle -32.14°\,\Omega \quad \text{or} \quad (71.69 - j45.04)\,\Omega$$

Thus the value of load resistance for maximum power transfer is **84.67 Ω** (i.e., $|z|$).

(b) With $z = (71.69 - j45.04)\,\Omega$ and $R = 84.67\,\Omega$ for maximum power transfer, the total circuit impedance

$$Z_T = 71.69 + 84.67 - j45.04 = (156.36 - j45.04)\,\Omega \quad \text{or} \quad 162.72\angle -16.07°\,\Omega$$

Current flowing in the load,

$$I = \frac{V}{Z_T} = \frac{200\angle 0°}{162.72\angle -16.07°} = 1.23\angle 16.07°\,A$$

Thus the maximum power transferred, $P = I^2 R = (1.23)^2(84.67) = $ **128 W**

Figure 31

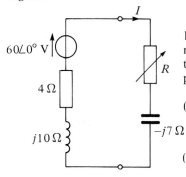

Problem 4. In the network shown in Fig. 31 the load consists of a fixed capacitive reactance of $7\,\Omega$ and a variable resistance R. Determine (a) the value of R for which the power transferred to the load is a maximum, and (b) the value of the maximum power.

(a) From condition (4), maximum power transfer is achieved when

$$R = \sqrt{[r^2 + (x + X)^2]} = \sqrt{[4^2 + (10 - 7)^2]} = \sqrt{(4^2 + 3^2)} = \textbf{5}\,\boldsymbol{\Omega}$$

(b) Current $I = \dfrac{60\angle 0°}{(4 + j10) + (5 - j7)} = \dfrac{60\angle 0°}{(9 + j3)} = \dfrac{60\angle 0°}{9.487\angle 18.43°}$

$$= 6.32\angle -18.43°\,A.$$

Thus the maximum power transferred, $P = I^2 R = (6.32)^2(5) = $ **200 W.**

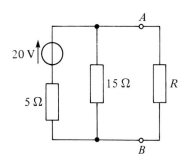

Figure 32

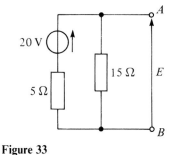

Figure 33

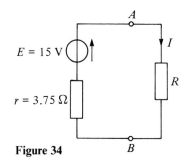

Figure 34

Problem 5. Determine the value of the load resistance R shown in Fig. 32 that gives maximum power dissipation and calculate the value of this power.

Using the procedure of Thévenin's theorem (see page 152):
(i) R is removed from the network as shown in Fig. 33.
(ii) P.d. across AB, $E = (15/(15 + 5)) (20) = 15\,\text{V}$.
(iii) Impedance "looking-in" at terminals AB with the 20 V source removed is given by $r = (5 \times 15)/(5 + 15) = 3.75\,\Omega$.
(iv) The equivalent Thévenin circuit supplying terminals AB is shown in Fig. 34. From condition (2), for maximum power transfer, $R = r$, i.e., $\textbf{R = 3.75\,\Omega}$.

$$\text{Current } I = \frac{E}{R + r} = \frac{15}{3.75 + 3.75} = 2\,\text{A}$$

Thus the maximum power dissipated in the load, $P = I^2 R = (2)^2(3.75) = \textbf{15\,W}$.

Figure 35

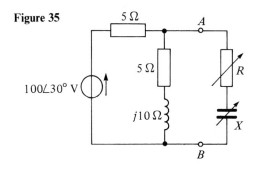

Figure 36

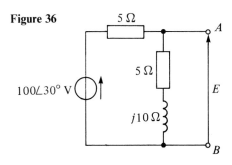

Problem 6. Determine, for the network shown in Fig. 35, (a) the values of R and X that will result in maximum power being transferred across terminals AB, and (b) the value of the maximum power.

(a) Using the procedure for Thévenin's theorem:
(i) Resistance R and reactance X are removed from the network as shown in Fig. 36.
(ii) P.d. across AB,

$$E = \left(\frac{5 + j10}{5 + j10 + 5}\right)(100\angle\,30°) = \frac{(11.18\angle\,63.43°)(100\angle\,30°)}{14.14\angle\,45°}$$

i.e.,

$$E = 79.07\angle\,48.43°\,\text{V}$$

(iii) With the $100\underline{/\,30°}$ V source removed the impedance, z, "looking in" at terminals AB is given by

$$z = \frac{(5)(5+j10)}{(5+5+j10)} = \frac{(5)(11.18\underline{/\,63.43°})}{(14.14\underline{/\,45°})} = 3.953\underline{/\,18.43°}\,\Omega$$
$$\text{or}\quad (3.75+j1.25)\,\Omega.$$

(iv) The equivalent Thévenin circuit is shown in Fig. 37. From condition 3, maximum power transfer is achieved when $X = -x$ and $R = r$, i.e., in this case when $X = -\mathbf{1.25\,\Omega}$ and $R = \mathbf{3.75\,\Omega}$.

(b) Current $I = \dfrac{E}{z+Z} = \dfrac{79.07\underline{/\,48.43°}}{(3.75+j1.25)+(3.75-j1.25)} = \dfrac{79.07\underline{/\,48.43°}}{7.5}$

$$= 10.54\underline{/\,48.43°}\ \text{A}.$$

Thus the maximum power transferred, $P = I^2R = (10.54)^2(3.75) = \mathbf{417\,W}$.

Further problems on the maximum power transfer theorem may be found in section 7, problems 11 to 20, page 207.

6. Impedance matching

It is seen from section 5 that when it is necessary to obtain the maximum possible amount of power from a source, it is advantageous if the circuit components can be adjusted to give equality of impedances. This adjustment is called **"impedance matching"** and is an important consideration in electronic and communications devices which normally involve small amounts of power. Examples where matching is important include coupling an aerial to a transmitter or receiver, or coupling a loudspeaker to an amplifier.

The mains power supply is considered as infinitely large compared with the demand upon it, and under such conditions it is unnecessary to consider the conditions for maximum power transfer. With transmission lines (see chapter 14), the lines are "matched", ideally, i.e., terminated in their characteristic impedance.

With d.c. generators, motors or secondary cells, for example, the internal impedance is usually very small and in such cases, if an attempt is made to make the load impedance as small as the source internal impedance, overloading of the source results.

A method of achieving maximum power transfer between a source and a load is to adjust the value of the load impedance to match the source impedance, which can be done using a **"matching-transformer"**.

A transformer is represented in Fig. 38 supplying a load impedance Z_L.

Figure 37

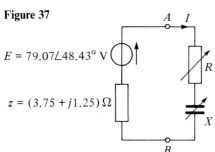

$E = 79.07\underline{/48.43°}$ V

$z = (3.75+j1.25)\,\Omega$

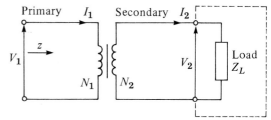

Figure 38 Matching impedance by means of a transformer

Small transformers used in low power networks are usually regarded as ideal (i.e., losses are negligible), such that

$$\frac{V_1}{V_2} = \frac{N_1}{N_2} = \frac{I_2}{I_1}$$

From Fig. 38, the primary input impedance $|z|$ is given by

$$|z| = \frac{V_1}{I_1} = \frac{(N_1/N_2)V_2}{(N_2/N_1)I_2} = \left(\frac{N_1}{N_2}\right)^2 \frac{V_2}{I_2}$$

Since the load impedance $|Z_L| = V_2/I_2$,

$$|\mathbf{z}| = \left(\frac{N_1}{N_2}\right)^2 |\mathbf{Z}_L| \qquad (15)$$

If the input impedance of Fig. 38 is purely resistive (say, r) and the load impedance is purely resistive (say, R_L) then equation (15) becomes

$$r = \left(\frac{N_1}{N_2}\right)^2 R_L \qquad (16)$$

Thus by varying the value of the transformer turns ratio, the equivalent input impedance of the transformer can be "matched" to the impedance of a source to achieve maximum power transfer.

Worked problems on impedance matching

Problem 1. Determine the optimum value of load resistance for maximum power transfer if the load is connected to an amplifier of output resistance $448\,\Omega$ through a transformer with a turns ratio of 8:1.

The equivalent input resistance r of the transformer must be $448\,\Omega$ for maximum power transfer. From equation (16), $r = (N_1/N_2)^2 R_L$, from which, load resistance $R_L = r(N_2/N_1)^2 = 448(\frac{1}{8})^2 = \mathbf{7\,\Omega.}$

Problem 2. A generator has an output impedance of $(450 + j60)\,\Omega$. Determine the turns ratio of an ideal transformer necessary to match the generator to a load of $(40 + j19)\,\Omega$ for maximum transfer of power.

Let the output impedance of the generator be z, where $z = (450 + j60)\,\Omega$ or $453.98\underline{/\,7.59°}\,\Omega$ and the load impedance be Z_L, where $Z_L = (40 + j19)\,\Omega$ or $44.28\underline{/\,25.41°}\,\Omega$. From Fig. 38 and equation (15), $z = (N_1/N_2)^2\,Z_L$. Hence

$$\text{transformer turns ratio } \left(\frac{N_1}{N_2}\right) = \sqrt{\left(\frac{z}{Z_L}\right)} = \sqrt{\left(\frac{453.98}{44.28}\right)} = \sqrt{(10.25)} = \mathbf{3.20}$$

Problem 3. A single-phase, 240/1920 V ideal transformer is supplied from a 240 V source through a cable of resistance $5\,\Omega$. If the load across the secondary winding is $1.60\,\text{k}\Omega$, determine (a) the primary current flowing, and (b) the power dissipated in the load resistance.

The network is shown in Fig. 39.

Figure 39

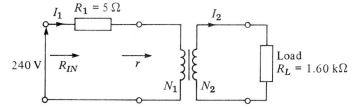

(a) Turns ratio $\dfrac{N_1}{N_2} = \dfrac{V_1}{V_2} = \dfrac{240}{1920} = \dfrac{1}{8}$

Equivalent input resistance of the transformer,

$$r = \left(\frac{N_1}{N_2}\right)^2 R_L = (\tfrac{1}{8})^2(1600) = 25\,\Omega$$

Total input resistance, $R_{IN} = R_1 + r = 5 + 25 = 30\,\Omega$. Hence the primary current, $I_1 = V_1/R_{IN} = 240/30 = \textbf{8 A}$.

(b) For an ideal transformer

$$\frac{V_1}{V_2} = \frac{I_2}{I_1},$$

from which

$$I_2 = I_1\left(\frac{V_1}{V_2}\right) = (8)\left(\frac{240}{1920}\right) = 1\,\text{A}$$

Power dissipated in the load resistance, $P = I_2^2 R_L = (1)^2(1600) \equiv \textbf{1.6 kW.}$

Problem 4. An a.c. source of $30\underline{/\,0^\circ}$ V and internal resistance 20 kΩ is matched to a load by a 20:1 ideal transformer. Determine for maximum power transfer (a) the value of the load resistance, and (b) the power dissipated in the load.

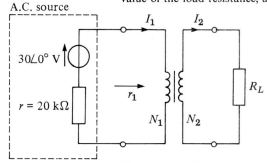

A.C. source

$30\angle 0^\circ$ V

$r = 20\ \text{k}\Omega$

Figure 40

The network diagram is shown in Fig. 40.

(a) For maximum power transfer, r_1 must be equal to 20 kΩ. From equation (16), $r_1 = (N_1/N_2)^2\, R_L$ from which

$$\text{load resistance } R_L = r_1\left(\frac{N_2}{N_1}\right)^2 = (20000)(\tfrac{1}{20})^2 = \textbf{50}\,\boldsymbol{\Omega}$$

(b) The total input resistance when the source is connected to the matching transformer is $(r + r_1)$. i.e., $20\,\text{k}\Omega + 20\,\text{k}\Omega = 40\,\text{k}\Omega$. Primary current, $I_1 = V/40000 = 30/40000 = 0.75\,\text{mA}$.

$$\frac{N_1}{N_2} = \frac{I_2}{I_1}$$

from which

$$I_2 = I_1\left(\frac{N_1}{N_2}\right) = (0.75 \times 10^{-3})(\tfrac{20}{1}) = 15\,\text{mA}$$

Power dissipated in load resistance R_L is given by

$$P = I_2^2 R_L = (15 \times 10^{-3})^2 (50)$$
$$= \textbf{0.01125 W}\quad \text{or}\quad \textbf{11.25 mW}$$

Further problems on impedance matching may be found in the following section (7), problems 21 to 25, page 208.

7. Further problems

Delta–star and star–delta transformations

1. Transform the delta connected networks shown in Fig. 41 to their equivalent star-connected networks.

$$\begin{bmatrix}\text{(a)} & Z_1 = 0.4\,\Omega, Z_2 = 2\,\Omega, Z_3 = 0.5\,\Omega \\ \text{(b)} & Z_1 = -j100\,\Omega, Z_2 = j100\,\Omega, Z_3 = 100\,\Omega\end{bmatrix}$$

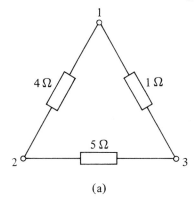

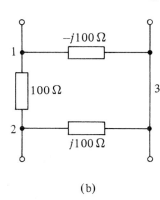

Figure 41   (a)   (b)

2. Determine the delta-connected equivalent networks for the star-connected impedances shown in Fig. 42.

$$\begin{bmatrix}\text{(a)} & Z_{12} = 18\,\Omega, Z_{23} = 9\,\Omega, Z_{31} = 13.5\,\Omega \\ \text{(b)} & Z_{12} = (10 + j0)\,\Omega, Z_{23} = (5 + j5)\,\Omega, \\ & Z_{31} = (0 - j10)\,\Omega.\end{bmatrix}$$

Figure 42

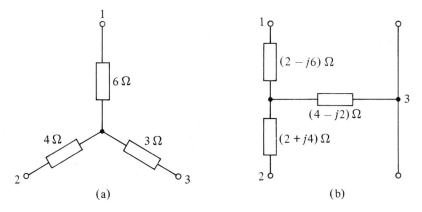

(a)   (b)

3. (a) Transform the π network shown in Fig. 43(a) to its equivalent star-connected network.

(b) Change the T-connected network shown in Fig. 43(b) to its equivalent delta-connected network.

$$
\left[
\begin{array}{l}
\text{(a)} \ \ Z_1 = 5.12\underline{/\,78.35°}\ \Omega, Z_2 = 6.82\underline{/\,-26.65°}\ \Omega \\
\qquad Z_3 = 10.23\underline{/\,-11.65°}\ \Omega \\
\text{(b)} \ \ Z_{12} = 35.93\underline{/\,40.50°}\ \Omega, Z_{23} = 53.89\underline{/\,-19.50°}\ \Omega \\
\qquad Z_{31} = 26.95\underline{/\,-49.50°}\ \Omega
\end{array}
\right]
$$

Figure 43

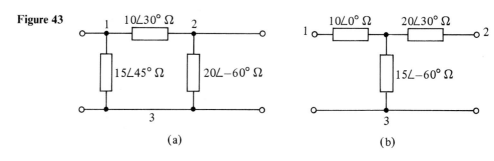

(a)   (b)

Figure 44

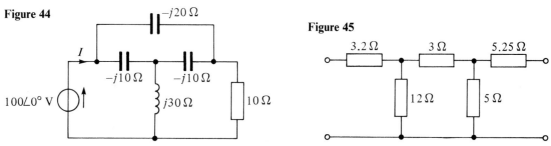

Figure 45

4. For the network shown in Fig. 44 determine (a) current I and (b) the power dissipated in the $10\,\Omega$ resistance.

[(a) $7.32\underline{/\,24.06°}$ A   (b) 668 W]

5. (a) A delta-connected network contains three $24\underline{/\,60°}\ \Omega$ impedances. Determine the impedances of the equivalent star-connected network.

(b) Three impedances, each of $(2 + j3)\,\Omega$, are connected in star. Determine the impedances of the equivalent delta-connected network.

[(a) Each impedance $= 8\underline{/\,60°}\ \Omega$ (b) Each impedance $= (6 + j9)\,\Omega$]

6. (a) Derive the star-connected network of three impedances equivalent to the network shown in Fig. 45.

(b) Obtain the delta-connected equivalent network for Fig. 45.

[(a) $5\,\Omega, 6\,\Omega, 3\,\Omega$ (b) $21\,\Omega, 12.6\,\Omega, 10.5\,\Omega$]

7. For the a.c. bridge network shown in Fig. 46, transform the delta-connected network ABC into an equivalent star, and hence determine the current flowing in the capacitor.

[130 mA]

8. For the network shown in Fig. 47 transform the delta-connected network ABC to an equivalent star-connected network, convert the 35 A, $2\,\Omega$ Norton circuit to an equivalent Thévenin circuit and hence determine the p.d. across the $12.5\,\Omega$ resistor.

[31.25 V]

9. Transform the delta-connected network ABC shown in Fig. 48 and hence determine the magnitude of the current flowing in the $20\,\Omega$ resistance.

[4.47 A]

10. For the network shown in Fig. 49 determine (a) the current supplied by the $80\underline{/\,0°}\,V$ source, and (b) the power dissipated in the $(2.00 - j0.916)\,\Omega$ impedance.

[(a) 9.73 A   (b) 98.6 W]

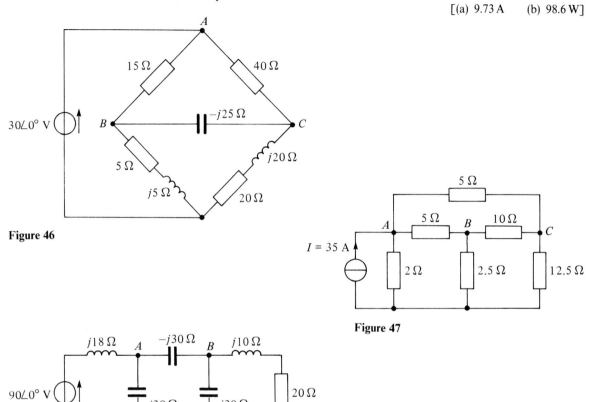

Figure 46

Figure 47

Figure 48

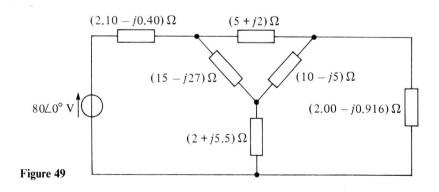

Figure 49

Maximum power transfer theorems

11. For the circuit shown in Fig. 50 determine the value of the source resistance *r* if the maximum power is to be dissipated in the 15 Ω load. Determine the value of this maximum power.

$$[r = 9\,\Omega, P = 208.3\ \text{W}]$$

12. In the circuit shown in Fig. 51 the load impedance Z_L is a pure resistance R. Determine (a) the value of R for maximum power to be transferred from the source to the load, and (b) the value of the maximum power delivered to R.

$$[(a)\ 11.18\,\Omega \quad (b)\ 151.1\ \text{W}]$$

13. If the load impedance Z_L in Fig. 51 of problem 12 consists of a variable resistance R and variable reactance X, determine (a) the value of Z_L which results in maximum power transfer, and (b) the value of the maximum power.

$$[(a)\ (10 + j5)\,\Omega \quad (b)\ 160\ \text{W}]$$

14. For the network shown in Fig. 52 determine (a) the value of the load resistance R_L required for maximum power transfer, and (b) the value of the maximum power.

$$[(a)\ 26.8\,\Omega \quad (b)\ 35.4\ \text{W}]$$

Figure 50 $E = 100\angle0°$ V
1 kHz

r

$R_L = 15\ \Omega$

1.91 mH

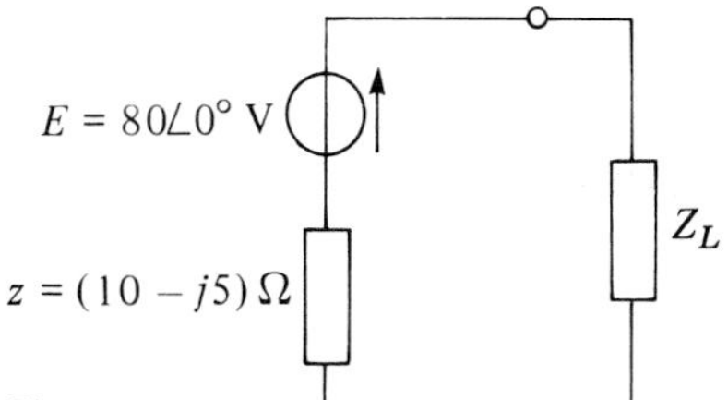

$E = 80\angle0°$ V

$z = (10 - j5)\ \Omega$

Z_L

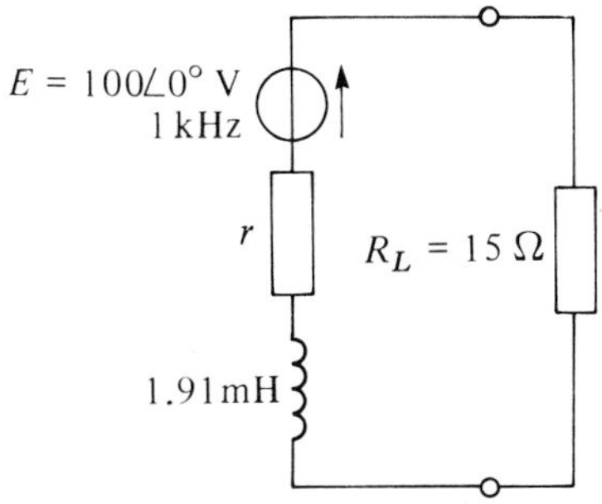

Figure 51

Figure 52 $E = 60\angle0°$ V
500 Hz

15 Ω

R_L

15.92 μF

4.77 mH

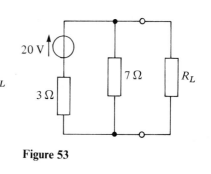

20 V

7 Ω

R_L

3 Ω

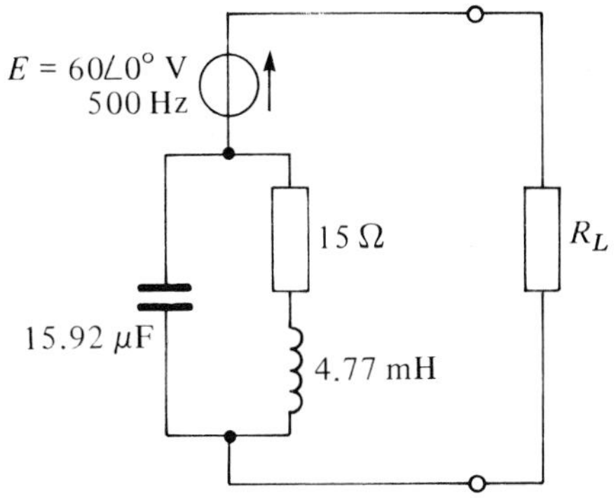

Figure 53

Figure 54

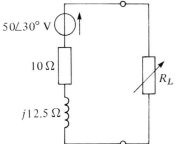

50∠30° V

10 Ω

R_L

$j12.5\ \Omega$

15. Find the value of the load resistance R_L shown in Fig. 53 that gives maximum power dissipation, and calculate the value of this power.

$$[R = 2.1\,\Omega, P = 23.3\ \text{W}]$$

16. For the circuit shown in Fig. 54 determine (a) the value of load resistance R_L which results in maximum power transfer, and (b) the value of the maximum power.

$$[(a)\ 16\,\Omega \quad (b)\ 48\ \text{W}]$$

17. Determine, for the network shown in Fig. 55, (a) the values of R and X which result in maximum power being transferred across terminals AB, and (b) the value of the maximum power.

$$[(a)\ R = 1.706\,\Omega, X = 0.177\,\Omega \quad (b)\ 269\ \text{W}]$$

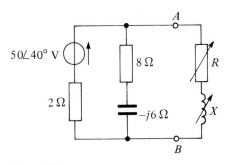

Figure 55

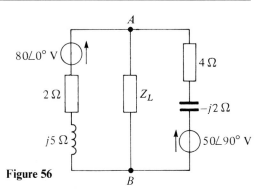

Figure 56

18. A source of $120\underline{/\,0°}$ V and impedance $(5+j3)\,\Omega$ supplies a load consisting of a variable resistor R in series with a fixed capacitive reactance of $8\,\Omega$. Determine (a) the value of R to give maximum power transfer, and (b) the value of the maximum power.

[(a) $7.07\,\Omega$ (b) $596.5\,\text{W}$]

19. If the load Z_L between terminals A and B of Fig. 56 is variable in both resistance and reactance determine the value of Z_L such that it will receive maximum power. Calculate the value of the maximum power.

[$R=3.47\,\Omega, X=-0.93\,\Omega, 13.6\,\text{W}$]

20. For the circuit of Fig. 57, determine the value of load impedance Z_L for maximum load power if (a) Z_L comprises a variable resistance R and variable reactance X, and (b) Z_L is a pure resistance R. Determine the values of load power in each case.

$$\begin{bmatrix}\text{(a)} & R=0.80\,\Omega, X=-1.4\,\Omega, P=225\,\text{W}\\ \text{(b)} & R=1.61\,\Omega, P=149.2\,\text{W}\end{bmatrix}$$

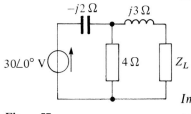

Figure 57

Impedance matching

21. The output stage of an amplifier has an output resistance of $144\,\Omega$. Determine the optimum turns ratio of a transformer that would match a load resistance of $9\,\Omega$ to the output resistance of the amplifier for maximum power transfer.

[$4:1$]

22. Find the optimum value of load resistance for maximum power transfer if a load is connected to an amplifier of output resistance $252\,\Omega$ through a transformer with a turns ratio of $6:1$.

[$7\,\Omega$]

23. A generator has an output impedance of $(300+j45)\,\Omega$. Determine the turns ratio of an ideal transformer necessary to match the generator to a load of $(37+j19)\,\Omega$ for maximum power transfer.

[2.70]

24. A single-phase, $240/2880\,\text{V}$ ideal transformer is supplied from a $240\,\text{V}$ source through a cable of resistance $3.5\,\Omega$. If the load across the secondary winding is $1.8\,\text{k}\Omega$, determine (a) the primary current flowing, and (b) the power dissipated in the load resistance.

[(a) $15\,\text{A}$ (b) $2.81\,\text{kW}$]

25. An a.c. source of $20\underline{/\,0°}$ V and internal resistance $10.24\,\text{k}\Omega$ is matched to a load for maximum power transfer by a $16:1$ ideal transformer. Determine (a) the value of the load resistance, and (b) the power dissipated in the load.

[(a) $40\,\Omega$ (b) $9.77\,\text{mW}$]

10 Complex waveforms

1. Introduction

In preceding chapters a.c. supplies have been assumed to be sinusoidal, this being a form of alternating quantity commonly encountered in electrical engineering. However, many supply waveforms are **not** sinusoidal. For example, sawtooth generators produce ramp waveforms, and rectangular waveforms may be produced by multivibrators. A waveform that is not sinusoidal is called a **complex wave**. Such a waveform may be shown to be composed of the sum of a series of sinusoidal waves having various interrelated periodic times.

A function $f(t)$ is said to be **periodic** if $f(t + T) = f(t)$ for all values of t, where T is some positive number. T is the interval between two successive repetitions and is called the **period** of the function $f(t)$. A sine wave having a period of $2\pi/\omega$ is a familiar example of a periodic function.

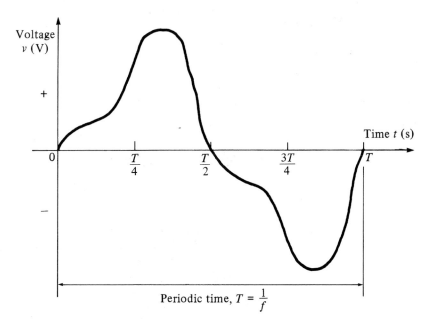

Figure 1 Typical complex periodic voltage waveform

A typical complex periodic-voltage waveform, shown in Fig. 1, has period T seconds and frequency f hertz. A complex wave such as this can be resolved into the sum of a number of sinusoidal waveforms, and each of the sine waves can have a different frequency, amplitude and phase.

The initial, major sine wave component has a frequency f equal to the frequency of the complex wave and this frequency is called the **fundamental frequency**. The other sine wave components are known as **harmonics**, these having frequencies which are integer multiples of frequency f. Hence the second harmonic has a frequency of $2f$, the third harmonic has a frequency of $3f$, and so on. Thus if the fundamental (i.e., supply) frequency of a complex wave is 50 Hz, then the third harmonic frequency is 150 Hz, the fourth harmonic frequency is 200 Hz, and so on.

2. The general equation for a complex waveform

The instantaneous value of a complex voltage wave v acting in a linear circuit may be represented by the general equation

$$v = V_{1m}\sin(\omega t + \psi_1) + V_{2m}\sin(2\omega t + \psi_2) + \cdots + V_{nm}\sin(n\omega t + \psi_n)\,\text{volts}$$

(1)

Here $V_{1m}\sin(\omega t + \psi_1)$ represents the fundamental component, of which V_{1m} is the maximum or peak value, frequency, $f = \omega/2\pi$ and ψ_1 is the phase angle with respect to time, $t = 0$.

Similarly, $V_{2m}\sin(2\omega t + \psi_2)$ represents the second harmonic component, and $V_{nm}\sin(n\omega t + \psi_n)$ represents the nth harmonic component, of which V_{nm} is the peak value, frequency $= n\omega/2\pi\ (= nf)$ and ψ_n is the phase angle.

In the same way, the instantaneous value of a complex current i may be represented by the general equation

$$i = I_{1m}\sin(\omega t + \theta_1) + I_{2m}\sin(2\omega t + \theta_2) + \cdots + I_{nm}\sin(n\omega t + \theta_n)\,\text{amperes}$$

(2)

Where equations (1) and (2) refer to the voltage across and the current flowing through a given linear circuit, the phase angle between the fundamental voltage and current is $\phi_1 = (\psi_1 - \theta_1)$, the phase angle between the second harmonic voltage and current is $\phi_2 = (\psi_2 - \theta_2)$, and so on.

It often occurs that not all harmonic components are present in a complex waveform. Sometimes only the fundamental and odd harmonics are present, and in others only the fundamental and even harmonics are present.

3. Harmonic synthesis

Harmonic analysis is the process of resolving a complex periodic waveform into a series of sinusoidal components of ascending order of frequency. Many of the waveforms met in practice can be represented by mathematical expressions similar to those of equations (1) and (2), and the magnitude of their harmonic components together with their phase may be calculated using **Fourier series. Numerical methods** are used to analyse waveforms for which simple mathematical expressions cannot be obtained. In a laboratory, waveform analysis may be performed using a **waveform analyser** which produces a direct readout of the component waves present in a complex wave.

By adding the instantaneous values of the fundamental and progressive harmonics of a complex wave for given instants in time, the shape of a

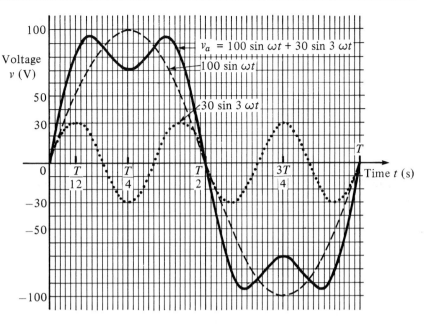

Figure 2

complex waveform can be gradually built up. This graphical procedure is known as **harmonic synthesis** (synthesis meaning "the putting together of parts or elements so as to make up a complex whole").

A number of examples of harmonic synthesis will now be considered.

Example 1

Consider the complex voltage expression given by

$$v_a = 100 \sin \omega t + 30 \sin 3\omega t \text{ volts}$$

The waveform is made up of a fundamental wave of maximum value 100 V and frequency, $f = \omega/2\pi$ hertz and a third harmonic component of maximum value 30 V and frequency = $3\omega/2\pi$ ($=3f$), the fundamental and third harmonics being initially in phase with each other.

Since the maximum value of the third harmonic is 30 V and that of the fundamental is 100 V, the resultant waveform v_a is said to contain 30/100, i.e., "30% third harmonic". In Fig. 2, the fundamental waveform is shown by the broken line plotted over one cycle, the periodic time being $2\pi/\omega$ seconds. On the same axis is plotted $30 \sin 3\omega t$, shown by the dotted line, having a maximum value of 30 V and for which three cycles are completed in time T seconds. At zero time, $30 \sin 3\omega t$ is in phase with $100 \sin \omega t$.

The fundamental and third harmonic are combined by adding ordinates at intervals to produce the waveform for v_a as shown. For example, at time $T/12$ seconds, the fundamental has a value of 50 V and the third harmonic a value of 30 V. Adding gives a value of 80 V for waveform v_a, at time $T/12$ seconds. Similarly, at time $T/4$ seconds, the fundamental has a value of 100 V and the third harmonic a value of -30 V. After addition, the resultant waveform v_a is 70 V at $T/4$. The procedure is continued between $t = 0$ and

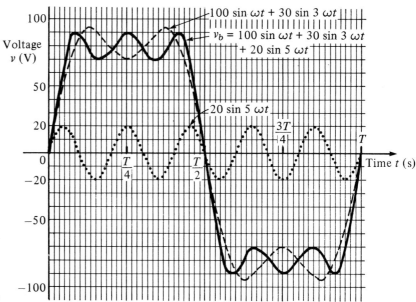

Figure 3

$t = T$ to produce the complex waveform for v_a. The negative half-cycle of waveform v_a is seen to be identical in shape to the positive half-cycle.

Example 2

Consider the addition of a fifth harmonic component to the complex waveform of Fig. 2, giving a resultant waveform expression

$$v_b = 100 \sin \omega t + 30 \sin 3\omega t + 20 \sin 5\omega t \text{ volts}$$

Figure 3 shows the effect of adding $(100 \sin \omega t + 30 \sin 3\omega t)$ obtained from Fig. 2 to $20 \sin 5\omega t$. The shapes of the negative and positive half-cycles are still identical. If further odd harmonics of the appropriate amplitude and phase were added to v_b, a good approximation to a square wave would result.

Example 3

Consider the complex voltage expression given by

$$v_c = 100 \sin \omega t + 30 \sin \left(3\omega t + \frac{\pi}{2} \right) \text{volts}$$

This expression is similar to voltage v_a in that the peak value of the fundamental and third harmonic are the same. However the third harmonic has a phase displacement of $\pi/2$ radian leading (i.e., leading $30 \sin 3\omega t$ by $\pi/2$ radian). Note that, since the periodic time of the fundamental is T seconds, the periodic time of the third harmonic is $T/3$ seconds, and a phase displacement of $\pi/2$ radian or $\frac{1}{4}$ cycle of the third harmonic represents a time interval of $(T/3) \div 4$, i.e., $T/12$ seconds.

Figure 4 shows graphs of 100 sin ωt and 30 sin $(3\,\omega t + (\pi/2))$ over the time for one cycle of the fundamental. When ordinates of the two graphs are added at intervals, the resultant waveform v_c is as shown. The shape of the

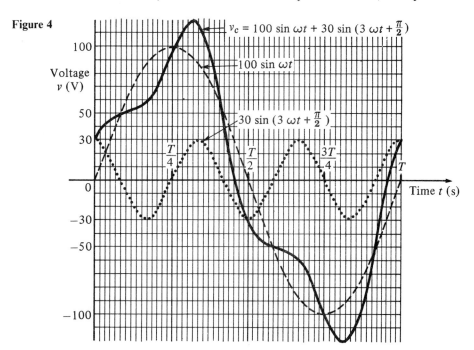

Figure 4

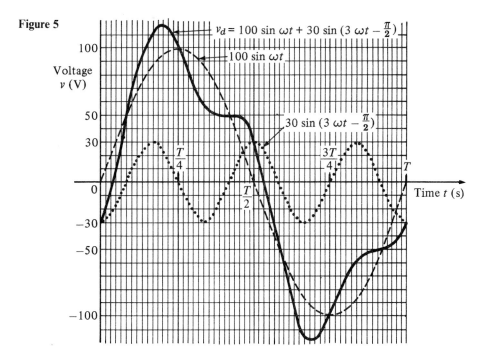

Figure 5

waveform v_c is quite different from that of waveform v_a shown in Fig. 2, even though the percentage third harmonic is the same. If the negative half-cycle in Fig. 4 is reversed it can be seen that the shape of the positive and negative half-cycles are identical.

Example 4

Consider the complex voltage expression given by

$$v_d = 100 \sin \omega t + 30 \sin \left(3\omega t - \frac{\pi}{2} \right) \text{volts}$$

The fundamental, $100 \sin \omega t$, and the third harmonic component, $30 \sin (3 \omega t - (\pi/2))$, are plotted in Fig. 5, the latter lagging $30 \sin 3\omega t$ by $\pi/2$ radian or $T/12$ seconds. Adding ordinates at intervals gives the resultant waveform v_d as shown. The negative half-cycle of v_d is identical in shape to the positive half-cycle.

Example 5

Consider the complex voltage expression given by

$$v_e = 100 \sin \omega t + 30 \sin (3\omega t + \pi) \text{volts}$$

The fundamental, $100 \sin \omega t$, and the third harmonic component, $30 \sin (3\omega t + \pi)$, are plotted as shown in Fig. 6, the latter leading $30 \sin 3 \omega t$

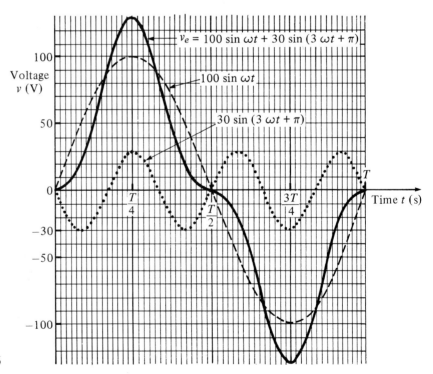

Figure 6

Figure 7

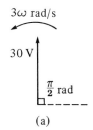

3ω rad/s

30 V

$\frac{\pi}{2}$ rad

(a)

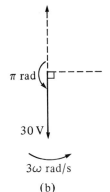

π rad

30 V

3ω rad/s

(b)

by π radian or $T/6$ seconds. Adding ordinates at intervals gives the resultant waveform v_e as shown. The negative half-cycle of v_e is identical in shape to the positive half-cycle.

Example 6

Consider the complex voltage expression given by

$$v_f = 100 \sin \omega t - 30 \sin \left(3\omega t + \frac{\pi}{2} \right) \text{volts}$$

The phasor representing $30 \sin (3\omega t + (\pi/2))$ is shown in Fig. 7(a) at time $t = 0$. The phasor representing $-30 \sin (3\omega t + (\pi/2))$ is shown in Fig. 7(b) where it is seen to be in the opposite direction to that shown in Fig. 7(a). $-30 \sin (3\omega t + (\pi/2))$ is the same as $30 \sin (3\omega t - (\pi/2))$. Thus

$$v_f = 100 \sin \omega t - 30 \sin \left(3\omega t + \frac{\pi}{2} \right) = 100 \sin \omega t + 30 \sin \left(3\omega t - \frac{\pi}{2} \right)$$

The waveform representing this expression has already been plotted in Fig. 5.

General conclusions on examples 1 to 6

Whenever odd harmonics are added to a fundamental waveform, whether initially in phase with each other or not, the positive and negative half-cycles of the resultant complex wave are identical in shape (i.e., in Figs 2 to 6, the values of voltage in the third quadrant—between $T/2$ seconds and $3T/4$ seconds—are identical to the voltage values in the first quadrant—between 0 and $T/4$ seconds, except that they are negative, and the values of voltage in the second and fourth quadrants are identical, except for the sign change). This is a feature of waveforms containing a fundamental and odd harmonics and is true whether harmonics are added or subtracted from the fundamental.

From Figs 2 to 6, it is seen that a waveform can change its shape considerably as a result of changes in both phase and magnitude of the harmonics.

Example 7

Consider the complex current expression given by

$$i_a = 10 \sin \omega t + 4 \sin 2\omega t \text{ amperes}$$

Current i_a consists of a fundamental component, $10 \sin \omega t$, and a second harmonic component, $4 \sin 2\omega t$, the components being initially in phase with each other. Current i_a contains 40% second harmonic. The fundamental and second harmonic are shown plotted separately in Fig. 8. By adding ordinates at intervals, the complex waveform representing i_a is produced as shown. It is noted that if all the values in the negative half-cycle were reversed then this half-cycle would appear as a mirror image of the positive half-cycle about a vertical line drawn through time, $t = T/2$.

Figure 8

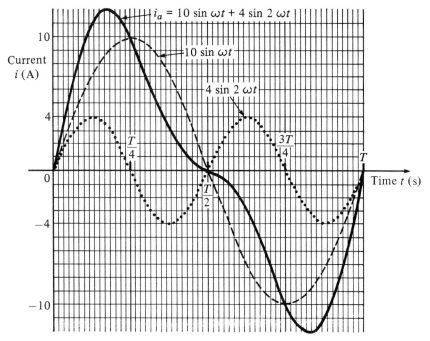

Figure 9

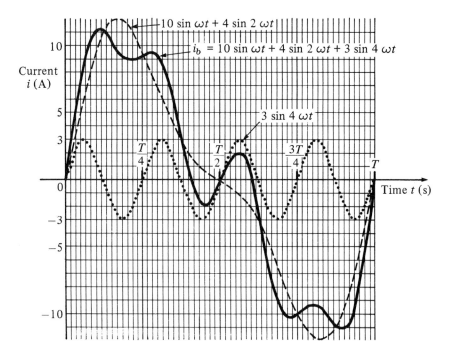

Example 8

Consider the complex current expression given by

$$i_b = 10 \sin \omega t + 4 \sin 2\omega t + 3 \sin 4\omega t \text{ amperes}$$

The waveforms representing $(10 \sin \omega t + 4 \sin 2\omega t)$ and the fourth harmonic component, $3 \sin 4\omega t$, are each shown separately in Fig. 9, the former waveform having been produced in Fig. 8. By adding ordinates at intervals, the complex waveform for i_b is produced as shown in Fig. 9. If the half-cycle between times $T/2$ and T is reversed then it is seen to be a mirror image of the half-cycle lying between 0 and $T/2$ about a vertical line drawn through the time, $t = T/2$.

Example 9

Consider the complex current expression given by

$$i_c = 10 \sin \omega t + 4 \sin \left(2\omega t + \frac{\pi}{2} \right) \text{amperes}$$

The fundamental component, $10 \sin \omega t$, and the second harmonic component, having an amplitude of 4A and a phase displacement of $\pi/2$ radian leading (i.e., leading $4 \sin 2\omega t$ by $\pi/2$ radian or $T/8$ seconds), are shown plotted separately in Fig. 10. By adding ordinates at intervals, the complex waveform for i_c is produced as shown. The positive and negative half-cycles of the resultant waveform i_c are seen to be quite dissimilar.

Figure 10

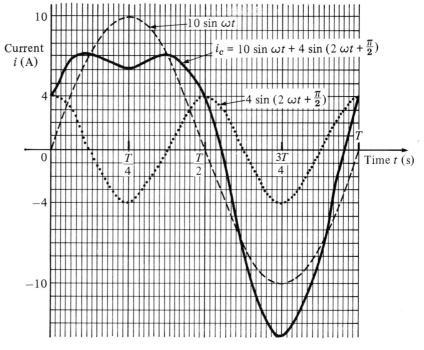

Figure 11

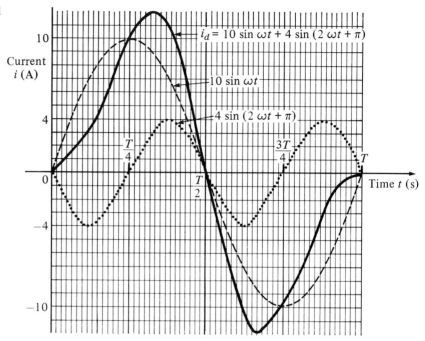

Example 10

Consider the complex current expression given by

$$i_d = 10 \sin \omega t + 4 \sin (2\omega t + \pi) \text{ amperes}$$

The fundamental, $10 \sin \omega t$, and the second harmonic component which leads $4 \sin 2\omega t$ by π rad are shown separately in Fig. 11. By adding ordinates at intervals, the resultant waveform i_d is produced as shown. If the negative half-cycle is reversed, it is seen to be a mirror image of the positive half-cycle about a line drawn vertically through time, $t = T/2$.

General conclusions on examples 7 to 10

Whenever even harmonics are added to a fundamental component:

(a) if the harmonics are initially in phase or if there is a phase-shift of π rad, the negative half-cycle, when reversed, is a mirror image of the positive half-cycle about a vertical line drawn through time, $t = T/2$;

(b) if the harmonics are initially out of phase with each other (i.e., other than π rad), the positive and negative half-cycles are dissimilar.

These are features of waveforms containing the fundamental and even harmonics.

Example 11

Consider the complex voltage expression given by

$$v_g = 50 \sin \omega t + 25 \sin 2\omega t + 15 \sin 3\omega t \text{ volts}$$

The fundamental and the second and third harmonics are each shown separately in Fig. 12. By adding ordinates at intervals, the resultant waveform v_g is produced as shown. If the negative half-cycle is reversed, it appears as a mirror image of the positive half-cycle about a vertical line drawn through time $t = T/2$.

Example 12

Consider the complex voltage expression given by

$$v_h = 50 \sin \omega t + 25 \sin(2\omega t - \pi) + 15 \sin\left(3\omega t + \frac{\pi}{2}\right) \text{volts}$$

The fundamental, the second harmonic lagging by π radian and the third harmonic leading by $\pi/2$ radian are initially plotted separately, as shown in Fig. 13. Adding ordinates at intervals gives the resultant waveform v_h as shown. The positive and negative half-cycles are seen to be quite dissimilar.

General conclusions on examples 11 and 12

Whenever a waveform contains both odd and even harmonics:

(a) if the harmonics are initially in phase with each other, the negative cycle, when reversed, is a mirror image of the positive half-cycle about a vertical line drawn through time, $t = T/2$;

Figure 12

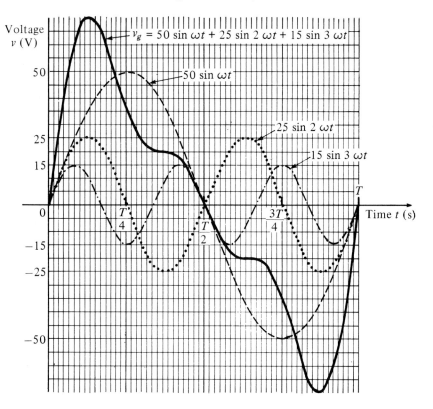

Figure 13

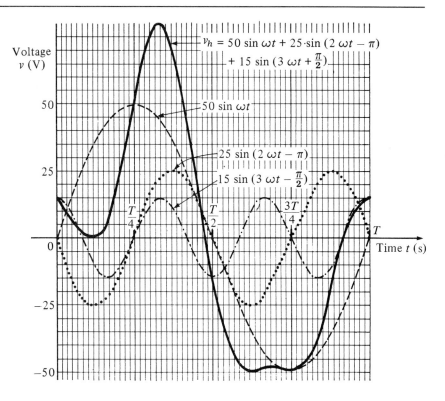

Voltage v (V)

$v_h = 50 \sin \omega t + 25 \cdot \sin (2\,\omega t - \pi)$
$+ 15 \sin (3\,\omega t + \frac{\pi}{2})$

50 sin ωt

25 sin $(2\,\omega t - \pi)$

15 sin $(3\,\omega t - \frac{\pi}{2})$

(b) if the harmonics are initially out of phase with each other, the positive and negative half-cycles are dissimilar.

Example 13

Consider the complex current expression given by

$$i = 32 + 50 \sin \omega t + 20 \sin \left(2\omega t - \frac{\pi}{2} \right) \text{mA}$$

The current i comprises three components—a 32 mA d.c. component, a fundamental of amplitude 50 mA and a second harmonic of amplitude 20 mA, lagging by $\pi/2$ radian. The fundamental and second harmonic are shown separately in Fig. 14. Adding ordinates at intervals gives the complex waveform $50 \sin \omega t + 20 \sin (2\omega t - (\pi/2))$.

This waveform is then added to the 32 mA d.c. component to produce the waveform i as shown. The effect of the d.c. component is seen to be to shift the whole wave 32 mA upward. The waveform approaches that expected from a half-wave rectifier (see section 8).

Worked problems on harmonic synthesis

Problem 1. A complex waveform v comprises a fundamental voltage of 240 V rms and frequency 50 Hz, together with a 20% third harmonic which has a phase angle

Figure 14

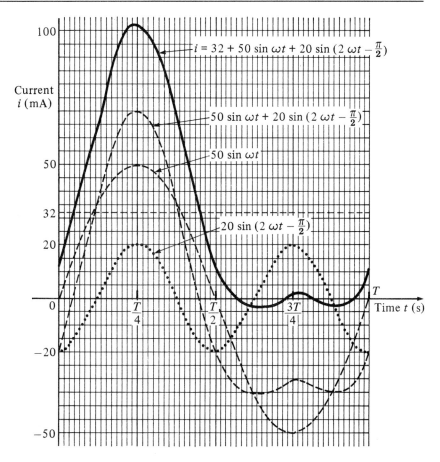

Current
i (mA)

$i = 32 + 50 \sin \omega t + 20 \sin (2 \omega t - \frac{\pi}{2})$

$50 \sin \omega t + 20 \sin (2 \omega t - \frac{\pi}{2})$

$50 \sin \omega t$

$20 \sin (2 \omega t - \frac{\pi}{2})$

Time t (s)

lagging by $3\pi/4$ rad at time $t = 0$. (a) Write down an expression to represent voltage v. (b) Use harmonic synthesis to sketch the complex waveform representing voltage v over one cycle of the fundamental component.

(a) A fundamental voltage having an rms value of 240 V has a maximum value, or amplitude of $(\sqrt{2})(240)$, i.e., 339.4 V.

   If the fundamental frequency is 50 Hz then angular velocity, $\omega = 2\pi f = 2\pi(50) = 100\pi$ rad/s. Hence the fundamental voltage is represented by $339.4 \sin 100\pi t$ volts. Since the fundamental frequency is 50 Hz, the time for one cycle of the fundamental is given by $T = 1/f = 1/50$ s or 20 ms.

   The third harmonic has an amplitude equal to 20% of 339.4 V, i.e. 67.9 V. The frequency of the third harmonic component is $3 \times 50 = 150$ Hz, thus the angular velocity is $2\pi(150)$, i.e., 300π rad/s. Hence the third harmonic voltage is represented by $67.9 \sin (300\pi t - (3\pi/4))$ volts. Thus

$$\textbf{voltage, } \mathbf{\mathit{v} = 339.4 \sin 100\pi \mathit{t} + 67.9 \sin\left(300\pi \mathit{t} - \frac{3\pi}{4} \right) \textbf{volts}}$$

(b) One cycle of the fundamental, $339.4 \sin 100\pi t$, is shown sketched in Fig. 15, together with three cycles of the third harmonic component,

$$67.9 \sin (300\pi t - (3\pi/4))$$

Figure 15

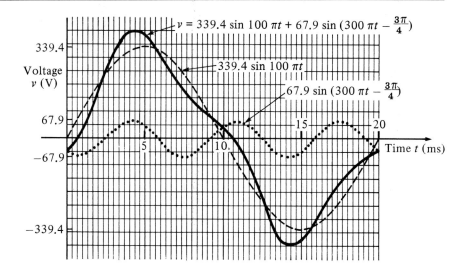

initially lagging by $3\pi/4$ rad. By adding ordinates at intervals, the complex waveform representing voltage is produced as shown. If the negative half-cycle is reversed it is seen to be identical to the positive half-cycle, which is a feature of waveforms containing the fundamental and odd harmonics.

Problem 2. For each of the periodic complex waveforms shown in Fig. 16, suggest whether odd or even harmonics (or both) are likely to be present.

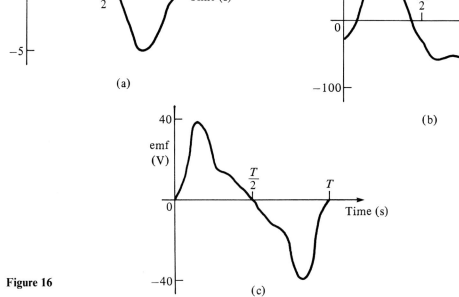

Figure 16

(a) If in Fig. 16(a) the negative half-cycle is reversed, it is seen to be identical to the positive half-cycle. This feature indicates that the complex current waveform is composed of a fundamental and odd harmonics only (see examples 1 to 6).
(b) In Fig. 16(b) the negative half-cycle is quite dissimilar to the positive half-cycle. This indicates that the complex voltage waveform comprises either:
(i) a fundamental and even harmonics, initially out of phase with each other (see example 9), or
(ii) a fundamental and odd and even harmonics, one or more of the harmonics being initially out of phase (see example 12).
(c) If in Fig. 16(c) the negative half-cycle is reversed, it is seen to be a mirror image of the positive half-cycle about a vertical line drawn through time $T/2$. This feature indicates that the complex emf waveform comprises either:
(i) a fundamental and even harmonics initially in phase with each other (see examples 7 and 8), or
(ii) a fundamental and odd and even harmonics, each initially in phase with each other (see example 11).

Further problems on harmonic synthesis may be found in section 9, problems 1 to 7, page 245.

4. Rms value, mean value and the form factor of a complex wave

Rms value

Let the instantaneous value of a complex current, i, be given by

$$i = I_{1m} \sin(\omega t + \theta_1) + I_{2m} \sin(2\omega t + \theta_2) + \cdots + I_{nm} \sin(n\omega t + \theta_n) \text{ amperes}$$

The effective or rms value of this current is given by

$$I = \sqrt{(\text{mean value of } i^2)}$$

$$i^2 = [I_{im} \sin(\omega t + \theta_1) + I_{2m} \sin(2\omega t + \theta_2) + \cdots + I_{nm} \sin(n\omega t + \theta_n)]^2$$

i.e.,

$$i^2 = I_{1m}^2 \sin^2(\omega t + \theta_1) + I_{2m}^2 \sin^2(2\omega t + \theta_2) + \cdots + I_{nm}^2 \sin^2(n\omega t + \theta_n)$$
$$+ 2I_{1m}I_{2m} \sin(\omega t + \theta_1) \sin(2\omega t + \theta_2) + \cdots \tag{3}$$

Without writing down all terms involved when squaring current i, it can be seen that two types of term result, these being:

(i) terms such as $I_{1m}^2 \sin^2(\omega t + \theta_1)$, $I_{2m}^2 \sin^2(2\omega t + \theta_2)$, and so on, and
(ii) terms such as $2I_{1m}I_{2m} \sin(\omega t + \theta_1) \sin(2\omega t + \theta_2)$, i.e., products of different harmonics.

The mean value of i^2 is the sum of the mean values of each term in equation (3).

Taking an example of the first type, say $I_{1m}^2 \sin^2(\omega t + \theta_1)$, the mean value over one cycle of the fundamental is determined using integral calculus:

Mean value of $I_{1m}^2 \sin^2(\omega t + \theta_1)$

$$= \frac{1}{2\pi} \int_0^{2\pi} I_{1m}^2 \sin^2(\omega t + \theta_1) \, \mathrm{d}(\omega t),$$

(since the mean value of $y = f(x)$ between $x = a$ and $x = b$ is given by $(1/(b-a)) \int_a^b y \, \mathrm{d}x$,

$$= \frac{I_{1m}^2}{2\pi} \int_0^{2\pi} \left\{ \frac{1 - \cos 2(\omega t + \theta_1)}{2} \right\} d(\omega t),$$

(since $\cos 2x = 1 - 2\sin^2 x$, from which $\sin^2 x = (1 - \cos 2x)/2$),

$$= \frac{I_{1m}^2}{4\pi} \left[\omega t - \frac{\sin 2(\omega t + \theta_1)}{2} \right]_0^{2\pi}$$

$$= \frac{I_{1m}^2}{4\pi} \left[\left(2\pi - \frac{\sin 2(2\pi + \theta_1)}{2} \right) - \left(0 - \frac{\sin 2(0 + \theta_1)}{2} \right) \right]$$

$$= \frac{I_{1m}^2}{4\pi} \left[2\pi - \frac{\sin 2(2\pi + \theta_1)}{2} + \frac{\sin 2\theta_1}{2} \right] = \frac{I_{1m}^2}{4\pi}(2\pi) = \frac{\boldsymbol{I_{1m}^2}}{\boldsymbol{2}}$$

Hence it follows that the mean value of $I_{nm}^2 \sin^2(n\omega t + \theta_n)$ is given by $I_{nm}^2/2$.
Taking an example of the second type, say,

$$2I_{1m}I_{2m} \sin(\omega t + \theta_1)\sin(2\omega t + \theta_2),$$

the mean value over one cycle of the fundamental is also determined using integration:

Mean value of $2I_{1m}I_{2m} \sin(\omega t + \theta_1)\sin(2\omega t + \theta_2)$

$$= \frac{1}{2\pi} \int_0^{2\pi} 2I_{1m}I_{2m} \sin(\omega t + \theta_1)\sin(2\omega t + \theta_2) \, d(\omega t)$$

$$= \frac{I_{1m}I_{2m}}{\pi} \int_0^{2\pi} \tfrac{1}{2} \{ \cos(\omega t + \theta_2 - \theta_1) - \cos(3\omega t + \theta_2 + \theta_1) \} \, d(\omega t),$$

(since $\sin A \sin B = \tfrac{1}{2}[\cos(A - B) - \cos(A + B)]$,
and taking $A = (2\omega t + \phi_2)$ and $B = (\omega t + \phi_1)$)

$$= \frac{I_{1m}I_{2m}}{2\pi} \left[\sin(\omega t + \theta_2 - \theta_1) - \frac{\sin(3\omega t + \theta_2 + \theta_1)}{3} \right]_0^{2\pi}$$

$$= \frac{I_{1m}I_{2m}}{2\pi} \left[\left(\sin(2\pi + \theta_2 - \theta_1) - \frac{\sin(6\pi + \theta_2 + \theta_1)}{3} \right) \right.$$

$$\left. - \left(\sin(\theta_2 - \theta_1) - \frac{\sin(\theta_2 + \theta_1)}{3} \right) \right]$$

$$= \frac{I_{1m}I_{2m}}{2\pi} [0] = \mathbf{0} \tag{4}$$

Hence it follows that all such products of different harmonics will have a mean value of zero. Thus

$$\text{mean value of } i^2 = \frac{I_{1m}^2}{2} + \frac{I_{2m}^2}{2} + \cdots + \frac{I_{nm}^2}{2}$$

Hence the rms value of current,

$$I = \sqrt{\left(\frac{I_{1m}^2}{2} + \frac{I_{2m}^2}{2} + \cdots + \frac{I_{nm}^2}{2} \right)}$$

i.e.,

$$I = \sqrt{\left(\frac{I_{1m}^2 + I_{2m}^2 + \cdots + I_{nm}^2}{2} \right)} \qquad (5)$$

For a sine wave, rms value $= (1/\sqrt{2}) \times$ maximum value, i.e., maximum value $= \sqrt{2} \times$ rms value. Hence, for example, $I_{1m} = \sqrt{2}I_1$, where I_1 is the rms value of the fundamental component, and $(I_{1m})^2 = (\sqrt{2}I_1)^2 = 2I_1^2$.

Thus, from equation (5), rms current

$$I = \sqrt{\left(\frac{2I_1^2 + 2I_2^2 + \cdots + 2I_n^2}{2} \right)}$$

i.e.,

$$I = \sqrt{(I_1^2 + I_2^2 + \cdots + I_n^2)} \qquad (6)$$

where $I_1, I_2, \ldots, I_n$ are the rms values of the respective harmonics.

By similar reasoning, for a complex voltage waveform represented by

$$v = V_{1m} \sin(\omega t + \psi_1) + V_{2m} \sin(2\omega t + \psi_2) + \cdots$$
$$+ V_{nm} \sin(n\omega t + \psi_n) \text{ volts},$$

the rms value of voltage, V, is given by

$$V = \sqrt{\left(\frac{V_{1m}^2 + V_{2m}^2 + \cdots + V_{nm}^2}{2} \right)} \qquad (7)$$

or

$$V = \sqrt{(V_1^2 + V_2^2 + \cdots + V_n^2)} \qquad (8)$$

where $V_1, V_2, \ldots, V_n$ are the rms values of the respective harmonics.

From equations (5) to (8) it is seen that the rms value of a complex wave is unaffected by the relative phase angles of the harmonic components. For a d.c. current or voltage, the instantaneous value, the mean value and the maximum value are equal. Thus, if a complex waveform should contain a d.c. component I_0, then the rms current I is given by

$$I = \sqrt{\left(I_0^2 + \frac{I_{1m}^2 + I_{2m}^2 + \cdots + I_{nm}^2}{2} \right)} \qquad (9)$$

or

$$I = \sqrt{(I_0^2 + I_1^2 + I_2^2 + \cdots + I_n^2)}$$

Mean value

The mean or average value of a complex quantity whose negative half-cycle is similar to its positive half-cycle is given, for current, by

$$i_{av} = \frac{1}{\pi} \int_0^\pi i \, \mathrm{d}(\omega t) \qquad (10)$$

and for voltage by

$$v_{av} = \frac{1}{\pi} \int_0^\pi v \, \mathrm{d}(\omega t) \qquad (11)$$

each waveform being taken over half a cycle.

Unlike rms values, mean values **are** affected by the relative phase angles of the harmonic components.

Form factor

The form factor of a complex waveform whose negative half-cycle is similar in shape to its positive half-cycle is defined as

$$\text{form factor} = \frac{\text{rms value of the waveform}}{\text{mean value}} \tag{12}$$

where the mean value is taken over half a cycle.

Changes in the phase displacement of the harmonics may appreciably alter the form factor of a complex waveform.

Worked problems on the rms value, mean value and form factor of a complex wave

Problem 1. Determine the rms value of the current waveform represented by

$$i = 100 \sin \omega t + 20 \sin(3\omega t + \pi/6) + 10 \sin(5\omega t + 2\pi/3)\, \text{mA}$$

From equation (5), the rms value of current is given by

$$I = \sqrt{\left(\frac{100^2 + 20^2 + 10^2}{2}\right)} = \sqrt{\left(\frac{10000 + 400 + 100}{2}\right)} = \textbf{72.46 mA}$$

Problem 2. A complex voltage is represented by

$$v = (10 \sin \omega t + 3 \sin 3\omega t + 2 \sin 5\omega t)\, \text{volts}$$

Determine for the voltage: (a) the rms value, (b) the mean value, and (c) the form factor.

(a) From equation (7), the rms value of voltage is given by

$$V = \sqrt{\left(\frac{10^2 + 3^2 + 2^2}{2}\right)} = \sqrt{\left(\frac{113}{2}\right)} = \textbf{7.52 V}$$

(b) From equation (11), the mean value of voltage is given by

$$v_{av} = \frac{1}{\pi} \int_0^\pi (10 \sin \omega t + 3 \sin 3\omega t + 2 \sin 5\omega t)\, \mathrm{d}(\omega t)$$

$$= \frac{1}{\pi}\left[-10 \cos \omega t - \frac{3 \cos 3\omega t}{3} - \frac{2 \cos 5\omega t}{5}\right]_0^\pi$$

$$= \frac{1}{\pi}\left[(-10 \cos \pi - \cos 3\pi - \tfrac{2}{5}\cos 5\pi) - (-10 \cos 0 - \cos 0 - \tfrac{2}{5}\cos 0)\right]$$

$$= \frac{1}{\pi}[(10 + 1 + \tfrac{2}{5}) - (-10 - 1 - \tfrac{2}{5})] = \frac{22.8}{\pi} = \textbf{7.26 V}$$

(c) From equation (12), form factor is given by

$$\text{form factor} = \frac{\text{rms value of the waveform}}{\text{mean value}} = \frac{7.52}{7.26} = \textbf{1.036}$$

Problem 3. A complex voltage waveform which has an rms value of 240 V contains 30% third harmonic and 10% fifth harmonic, both of the harmonics being initially in phase with each other. (a) Determine the rms value of the fundamental and each harmonic. (b) Write down an expression to represent the complex voltage waveform if the frequency of the fundamental is 31.83 Hz.

(a) From equation (8), rms voltage $V = \sqrt{(V_1^2 + V_3^2 + V_5^2)}$. Since $V_3 = 0.30\,V_1$, $V_5 = 0.10\,V_1$ and $V = 240\,V$, then

$$240 = \sqrt{[V_1^2 + (0.30\,V_1)^2 + (0.10\,V_1)^2]}$$

i.e.,

$$240 = \sqrt{(1.10\,V_1^2)} = 1.049\,V_1$$

from which the rms value of the fundamental, $V_1 = 240/1.049 = \mathbf{22.88\ V}$. Rms value of the third harmonic,

$$V_3 = 0.30\,V_1 = (0.30)(228.8) = \mathbf{68.64\ V}$$

and the rms value of the fifth harmonic,

$$V_5 = 0.10\,V_1 = (0.10)(228.8) = \mathbf{22.88\ V}$$

(b) Maximum value of the fundamental,

$$V_{1m} = \sqrt{2}\,V_1 = \sqrt{2}(228.8) = 323.6\ \text{V}$$

Maximum value of the third harmonic,

$$V_{3m} = \sqrt{2}\,V_3 = \sqrt{2}(68.64) = 97.07\ \text{V}$$

Maximum value of the fifth harmonic,

$$V_{5m} = \sqrt{2}\,V_5 = \sqrt{2}(22.88) = 32.36\ \text{V}$$

Since the fundamental frequency is 31.83 Hz, the fundamental voltage may be written as $323.6 \sin 2\pi(31.83)t$, i.e., $323.6 \sin 200\,t$ volts.

The third harmonic component is $97.07 \sin 600\,t$ volts and the fifth harmonic component is $32.36 \sin 1000\,t$ volts. Hence an expression representing the complex voltage waveform is given by

$$v = (\mathbf{323.6 \sin 200\,t + 97.07 \sin 600\,t + 32.36 \sin 1000\,t})\ \textbf{volts}$$

Further problems on rms values, mean values and form factor of complex waves may be found in section 9, problems 8 to 13, page 247.

5. Power associated with complex waves

Let a complex voltage wave be represented by

$$v = V_{1m} \sin \omega t + V_{2m} \sin 2\omega t + V_{3m} \sin 3\omega t + \cdots,$$

and when this is applied to a circuit let the resulting current be represented by

$$i = I_{1m} \sin(\omega t - \phi_1) + I_{2m} \sin(2\omega t - \phi_2) + I_{3m} \sin(3\omega t - \phi_3) + \cdots$$

(Since the phase angles are lagging, the circuit in this case is inductive.)

At any instant in time the power p supplied to the circuit is given by $p = vi$ i.e.,

$$p = (V_{1m} \sin \omega t + V_{2m} \sin 2\omega t + \cdots)(I_{1m} \sin(\omega t - \phi_1)$$
$$+ I_{2m} \sin(2\omega t - \phi_2) + \cdots)$$
$$= V_{1m}I_{1m} \sin \omega t \sin(\omega t - \phi_1) + V_{1m}I_{2m} \sin \omega t \sin(2\omega t - \phi_2) + \cdots \quad (13)$$

The average or active power supplied over one cycle is given by the sum of the average values of each individual product term taken over one cycle. It is seen from equation (4) that the average value of product terms involving harmonics of different frequencies is always zero. This means therefore that only products of voltage and current harmonics of the same frequency need be considered in equation (13).

Taking the first term, for example, the average power P_1 over one cycle of the fundamental is given by

$$P_1 = \frac{1}{2\pi} \int_0^{2\pi} V_{1m} I_{1m} \sin \omega t \sin(\omega t - \phi_1) \, d(\omega t)$$

$$= \frac{V_{1m} I_{1m}}{2\pi} \int_0^{2\pi} \tfrac{1}{2} \{\cos \phi_1 - \cos(2\omega t - \phi_1)\} \, d(\omega t)$$

since $\sin A \sin B = \tfrac{1}{2} \{\cos(A - B) - \cos(A + B)\}$,

$$= \frac{V_{1m} I_{1m}}{4\pi} \left[(\omega t) \cos \phi_1 - \frac{\sin(2\omega t - \phi_1)}{2} \right]_0^{2\pi}$$

$$= \frac{V_{1m} I_{1m}}{4\pi} \left[\left(2\pi \cos \phi_1 - \frac{\sin(4\pi - \phi_1)}{2} \right) - \left(0 - \frac{\sin(-\phi_1)}{2} \right) \right]$$

$$= \frac{V_{1m} I_{1m}}{4\pi} [2\pi \cos \phi_1] = \frac{V_{1m} I_{1m}}{2} \cos \phi_1$$

$V_{1m} = \sqrt{2} V_1$ and $I_{1m} = \sqrt{2} I_1$, where V_1 and I_1 are rms values, hence

$$P_1 = \frac{(\sqrt{2} V_1)(\sqrt{2} I_1)}{2} \cos \phi_1$$

i.e.,

$$P_1 = V_1 I_1 \cos \phi_1 \text{ watts}$$

Similarly, the average power supplied over one cycle of the fundamental for the second harmonic is $V_2 I_2 \cos \phi_2$, and so on. Hence the total power supplied by complex voltages and currents is the sum of the powers supplied by each harmonic component acting on its own. The average power P supplied for one cycle of the fundamental is given by

$$P = V_1 I_1 \cos \phi_1 + V_2 I_2 \cos \phi_2 + \cdots + V_n I_n \cos \phi_n \qquad (14)$$

If the voltage waveform contains a d.c. component V_0 which causes a direct current component I_0, then the average power supplied by the d.c. component is $V_0 I_0$ and the total average power P supplied is given by

$$P = V_0 I_0 + V_1 I_1 \cos \phi_1 + V_2 I_2 \cos \phi_2 + \cdots + V_n I_n \cos \phi_n \qquad (15)$$

Alternatively, if R is the equivalent series resistance of a circuit then the total power is given by

$$P = I_0^2 R + I_1^2 R + I_2^2 R + I_3^2 R + \cdots$$

i.e.,

$$P = I^2 R \qquad (16)$$

where I is the rms value of current i.

Power factor

When harmonics are present in a waveform the overall circuit power factor is defined as

$$\text{overall power factor} = \frac{\text{total power supplied}}{\text{total rms voltage} \times \text{total rms current}}$$
$$= \frac{\text{total power}}{\text{volt amperes}}$$

i.e.,

$$\textbf{p.f.} = \frac{V_1 I_1 \cos \phi_1 + V_2 I_2 \cos \phi_2 + \cdots}{V I} \tag{17}$$

Worked problems on power associated with complex waves

Problem 1. Determine the average power in a $20\,\Omega$ resistance if the current i flowing through it is of the form

$$i = (12 \sin \omega t + 5 \sin 3\omega t + 2 \sin 5\omega t)\,\text{amperes}$$

From equation (5), rms current,

$$I = \sqrt{\left(\frac{12^2 + 5^2 + 2^2}{2}\right)} = 9.30\,\text{A}$$

From equation (16), average power,

$$P = I^2 R = (9.30)^2(20) = \textbf{1730 W or 1.73 kW}$$

Problem 2. A complex voltage v given by

$$v = 60 \sin \omega t + 15 \sin\left(3\omega t + \frac{\pi}{4}\right) + 10 \sin\left(5\omega t - \frac{\pi}{2}\right)\,\text{volts}$$

is applied to a circuit and the resulting current i is given by

$$i = 2 \sin\left(\omega t - \frac{\pi}{6}\right) + 0.3 \sin\left(3\omega t - \frac{\pi}{12}\right) + 0.1 \sin\left(5\omega t - \frac{8\pi}{9}\right)\,\text{amperes}$$

Determine (a) the total active power supplied to the circuit, and (b) the overall power factor.

(a) From equation (14), total power supplied,

$$P = V_1 I_1 \cos \phi_1 + V_3 I_3 \cos \phi_3 + V_5 I_5 \cos \phi_5$$
$$= \left(\frac{60}{\sqrt{2}}\right)\left(\frac{2}{\sqrt{2}}\right)\cos\left(0 - \left(-\frac{\pi}{6}\right)\right) + \left(\frac{15}{\sqrt{2}}\right)\left(\frac{0.3}{\sqrt{2}}\right)\cos\left(\frac{\pi}{4} - \left(-\frac{\pi}{12}\right)\right)$$
$$+ \left(\frac{10}{\sqrt{2}}\right)\left(\frac{0.1}{\sqrt{2}}\right)\cos\left(-\frac{\pi}{2} - \left(-\frac{8\pi}{9}\right)\right)$$
$$= 51.96 + 1.125 + 0.171 = \textbf{53.26 W}$$

(b) From equation (5), rms current,

$$I = \sqrt{\left(\frac{2^2 + 0.3^3 + 0.1^2}{2}\right)} = 1.43\,\text{A}$$

and from equation (7), rms voltage,

$$V = \sqrt{\left(\frac{60^2 + 15^2 + 10^2}{2}\right)} = 44.30 \text{ V}$$

From equation (17),

$$\text{overall power factor} = \frac{53.26}{(44.30)(1.43)} = \textbf{0.841}$$

(With a sinusoidal waveform,

$$\text{power factor} = \frac{\text{power}}{\text{volt-amperes}} = \frac{VI \cos \phi}{VI} = \cos \phi$$

Thus power factor depends upon the value of phase angle ϕ, and is lagging for an inductive circuit and leading for a capacitive circuit. However, with a complex waveform, power factor is not given by $\cos \phi$. In the expression for power in equation (14) there are n phase-angle terms, $\phi_1, \phi_2, \ldots, \phi_n$, all of which may be different. It is for this reason that is is not possible to state whether the overall power factor is lagging or leading when harmonics are present.)

Further problems on power associated with complex waves may be found in section 9, problems 14 to 17, page 247.

6. Harmonics in single-phase circuits

When a complex alternating voltage wave, i.e., one containing harmonics, is applied to a single-phase circuit containing resistance, inductance and/or capacitance (i.e., linear circuit elements), then the resulting current will also be complex and contain harmonics.

Let a complex voltage v be represented by

$$v = V_{1m} \sin \omega t + V_{2m} \sin 2\omega t + V_{3m} \sin 3\omega t + \cdots$$

(a) Pure resistance

The impedance of a pure resistance R is independent of frequency and the current and voltage are in phase for each harmonic. Thus the general expression for current i is given by

$$i = \frac{v}{R} = \frac{V_{1m}}{R} \sin \omega t + \frac{V_{2m}}{R} \sin 2\omega t + \frac{V_{3m}}{R} \sin 3\omega t + \cdots \qquad (18)$$

The percentage harmonic content in the current wave is the same as that in the voltage wave. For example, the percentage second harmonic content from equation (18) is

$$\frac{V_{2m}/R}{V_{1m}/R} \times 100\%, \quad \text{i.e.,} \quad \frac{V_{2m}}{V_{1m}} \times 100\%$$

the same as for the voltage wave. The current and voltage waveforms will therefore be identical in shape.

(b) Pure inductance

The impedance of a pure inductance L, i.e., inductive reactance X_L ($= 2\pi f L$), varies with the harmonic frequency when voltage v is applied to it. Also, for every harmonic term, the current will lag the voltage by 90° or $\pi/2$ rad. The current i is given by

$$i = \frac{v}{X_L} = \frac{V_{1m}}{\omega L}\sin\left(\omega t - \frac{\pi}{2}\right) + \frac{V_{2m}}{2\omega L}\sin\left(2\omega t - \frac{\pi}{2}\right) + \frac{V_{3m}}{3\omega L}\sin\left(3\omega t - \frac{\pi}{2}\right)$$

$$(19)$$

since for the nth harmonic the reactance is $n\omega L$.

Equation (19) shows that for, say, the nth harmonic, the percentage harmonic content in the current waveform is only $1/n$ of the corresponding harmonic content in the voltage waveform.

If a complex current contains a d.c. component then the direct voltage drop across a pure inductance is zero.

(c) Pure capacitance

The impedance of a pure capacitance C, i.e., capacitive reactance X_C ($= 1/(2\pi f C)$), varies with the harmonic frequency when voltage v is applied to it. Also, for every harmonic term the current will lead the voltage by 90° or $\pi/2$ rad. The current i is given by

$$i = \frac{v}{X_C} = \frac{V_{1m}}{1/\omega C}\sin\left(\omega t + \frac{\pi}{2}\right) + \frac{V_{2m}}{1/2\omega C}\sin\left(2\omega t + \frac{\pi}{2}\right)$$

$$+ \frac{V_{3m}}{1/3\omega C_1}\sin\left(3\omega t + \frac{\pi}{2}\right) + \cdots,$$

since for the nth harmonic the reactance is $1/(n\omega C)$. Hence current,

$$i = V_{1m}(\omega C)\sin\left(\omega t + \frac{\pi}{2}\right) + V_{2m}(2\omega C)\sin\left(2\omega t + \frac{\pi}{2}\right)$$

$$+ V_{3m}(3\omega C)\sin\left(3\omega t + \frac{\pi}{2}\right) + \cdots \quad (20)$$

Equation (20) shows that the percentage harmonic content of the current waveform is n times larger for the nth harmonic than that of the corresponding harmonic voltage.

If a complex current contains a d.c. component then none of this direct current will flow through a pure capacitor, although the alternating components of the supply still operate.

Worked problems on harmonics in single-phase circuits

Problem 1. A complex voltage waveform represented by

$$v = 100\sin\omega t + 30\sin\left(3\omega t + \frac{\pi}{3}\right) + 10\sin\left(5\omega t - \frac{\pi}{6}\right)\text{volts}$$

is applied across (a) a pure $40\,\Omega$ resistance, (b) a pure 7.96 mH inductance, and (c) a pure $25\,\mu$F capacitor. Determine for each case an expression for the current flowing if the fundamental frequency is 1 kHz.

(a) From equation (18),

$$\text{current } i = \frac{v}{R} = \frac{100}{40}\sin \omega t + \frac{30}{40}\sin\left(3\omega t + \frac{\pi}{3}\right) + \frac{10}{40}\sin\left(5\omega t - \frac{\pi}{6}\right)$$

i.e.,

$$i = 2.5\sin \omega t + 0.75\sin\left(3\omega t + \frac{\pi}{3}\right) + 0.25\sin\left(5\omega t - \frac{\pi}{6}\right)\text{ amperes}$$

(b) At the fundamental frequency, $\omega L = 2\pi(1000)(7.96 \times 10^{-3}) = 50\,\Omega$. From equation (19),

$$\text{current } i = \frac{100}{50}\sin\left(\omega t - \frac{\pi}{2}\right) + \frac{30}{3 \times 50}\sin\left(3\omega t + \frac{\pi}{3} - \frac{\pi}{2}\right)$$

$$+ \frac{10}{5 \times 50}\sin\left(5\omega t - \frac{\pi}{6} - \frac{\pi}{2}\right)$$

i.e.,

$$\text{current } i = 2\sin\left(\omega t - \frac{\pi}{2}\right) + 0.20\sin\left(3\omega t - \frac{\pi}{6}\right)$$

$$+ 0.04\sin\left(5\omega t - \frac{2\pi}{3}\right)\text{ amperes}$$

(c) At the fundamental frequency, $\omega C = 2\pi(1000)(25 \times 10^{-6}) = 0.157$. From equation (20),

$$\text{current } i = 100(0.157)\sin\left(\omega t + \frac{\pi}{2}\right) + 30(3 \times 0.157)\sin\left(3\omega t + \frac{\pi}{3} + \frac{\pi}{2}\right)$$

$$+ 10(5 \times 0.157)\sin\left(5\omega t - \frac{\pi}{6} + \frac{\pi}{2}\right)$$

i.e.,

$$i = 15.70\sin\left(\omega t + \frac{\pi}{2}\right) + 14.13\sin\left(3\omega t + \frac{5\pi}{6}\right) + 7.85\sin\left(5\omega t + \frac{\pi}{3}\right)\text{ amperes}$$

Problem 2. A supply voltage v given by

$$v = (240\sin 314\,t + 40\sin 942\,t + 30\sin 1570\,t)\text{ volts}$$

is applied to a circuit comprising a resistance of $12\,\Omega$ connected in series with a coil of inductance 9.55 mH. Determine (a) an expression to represent the instantaneous value of the current, (b) the rms voltage, (c) the rms current, (d) the power dissipated, and (e) the overall power factor.

(a) The supply voltage comprises a fundamental, $240\sin 314\,t$, a third harmonic, $40\sin 942\,t$ (third harmonic since 942 is 3×314) and a fifth harmonic, $30\sin 1570\,t$.

Fundamental

Since the fundamental frequency, $\omega_1 = 314$ rad/s, inductive reactance,

$X_{L1} = \omega_1 L = (314)(9.55 \times 10^{-3}) = 3.0\,\Omega$. Hence impedance at the fundamental frequency,

$$Z_1 = (12 + j3.0)\,\Omega = 12.37\underline{/\,14.04°}\,\Omega$$

Maximum current at fundamental frequency,

$$I_{1m} = \frac{V_{1m}}{Z_1} = \frac{240\underline{/\,0°}}{12.37\underline{/\,14.04°}} = 19.40\underline{/\,-14.04°}\,\text{A}$$

$14.04° = 14.04 \times (\pi/180)\,\text{rad} = 0.245\,\text{rad}$, thus $I_{1m} = 19.40\underline{/\,-0.245}\,\text{A}$.

Hence the fundamental current $i_1 = 19.40\,(\sin 314t - 0.245)\,\text{A}$.
(Note that with an expression of the form $R\sin(\omega t \pm \alpha)$, ωt is an angle measured in radians, thus the phase displacement, α, should also be expressed in radians.)

Third harmonic

Since the third harmonic frequency, $\omega_3 = 942\,\text{rad/s}$, inductive reactance, $X_{L3} = 3X_{L1} = 9.0\,\Omega$. Hence impedance at the third harmonic frequency,

$$Z_3 = (12 + j9.0)\,\Omega = 15\underline{/\,36.87°}\,\Omega$$

Maximum current at the third harmonic frequency,

$$I_{3m} = \frac{V_{3m}}{Z_3} = \frac{40\underline{/\,0°}}{15\underline{/\,36.87°}}$$

$$= 2.67\underline{/\,-36.87°}\,\text{A}$$

$$= 2.67\underline{/\,-0.644}\,\text{A}$$

Hence the third harmonic current, $i_3 = 2.67\sin(942t - 0.644)\,\text{A}$.

Fifth harmonic

Inductive reactance, $X_{L5} = 5X_{L1} = 15\,\Omega$.
Impedance $Z_5 = (12 + j15)\,\Omega = 19.21\underline{/\,51.34°}\,\Omega$.

Current, $I_{5m} = \dfrac{V_{5m}}{Z_5} = \dfrac{30\underline{/\,0°}}{19.21\underline{/\,51.34°}} = 1.56\underline{/\,-51.34°}\,\text{A} = 1.56\underline{/\,-0.896}\,\text{A}$

Hence the fifth harmonic current, $i_5 = 1.56\sin(1570t - 0.896)\,\text{A}$.

Thus an expression to represent the instantaneous current, i, is given by $i = i_1 + i_3 + i_5$ i.e.,

$$i = \mathbf{19.40\sin(314t - 0.245) + 2.67\sin(942t - 0.644)}$$
$$\mathbf{+\ 1.56\sin(1570t - 0.896)\ amperes}$$

(b) From equation (7), rms voltage,

$$V = \sqrt{\left(\frac{240^2 + 40^2 + 30^2}{2}\right)} = \mathbf{173.35\ V}$$

(c) From equation (5), rms current,

$$I = \sqrt{\left(\frac{19.40^2 + 2.67^2 + 1.56^2}{2}\right)} = \mathbf{13.89\ A}$$

(d) From equation (16), power dissipated,

$$P = I^2 R = (13.89)^2(12) = \mathbf{2315\ W}\quad\text{or}\quad\mathbf{2.315\ kW}$$

(Alternatively, equation (14) may be used to determine power.)

(e) From equation (17),

$$\text{overall power factor} = \frac{2315}{(173.35)(13.89)} = \textbf{0.961}$$

Problem 3. A complex current i of fundamental frequency 50 Hz is represented by

$$i = (20 \sin \omega t + 4 \sin 3\omega t + 3 \sin 5\omega t) \text{ amperes}$$

This current flows through a pure inductance L and the p.d. across the inductance is measured by a voltmeter as 50 V. Determine the value of inductance L.

The 50 V indicated by the voltmeter means "50 V rms". From equation (7), rms voltage,

$$V = \sqrt{\left(\frac{V_{1m}^2 + V_{3m}^2 + V_{5m}^2}{2} \right)}$$

Voltage $V_{1m} = I_{1m}X_{L1} = (I_{1m})(2\pi fL) = (20)(2\pi 50L) = (20)(314L)$
Voltage $V_{3m} = I_{3m}X_{L3} = (I_{3m})(2\pi f_3 L) = (4)(2\pi 150L) = (4)(942L)$
Voltage $V_{5m} = I_{5m}X_{L5} = (I_{5m})(2\pi f_5 L) = (3)(2\pi 250L) = (3)(1570L)$

Hence

$$50 = \sqrt{\left\{ \frac{[(20)(314L)]^2 + [(4)(942L)]^2 + [(3)(1570L)]^2}{2} \right\}}$$

from which

$$50^2 = \frac{(314)^2 L^2 [20^2 + 12^2 + 15^2]}{2}$$

and

$$L^2 = \frac{2(50)^2}{(314)^2 [400 + 144 + 225]} = 6.5945 \times 10^{-5}$$

Thus **inductance, $L = \sqrt{(6.5945 \times 10^{-5})} = \textbf{8.12 mH}$.**

Problem 4. An emf is represented by

$$e = 50 + 200 \sin \omega t + 40 \sin \left(2\omega t - \frac{\pi}{2} \right) + 5 \sin \left(4\omega t + \frac{\pi}{4} \right) \text{volts,}$$

the fundamental frequency being 50 Hz. The emf is applied across a circuit comprising a 100 μF capacitor connected in series with a 50 Ω resistor. Obtain an expression for the current flowing and hence determine the rms value of current.

D.C. component

In a d.c. circuit no current will flow through a capacitor. The current waveform will not possess a d.c. component even though the emf waveform has a 50 V d.c. component. Hence $i_0 = 0$.

Fundamental

Capacitive reactance,

$$X_{C1} = \frac{1}{2\pi fC} = \frac{1}{2\pi(50)(100 \times 10^{-6})} = 31.83 \, \Omega$$

Impedance $Z_1 = (50 - j31.83) \, \Omega = 59.27 \underline{/-32.48°} \, \Omega$

$$I_{1m} = \frac{V_{1m}}{Z_1} = \frac{200 \underline{/\ 0°}}{59.27 \underline{/\ -32.48°}} = 3.374 \underline{/\ 32.48°} \text{ A} = 3.374 \underline{/\ 0.567} \text{ A}$$

Hence the fundamental current, $i_1 = 3.374 \sin(\omega t + 0.567)$ A.

Second harmonic

Capacitive reactance,

$$X_{C2} = \frac{1}{2(2\pi f C)} = \frac{31.83}{2} = 15.92 \,\Omega$$

Impedance, $Z_2 = (50 - j15.92)\,\Omega = 52.47 \underline{/\ -17.66°}\,\Omega.$

$$I_{2m} = \frac{V_{2m}}{Z_2} = \frac{40 \underline{/\ -\pi/2}}{52.47 \underline{/\ -17.66°}} = 0.762 \underline{/\ \left(\left(-\frac{\pi}{2} - (-17.66°) \right) \right.} = 0.762 \underline{/\ -72.34°} \text{ A}$$

Hence the second harmonic current, $i_2 = 0.762 \sin(2\omega t - 72.34°)$ A
$$= 0.762 \sin(2\omega t - 1.263) \text{ A}$$

Fourth harmonic

Capacitive reactance, $X_{C4} = \frac{1}{4}X_{C1} = \frac{31.83}{4} = 7.958 \,\Omega$
Impedance, $Z_4 = (50 - j7.958)\,\Omega = 50.63 \underline{/\ -9.04°}\,\Omega$

$$I_{4m} = \frac{V_{4m}}{Z_4} = \frac{5 \underline{/\ \pi/4}}{50.63 \underline{/\ -9.04°}} = 0.099 \underline{/\ (\pi/4 - (-9.04°))} = 0.099 \underline{/\ 54.04°} \text{ A}$$

Hence the fourth harmonic current, $i_4 = 0.099 \sin(4\omega t + 54.04°)$ A
$$= 0.099 \sin(4\omega t + 0.943) \text{ A}$$

An expression for current flowing is therefore given by $i = i_0 + i_1 + i_2 + i_4$ i.e.,

$$i = 3.374 \sin(\omega t + 0.567) + 0.762 \sin(2\omega t - 1.263)$$
$$+ 0.099 \sin(4\omega t + 0.943) \text{ amperes}$$

From equation (5), rms current,

$$I = \sqrt{\left(\frac{3.374^2 + 0.762^2 + 0.099^2}{2} \right)} = 2.45 \text{ A}$$

Problem 5. A complex voltage v is represented by:

$$v = 25 + 100 \sin \omega t + 40 \sin\left(3\omega t + \frac{\pi}{6} \right) + 20 \sin\left(5\omega t + \frac{\pi}{12} \right) \text{volts}$$

where $\omega = 10^4$ rad/s. The voltage is applied to a series circuit comprising a $5.0\,\Omega$ resistance and a $500\,\mu H$ inductance.

Determine (a) an expression to represent the current flowing in the circuit, (b) the rms value of current, correct to two decimal places, and (c) the power dissipated in the circuit, correct to three significant figures.

(a) D.C. component

Inductance has no effect on a steady current. Hence the d.c. component of the current, i_0, is given by

$$i_0 = \frac{v_0}{R} = \frac{25}{5.0} = 5.0 \text{ A}$$

Fundamental

Inductive reactance, $X_{L1} = \omega L = (10^4)(500 \times 10^{-6}) = 5\,\Omega$.
Impedance, $Z_1 = (5 + j5)\,\Omega = 7.071\underline{/\ 45°}\,\Omega$.

$$I_{1m} = \frac{V_{1m}}{Z_1} = \frac{100\underline{/\ 0°}}{7.071\underline{/\ 45°}} = 14.14\underline{/\ -45°}\,\text{A}$$

$$= 14.14\underline{/\ -\pi/4}\,\text{A or } 14.14\underline{/\ -0.785}\,\text{A}$$

Hence fundamental current, $i_1 = 14.14\sin(\omega t - 0.785)\,\text{A}$

Third harmonic

Inductive reactance at third harmonic frequency, $X_{L3} = 3X_{L1} = 15\,\Omega$.
Impedance, $Z_3 = (5 + j15)\,\Omega = 15.81\underline{/\ 71.57°}\,\Omega$.

$$I_{3m} = \frac{V_{3m}}{Z_3} = \frac{40\underline{/\ \pi/6}}{15.81\underline{/\ 71.57°}} = 2.53\underline{/\ -41.57°}\,\text{A} = 2.53\underline{/\ -0.726}\,\text{A}$$

Hence the third harmonic current, $i_3 = 2.53\sin(3\omega t - 41.57°)\,\text{A}$
$$= 2.53\sin(3\omega t - 0.726)\,\text{A}$$

Fifth harmonic

Inductive reactance at fifth harmonic frequency, $X_{L5} = 5X_{L1} = 25\,\Omega$.
Impedance, $Z_5 = (5 + j25)\,\Omega = 25.495\underline{/\ 78.69°}\,\Omega$.

$$I_{5m} = \frac{V_{5m}}{Z_5} = \frac{20\underline{/\ \pi/12}}{25.495\underline{/\ 78.69°}} = 0.784\underline{/\ -63.69°}\,\text{A} = 0.784\underline{/\ -1.112}\,\text{A}$$

Hence the fifth harmonic current, $i_5 = 0.784\sin(5\omega t - 63.69°)\,\text{A}$
$$= 0.784\sin(5\omega t - 1.112)\,\text{A}$$

Thus current, $i = i_0 + i_1 + i_3 + i_5$, i.e.,

$$i = 5 + 14.14\sin(\omega t - 0.785) + 2.53\sin(3\omega t - 0.726) + 0.784\sin(5\omega t - 1.112)\,\text{A}$$

(b) From equation (9), rms current,

$$I = \sqrt{\left(5.0^2 + \frac{14.14^2 + 2.53^2 + 0.784^2}{2}\right)}$$

$$= 11.3348\,\text{A} = \mathbf{11.33\,A}, \text{ correct to two decimal places.}$$

(c) From equation (16), power dissipated,

$$P = I^2 R = (11.3348)^2(5.0) = 642.4\,\text{W}$$
$$= \mathbf{642\,W}, \text{ correct to three significant figures.}$$

(Alternatively, from equation (15),

$$\text{power } P = (25)(5.0) + \left(\frac{100}{\sqrt{2}}\right)\left(\frac{14.14}{\sqrt{2}}\right)\cos 45° + \left(\frac{40}{\sqrt{2}}\right)\left(\frac{2.53}{\sqrt{2}}\right)\cos 71.57°$$

$$+ \left(\frac{20}{\sqrt{2}}\right)\left(\frac{0.784}{\sqrt{2}}\right)\cos 78.69°$$

$$= 125 + 499.92 + 16.00 + 1.54$$

$$= 642.46\,\text{W or } \mathbf{642\,W}, \text{ correct to three significant figures, as above.)}$$

Problem 6. The voltage applied to a particular circuit comprising two components connected in series is given by

$$v = (30 + 40 \sin 10^3 t + 25 \sin 2 \times 10^3 t + 15 \sin 4 \times 10^3 t) \text{ volts}$$

and the resulting current is given by

$$i = 0.743 \sin(10^3 t + 1.190) + 0.781 \sin(2 \times 10^3 t + 0.896)$$
$$+ 0.636 \sin(4 \times 10^3 t + 0.559) \text{ A}$$

Determine (a) the average power supplied, (b) the type of components present, and (c) the values of the components.

(a) From equation (15), the average power P is given by

$$P = (30)(0) + \left(\frac{40}{\sqrt{2}}\right)\left(\frac{0.743}{\sqrt{2}}\right)\cos 1.190 + \left(\frac{25}{\sqrt{2}}\right)\left(\frac{0.781}{\sqrt{2}}\right)\cos 0.896$$
$$+ \left(\frac{15}{\sqrt{2}}\right)\left(\frac{0.636}{\sqrt{2}}\right)\cos 0.559$$

i.e., $P = 0 + 5.523 + 6.099 + 4.044 = \mathbf{15.67\,W.}$

(b) The expression for the voltage contains a d.c. component of 30 V. However there is no corresponding term in the expression for current. This indicates that one of the components is a **capacitor** (since in a d.c. circuit a capacitor offers an infinite impedance to a direct current). Since power is delivered to the circuit the other component is a **resistor**.

(c) From equation (5), rms current,

$$I = \sqrt{\left(\frac{0.743^2 + 0.781^2 + 0.636^2}{2}\right)} = 0.885 \text{ A}$$

Average power $P = I^2 R$, from which,

$$\text{resistance } R = \frac{P}{I^2} = \frac{15.67}{(0.885)^2} = \mathbf{20\,\Omega}$$

At the fundamental frequency, $\omega = 10^3$ rad/s,

$$\text{impedance } |Z_1| = \frac{V_{1m}}{I_{1m}} = \frac{40}{0.743} = 53.84\,\Omega$$

Impedance $|Z_1| = \sqrt{(R^2 + X_{C1}^2)}$, from which

$$X_{C1} = \sqrt{(Z_1^2 - R^2)} = \sqrt{(53.84^2 - 20^2)} = 50\,\Omega$$

Hence $1/\omega C = 50$, from which,

$$\mathbf{capacitance\ } C = \frac{1}{\omega(50)} = \frac{1}{10^3(50)} = \mathbf{20\,\mu F}$$

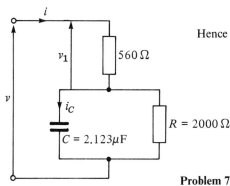

Figure 17

Problem 7. In the circuit shown in Fig. 17 the supply voltage v is given by $v = 300 \sin 314t + 120 \sin(942t + 0.698)$ volts. Determine (a) an expression for the supply current, i, (b) the percentage harmonic content of the supply current, (c) the total power dissipated, (d) an expression for the p.d. shown as v_1, and (e) an expression for current i_c.

(a) Capacitive reactance of the 2.123 μF capacitor at the fundamental frequency is given by

$$X_{C1} = \frac{1}{(314)(2.123 \times 10^{-6})} = 1500 \, \Omega$$

At the fundamental frequency the total circuit impedance, Z_1, is given by

$$Z_1 = 560 + \frac{(2000)(-j1500)}{(2000 - j1500)} = 560 + \frac{3 \times 10^6 \angle -90°}{2500 \angle -36.87°} = 560 + 1200 \angle -53.13°$$

$$= 560 + 720 - j960 = (1280 - j960) \, \Omega$$

$$= 1600 \angle -36.87° \, \Omega = 1600 \angle -0.644 \, \Omega$$

Since for the nth harmonic the capacitive reactance is $1/(n\omega C)$, the capacitive reactance of the third harmonic is $\frac{1}{3} X_{C1} = \frac{1}{3}(1500) = 500 \, \Omega$. Hence at the third harmonic frequency the total circuit impedance, Z_3, is given by

$$Z_3 = 560 + \frac{(2000)(-j500)}{(2000 - j500)} = 560 + \frac{10^6 \angle -90°}{2061.55 \angle -14.04°} = 560 + 485.07 \angle -75.96°$$

$$= 560 + 117.68 - j470.58 = (677.68 - j470.58) \, \Omega$$

$$= 825 \angle -34.78° \, \Omega$$

$$= 825 \angle -0.607 \, \Omega$$

The fundamental current

$$i_1 = \frac{v_1}{Z_1} = \frac{300 \angle 0}{1600 \angle -0.644} = 0.188 \angle 0.644 \, A$$

The third harmonic current

$$i_3 = \frac{v_3}{Z_3} = \frac{120 \angle 0.698}{825 \angle -0.607} = 0.145 \angle 1.305 \, A$$

Thus, supply current, $i = 0.188 \sin(314t + 0.644) + 0.145 \sin(942t + 1.305)$ A.

(b) Percentage harmonic content of the supply current is given by

$$\frac{0.145}{0.188} \times 100\% = \mathbf{77\%}$$

(c) From equation (14), total active power

$$P = \left(\frac{300}{\sqrt{2}}\right)\left(\frac{0.188}{\sqrt{2}}\right)\cos 0.644 + \left(\frac{120}{\sqrt{2}}\right)\left(\frac{0.145}{\sqrt{2}}\right)\cos 0.607$$

i.e.,

$$P = 22.56 + 7.15 = \mathbf{29.71 \, W}$$

(d) Voltage $v_1 = iR = 560[0.188 \sin(314t + 0.644) + 0.145 \sin(942t + 1.305)]$
i.e.,

$$v_1 = \mathbf{105.3 \sin(314t + 0.644) + 81.2 \sin(942t + 1.305) \, volts.}$$

(e) Current $i_c = i_1 \left(\dfrac{R}{R - jX_{C1}} \right) + i_3 \left(\dfrac{R}{R - jX_{C3}} \right)$ by current division

$$= (0.188 \underline{/\ 36.87°}) \left(\dfrac{2000}{2000 - j1500} \right)$$

$$+ (0.145 \underline{/\ 74.78°}) \left(\dfrac{2000}{2000 - j500} \right)$$

$$= (0.188 \underline{/\ 36.87°}) \left(\dfrac{2000}{2500 \underline{/\ -36.87°}} \right)$$

$$+ (0.145 \underline{/\ 74.78°}) \left(\dfrac{2000}{2061.55 \underline{/\ -14.04°}} \right)$$

$$= 0.150 \underline{/\ 73.74°} + 0.141 \underline{/\ 88.82°} = 0.150 \underline{/\ 1.287} + 0.141 \underline{/\ 1.550}$$

Hence $i_c = 0.150 \sin(314t + 1.287) + 0.141 \sin(942t + 1.550)$ A.

Further problems on harmonics in single phase circuits may be found in section 9, problems 18 to 29, page 248.

7. Resonance due to harmonics

In industrial circuits at power frequencies the typical values of L and C involved make resonance at the fundamental frequency very unlikely. (An exception to this is with the capacitor-start induction motor where the start-winding can achieve unity power factor during run-up.)

However, if the voltage waveform is not a pure sine wave it is quite possible for the resonant frequency to be near the frequency of one of the harmonics. In this case the magnitude of the particular harmonic in the current waveform is greatly increased and may even exceed that of the fundamental. The effect of this is a great distortion of the resultant current waveform so that dangerous volt drops may occur across the inductance and capacitance in the circuit.

When a circuit resonates at one of the harmonic frequencies of the supply voltage, the effect is called **selective or harmonic resonance.**

For resonance with the fundamental, the condition is $\omega L = 1/(\omega C)$; for resonance at, say, the third harmonic, the condition is $3\omega L = 1/(3\omega C)$; for resonance at the nth harmonic, the condition is $\boldsymbol{n\omega L = 1/(n\omega C)}$.

Worked problems on harmonic resonance

Problem 1. A voltage waveform having a fundamental of maximum value 400 V and a third harmonic of maximum value 10 V is applied to the circuit shown in Fig. 18. Determine (a) the frequency for resonance with the third harmonic, and (b) the maximum value of the fundamental and third harmonic components of current.

(a) Resonance with the third harmonic means that $3\omega L = 1/(3\omega C)$, i.e.,

$$\omega = \sqrt{\left(\dfrac{1}{9LC} \right)} = \dfrac{1}{3\sqrt{(0.5)(0.2 \times 10^{-6})}} = 1054 \text{ rad/s}$$

from which, **frequency,** $f = \dfrac{\omega}{2\pi} = \dfrac{1054}{2\pi} = \boldsymbol{167.7\ Hz}$

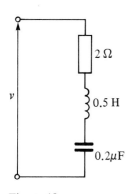

$2\,\Omega$

v

$0.5\ H$

$0.2\mu F$

Figure 18

(b) At the fundamental frequency,

$$\text{impedance } Z_1 = R + j\left(\omega L - \frac{1}{\omega C}\right)$$

$$= 2 + j\left[(1054)(0.5) - \frac{1}{(1054)(0.2 \times 10^{-6})}\right]$$

$$= (2 - j4217)\,\Omega$$

i.e.,
$$Z_1 = 4217\underline{/-89.97^\circ}\,\Omega$$

Maximum value of current at the fundamental frequency,

$$I_{1m} = \frac{V_{1m}}{Z_1} = \frac{400}{4217} = 0.095\,\text{A}$$

At the third harmonic frequency,

$$Z_3 = R + j\left(3\omega L - \frac{1}{3\omega C}\right) = R$$

since resonance occurs at the third harmonic, i.e., $Z_3 = 2\,\Omega$.
Maximum value of current at the third harmonic frequency,

$$I_{3m} = \frac{V_{3m}}{Z_3} = \frac{10}{2} = 5\,\text{A}$$

(Note that the magnitude of I_{3m} compared with I_{1m} is $5/0.095$, i.e. $\times$ **52.6 greater**.)

Problem 2. A voltage wave has an amplitude of 800 V at the fundamental frequency of 50 Hz and its nth harmonic has an amplitude 1.5% of the fundamental. The voltage is applied to a series circuit containing resistance $5\,\Omega$, inductance 0.369 H and capacitance 0.122 μF. Resonance occurs at the nth harmonic. Determine (a) the value of n, (b) the maximum value of current at the nth harmonic, (c) the p.d. across the capacitor at the nth harmonic and (d) the maximum value of the fundamental current.

(a) For resonance at the nth harmonic, $n\omega L = 1/(n\omega C)$, from which

$$n^2 = \frac{1}{\omega^2 LC} \qquad \text{and} \qquad n = \frac{1}{\omega \sqrt{(LC)}}$$

Hence

$$n = \frac{1}{2\pi 50\sqrt{(0.369)(0.122 \times 10^{-6})}} = 15$$

Thus resonance occurs at the 15th harmonic.

(b) At resonance, impedance $Z_{15} = R = 5\,\Omega$. Hence the maximum value of current at the 15th harmonic,

$$I_{15m} = \frac{V_{15m}}{R} = \frac{(1.5/100) \times 800}{5} = 2.4\,\text{A}$$

(c) At the 15th harmonic, capacitive reactance, .

$$X_{C15} = \frac{1}{15\omega C} = \frac{1}{15(2\pi 50)(0.122 \times 10^{-6})} = 1739\,\Omega$$

Hence the p.d. across the capacitor at the 15th harmonic $= (I_{15m})(X_{C15}) = (2.4)(1739) = $ **4.174 kV.**

(d) At the fundamental frequency, inductive reactance, $X_{L1} = \omega L = (2\pi 50)(0.369) = 115.9\,\Omega$, and capacitive reactance,

$$X_{C1} = \frac{1}{\omega C} = \frac{1}{(2\pi 50)(0.122 \times 10^{-6})} = 26091\,\Omega$$

Impedance at the fundamental frequency, $|Z_1| = \sqrt{[R^2 + (X_C - X_L)^2]} = 25975\,\Omega$.

Maximum value of current at the fundamental frequency,

$$I_{1m} = \frac{V_{1m}}{Z_1} = \frac{800}{25975} = \textbf{0.031 A} \quad \text{or} \quad \textbf{31 mA}$$

Further problems on harmonic resonance may be found in section 9, problems 30 to 34, page 250.

8. Sources of harmonics

Harmonics may be produced in the output waveform of an a.c. generator. This may be due either to "tooth-ripple", caused by the effect of the slots that accommodate the windings, or to the nonsinusoidal airgap flux distribution.

Great care is taken to ensure a sinusoidal output from generators in large supply systems; however, nonlinear loads will cause harmonics to appear in the load current waveform. Thus harmonics are produced in devices that have a nonlinear response to their inputs. Nonlinear circuit elements (i.e., those in which the current flowing through them is not proportional to the applied voltage) include rectifiers and any large-signal electronic amplifier in which diodes, transistors, valves or iron-cored inductors are used.

A **rectifier** is a device for converting an alternating or an oscillating current into a unidirectional or approximate direct current. A rectifier has a low impedance to current flow in one direction and a nearly infinite impedance to current flow in the opposite direction. Thus when an alternating current is applied to a rectifier, current will flow through it during the positive half-cycles only; the current is zero during the negative half-cycles. A typical current waveform is shown in Fig. 19. This "half-wave rectification" is produced by using a single diode. The waveform is similar in shape to that shown in Fig. 14, page 221, where the d.c. component brought the negative half-cycle up to the zero current point. The waveform shown in Fig. 19 is typical of one containing a fairly large second harmonic.

Transistors and **valves** are nonlinear devices in that sinusoidal input results in different positive and negative half-cycle amplifications. This means that the output half-cycles have different amplitudes. Since they have a different shape, even harmonic distortion is suggested (see section 3).

Figure 19 Typical current waveform containing a fairly large second harmonic

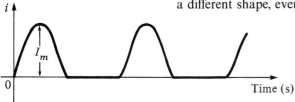

Time (s)

Ferromagnetic-cored coils are a source of harmonic generation in a.c. circuits became of the nonlinearity of the B/H curve and the hysteresis loop, especially if saturation occurs.

Let a sinusoidal voltage $v = V_m \sin \omega t$ be applied to a ferromagnetic-cored coil (having low resistance relative to inductive reactance) of cross-sectional area A square metres and possessing N turns.

If Φ is the flux produced in the core then the instantaneous voltage is given by $v = N(\mathrm{d}\Phi/\mathrm{d}t)$.

If B is the flux density of the core, then, since $\Phi = BA$,

$$v = N \frac{\mathrm{d}}{\mathrm{d}t}(BA) = NA \frac{\mathrm{d}B}{\mathrm{d}t},$$

since area A is a constant for a particular core.

Separating the variables gives

$$\int \mathrm{d}B = \frac{1}{NA} \int v \, \mathrm{d}t$$

i.e.,

$$B = \frac{1}{NA} \int V_m \sin \omega t \, \mathrm{d}t = \frac{-V_m}{\omega NA} \cos \omega t$$

Since $-\cos \omega t = \sin(\omega t - 90°)$,

$$B = \frac{V_m}{\omega NA} \sin(\omega t - 90°) \tag{21}$$

Equation (21) shows that if the applied voltage is sinusoidal, the flux density B in the iron core must also be sinusoidal but lagging by 90°.

The condition of low resistance relative to inductive reactance, giving a sinusoidal flux from a sinusoidal supply voltage, is called **free magnetisation**.

Consider the application of a sinusoidal voltage to a coil wound on a core with a hysteresis loop as shown in Fig. 20(a). The horizontal axis of a hysteresis loop is magnetic field strength H, but since $H = Ni/l$ and N and l (the length of the flux path) are constant, the axis may be directly scaled as current i (i.e., $i = Hl/N$). Figure 20(b) shows sinusoidal voltage v and flux density B waveforms, B lagging v by 90°.

The current waveform is shown in Fig. 20(c) and is derived as follows. At time t_1, point a on the voltage curve corresponds to point b on the flux density curve and the point c on the hysteresis loop. The current at time t_1 is given by the distance dc. Plotting this current on a vertical time-scale gives the derived point e on the current curve. A similar procedure is adopted for times t_2, t_3 and so on over one cycle of the voltage. (Note that it is important to move around the hysteresis loop in the correct direction.) It is seen from the current curve that it is nonsinusoidal and that the positive and negative half-cycles are identical. This indicates that the waveform contains only odd harmonics (see section 3).

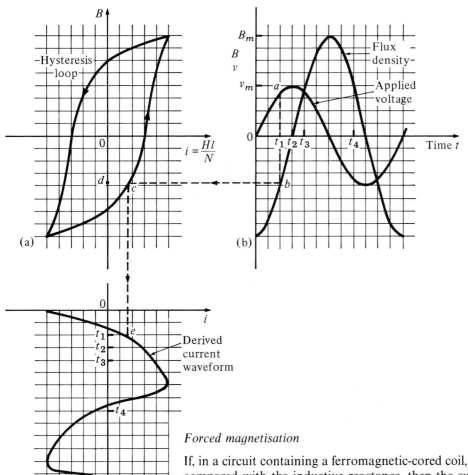

Figure 20

Forced magnetisation

If, in a circuit containing a ferromagnetic-cored coil, the resistance is high compared with the inductive reactance, then the current flowing from a sinusoidal supply will tend to be sinusoidal. This means that the flux density B of the core cannot be sinusoidal since it is related to the current by the hysteresis loop. This means, in turn, that the induced voltage due to the alternating flux (i.e., $v = NA(dB/dt)$) will not be sinusoidal. This condition is called forced magnetisation.

The shape of the induced voltage waveform under forced magnetisation is obtained as follows. The current waveform is shown on a vertical axis in Fig. 21(a). The hysteresis loop corresponding to the maximum value of circuit current is drawn as shown in Fig. 21(b). The flux density curve which is derived from the sinusoidal current waveform is shown in Fig. 21(c). Point a on the current wave at time t_1 corresponds to point b on the hysteresis loop and to point c on the flux density curve. By taking other points throughout the current cycle the flux density curve is derived as shown.

The relationship between the induced voltage v and the flux density B is given by $v = NA(dB/dt)$. Here dB/dt represents the rate of change of flux density with respect to time, i.e., the gradient of the B/t curve. At point d the gradient of the B/t curve is a maximum in the positive direction. Thus v will be maximum positive as shown by point d' in Fig. 21(d). At point e the gradient (i.e., dB/dt) is zero, thus v is zero, as shown by point e'. At point f the gradient is maximum in a negative direction, thus v is maximum negative as shown by point f'. If all such points are taken around the B/t curve, the curve representing induced voltage, shown in Fig. 21(d), is produced. The resulting voltage waveform is nonsinusoidal. The positive and negative half-cycles are identical in shape, indicating that the waveform contains a fundamental and a prominent third harmonic.

The amount of power delivered to a load can be controlled using a **thyristor**, which is a semiconductor device. Examples of applications of controlled rectification include lamp and heater controls and the control of

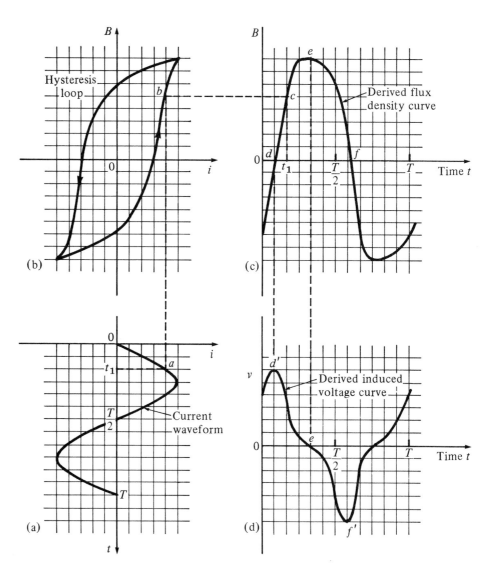

Figure 21

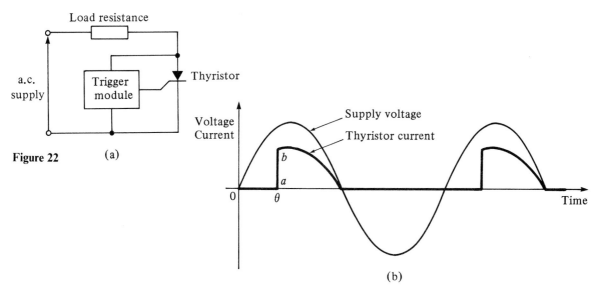

Figure 22   (a)

(b)

motor speeds. A basic circuit used for single-phase power control is shown in Fig. 22(a). The trigger module contains circuitry to produce the necessary gate current to turn the thyristor on. If the pulse is applied at time θ/ω, where θ is the firing or triggering angle, then the current flowing in the load resistor has a waveform as shown in Fig. 22(b). The sharp rise-time (shown as ab in Fig. 22(b)), however, gives rise to harmonics.

In **microelectronic systems** rectangular waveforms are common. Again, fast rise-times give rise to harmonics, especially at high frequency. These harmonics can be fed back to the mains if not filtered.

There are thus a large number of sources of harmonics.

9. Further problems

Harmonic synthesis

1. A complex current waveform i comprises a fundamental current of 50 A rms and frequency 100 Hz, together with a 24% third harmonic, both being in phase with each other at zero time. (a) Write down an expression to represent current i. (b) Sketch the complex waveform of current using harmonic synthesis over one cycle of the fundamental.

$$[\text{(a) } i = (70.71 \sin 628.3t + 16.97 \sin 1885t)\, \text{A.}]$$

2. A complex current wave i is described by

$$i = 3 \sin \omega t + \sin(3\omega t - \pi)\, \text{A}$$

Sketch one cycle of the fundamental wave. On the same axis, sketch the third harmonic. Use harmonic synthesis to sketch the complex current waveform.

3. A complex voltage waveform v is comprised of a 212.1 V rms fundamental voltage at a frequency of 50 Hz, a 30% second harmonic component lagging the fundamental voltage at zero time by $\pi/2$ rad, and a 10% fourth harmonic component leading the fundamental at zero time by $\pi/3$ rad. (a) Write down an expression to represent voltage v. (b) Sketch the complex voltage waveform using harmonic synthesis over one cycle of the fundamental waveform.

$$\left[\begin{array}{l} \text{(a) } v = 300 \sin 314.2t + 90 \sin(628.4t - (\pi/2)) \\ \qquad\qquad + 30 \sin(1256.8t + (\pi/3))\, \text{volts.} \end{array} \right]$$

4. A voltage waveform is represented by

$$v = 20 + 50 \sin \omega t + 20 \sin(2\omega t - \pi/2) \text{ volts.}$$

Draw the complex waveform over one cycle of the fundamental by using harmonic synthesis.

5. Write down an expression representing a current having a fundamental component of amplitude 16 A and frequency 1 kHz, together with its third and fifth harmonics being respectively one-eighth and one-tenth the amplitude of the fundamental, all components being in phase at zero time. Sketch the complex current waveform for one cycle of the fundamental using harmonic synthesis.

$$[i = (16 \sin 2\pi 10^3 t + 2 \sin 6\pi 10^3 t + 1.6 \sin \pi 10^4 t) \text{ A}]$$

6. For each of the waveforms shown in Fig. 23, state which harmonics are likely to be present.

⎡ (a) Fundamental and even harmonics, initially in phase with each other.
(b) Fundamental and odd harmonics only. (c) Fundamental and even harmonics, initially out of phase with each other (or all harmonics present, some being initially out of phase with each other. ⎤

7. A voltage waveform is described by

$$v = 200 \sin 377t + 80 \sin(1131t + (\pi/4)) + 20 \sin(1885t - (\pi/3)) \text{ volts}$$

Determine (a) the fundamental and harmonic frequencies of the waveform (b) the percentage third harmonic and (c) the percentage fifth harmonic. Sketch the voltage waveform using harmonic synthesis over one cycle of the fundamental.

⎡ (a) 60 Hz, 180 Hz, 300 Hz.
(b) 40%   (c) 10%. ⎤

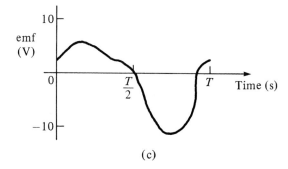

(a)

(b)

Figure 23

(c)

Rms values, mean values and form factor of complex waves

8. Determine the rms value of a complex current wave represented by

$$i = 3.5 \sin \omega t + 0.8 \sin\left(3\omega t - \frac{\pi}{3}\right) + 0.2 \sin\left(5\omega t + \frac{\pi}{2}\right) A$$

[2.54 A]

9. Derive an expression for the rms value of a complex voltage waveform represented by

$$v = V_0 + V_{1m}\sin(\omega t + \phi_1) + V_{3m}\sin(3\omega t + \phi_3) \text{ volts}$$

Calculate the rms value of a voltage waveform given by

$$v = 80 + 240 \sin \omega t + 50 \sin\left(2\omega t + \frac{\pi}{4}\right) + 20 \sin\left(4\omega t - \frac{\pi}{3}\right) \text{ volts}$$

[191.4 V]

10. A voltage waveform comprising a fundamental and a third harmonic has an rms value of 120 V. The maximum value of the fundamental is 158.4 V. Determine the rms value of the third harmonic.

[43.1 V]

11. A complex voltage waveform is given by

$$v = 150 \sin 314t + 40 \sin\left(942t - \frac{\pi}{2}\right) + 30 \sin(1570t + \pi) \text{ volts}$$

Determine for the voltage (a) the third harmonic frequency, (b) its rms value, (c) its mean value and (d) the form factor.

[(a) 150 Hz   (b) 111.8 V   (c) 91.7 V   (d) 1.22]

12. A complex voltage waveform has an rms value of 220 V and it contains 25% third harmonic and 15% fifth harmonic. (a) Determine the rms value of the fundamental and each harmonic. (b) Write down an expression to represent the complex current waveform if the frequency of the fundamental is 60 Hz.

$$\left[\begin{array}{l} \text{(a) } 211.2 \text{ V, } 52.8 \text{ V, } 31.7 \text{ V;} \\ \text{(b) } v = 298.7 \sin 377t + 74.7 \sin 1131t + 44.8 \sin(1885t) \text{ V} \end{array}\right]$$

13. Define the term "form factor" when applied to a symmetrical complex waveform. Calculate the form factor of an alternating voltage which is represented by

$$v = (50 \sin 314t + 15 \sin 942t + 6 \sin 1570t) \text{ volts}$$

[1.038]

Power associated with complex waves

14. Determine the average power in a 50 Ω resistor if the current i flowing through it is represented by

$$i = (140 \sin \omega t + 40 \sin 3\omega t + 20 \sin 5\omega t) \text{ mA}$$

[0.54 W]

15. A voltage waveform represented by

$$v = 100 \sin \omega t + 22 \sin\left(3\omega t - \frac{\pi}{6}\right) + 8 \sin\left(5\omega t - \frac{\pi}{4}\right) \text{ volts}$$

is applied to a circuit and the resulting current i is given by

$$i = 5 \sin\left(\omega t + \frac{\pi}{3}\right) + 1.91 \sin 3\omega t + 0.76 \sin (5\omega t - 0.452) \text{ amperes}$$

Calculate (a) the total active power supplied to the circuit, and (b) the overall power factor.

$$[(a)\ 146.1\,\text{W}; \quad (b)\ 0.526]$$

16. Determine the rms voltage, rms current and average power supplied to a network if the applied voltage is given by

$$v = 100 + 50\sin\left(400t - \frac{\pi}{3}\right) + 40\sin\left(1200t - \frac{\pi}{6}\right)\text{volts}$$

and the resulting current is given by

$$i = 0.928\sin(400t + 0.424) + 2.14\sin(1200t + 0.756)\,\text{amperes}$$

$$[109.8\,\text{V};\ 1.65\,\text{A};\ 14.60\,\text{W}]$$

17. A voltage $v = 40 + 20\sin 300t + 8\sin 900t + 3\sin 1500t$ volts is applied to the terminals of a circuit and the resulting current is given by

$$i = 4 + 1.715\sin(300t - 0.540) + 0.389\sin(900t - 1.064)$$
$$+ 0.095\sin(1500t - 1.249)\,\text{A}$$

Determine (a) the rms voltage, (b) the rms current and (c) the average power.

$$[(a)\ 42.85\,\text{V} \quad (b)\ 4.189\,\text{A} \quad (c)\ 175.5\,\text{W}]$$

Harmonics in single-phase circuits

18. A complex voltage waveform represented by

$$v = 240\sin \omega t + 60\sin\left(3\omega t - \frac{\pi}{4}\right) + 30\sin\left(5\omega t + \frac{\pi}{3}\right)\text{volts}$$

is applied across (a) a pure $50\,\Omega$ resistance, (b) a pure $4.974\,\mu\text{F}$ capacitor, and (c) a pure $15.92\,\text{mH}$ inductance. Determine for each case an expression for the current flowing if the fundamental frequency is 400 Hz.

$$\begin{bmatrix}
\text{(a)}\ i = 4.8\sin \omega t + 1.2\sin\left(3\omega t - \frac{\pi}{4}\right) + 0.6\sin\left(5\omega t + \frac{\pi}{3}\right)\text{A} \\[2mm]
\text{(b)}\ i = 3\sin\left(\omega t + \frac{\pi}{2}\right) + 2.25\sin\left(3\omega t + \frac{\pi}{4}\right) + 1.875\sin\left(5\omega t + \frac{5\pi}{6}\right)\text{A} \\[2mm]
\text{(c)}\ i = 6\sin\left(\omega t - \frac{\pi}{2}\right) + 0.5\sin\left(3\omega t - \frac{3\pi}{4}\right) + 0.15\sin\left(5\omega t - \frac{\pi}{6}\right)\text{A}
\end{bmatrix}$$

19. An rms current of 5.0 A containing a third harmonic component flows in a coil which has a resistance of $8.0\,\Omega$ and an inductance of 40 mH. The fundamental frequency is 50 Hz and the rms voltage across the coil is 75 V. Determine the magnitudes of the fundamental and harmonic components of the current.

$$[7.062\,\text{A};\ 0.349\,\text{A}]$$

20. A complex current given by

$$i = 5\sin\left(\omega t + \frac{\pi}{3}\right) + 8\sin\left(3\omega t + \frac{2\pi}{3}\right)\text{mA}$$

flows through a pure 2000 pF capacitor. If the frequency of the fundamental component is 4 kHz, determine (a) the rms value of current, (b) an expression for the p.d. across the capacitor and (c) the rms value of voltage.

$$\begin{bmatrix}\text{(a)}\ 6.671\,\text{mA} \quad \text{(b)}\ v = 99.47\sin(\omega t - (\pi/6)) + 53.05\sin(3\omega t + (\pi/6))\,\text{V} \\ \text{(c)}\,79.71\,\text{V}\end{bmatrix}$$

21. A complex voltage, v, given by

$$v = 200 \sin \omega t + 42 \sin 3\omega t + 25 \sin 5\omega t \text{ volts}$$

is applied to a circuit comprising a $6\,\Omega$ resistance in series with a coil of inductance $5\,\text{mH}$. Determine, for a fundamental frequency of $50\,\text{Hz}$, (a) an expression to represent the instantaneous value of the current flowing, (b) the rms voltage, (c) the rms current, (d) the power dissipated, and (e) the overall power factor.

$$\left[\begin{array}{l} \text{(a)} \; i = 32.25 \sin(314t - 0.256) + 5.50 \sin(942t - 0.666) \\ \qquad + 2.53 \sin(1570t - 0.919)\,\text{A} \\ \text{(b)} \; 145.6\,\text{V} \qquad \text{(c)} \; 23.20\,\text{A} \qquad \text{(d)} \; 3.23\,\text{kW} \qquad \text{(e)} \; 0.956 \end{array}\right]$$

22. A complex current i is represented by

$$i = 15 \sin 377t + 3 \sin 1131t + \sin 1885t \text{ amperes}$$

and the fundamental frequency is $60\,\text{Hz}$. The current flows through a pure inductance L and p.d. across the inductance is measured by a voltmeter as $100\,\text{V}$. Determine the value of inductance L.

$$[20.62\,\text{mH}]$$

23. (a) Explain why a complex current flowing through a capacitor has a larger harmonic content than the complex voltage across the capacitor.
    (b) An emf e is given by

$$e = 40 + 150 \sin \omega t + 30 \sin\left(2\omega t - \frac{\pi}{4}\right) + 10 \sin\left(4\omega t - \frac{\pi}{3}\right) \text{volts}$$

the fundamental frequency being $50\,\text{Hz}$. The emf is applied across a circuit comprising a $100\,\Omega$ resistance in series with a $15\,\mu\text{F}$ capacitor. Determine (i) the rms value of voltage, (ii) an expression for the current flowing and (iii) the rms value of current.

$$\left[\begin{array}{ll} \text{(i)} \; 115.5\,\text{V} & \text{(ii)} \; i = 0.639 \sin(\omega t + 1.130) + 0.206 \sin(2\omega t + 0.030) \\ \text{(iii)} \; 0.479\,\text{A} & \qquad + 0.088 \sin(4\omega t - 0.559)\,\text{A} \end{array}\right]$$

24. A circuit comprises a $100\,\Omega$ resistance in series with a $1\,\text{mH}$ inductance. The supply voltage is given by

$$v = 40 + 200 \sin \omega t + 50 \sin\left(3\omega t + \frac{\pi}{4}\right) + 15 \sin\left(5\omega t + \frac{\pi}{6}\right) \text{volts}$$

where $\omega = 10^5\,\text{rad/s}$. Determine for the circuit (a) an expression to represent the current flowing, (b) the rms value of current and (c) the power dissipated.

$$\left[\begin{array}{l} \text{(a)} \; i = 0.40 + 1.414 \sin(\omega t - (\pi/4)) + 0.158 \sin(3\omega t - 0.464) \\ \qquad + 0.029 \sin(5\omega t - 0.850)\,\text{A} \\ \text{(b)} \; 1.08\,\text{A} \qquad \text{(c)} \; 117\,\text{W} \end{array}\right]$$

25. The emf applied to a circuit comprising two components connected in series is given by

$$v = 50 + 150 \sin(2 \times 10^3 t) + 40 \sin(4 \times 10^3 t) + 20 \sin(8 \times 10^3 t) \text{ volts}$$

and the resulting current is given by

$$i = 1.011 \sin(2 \times 10^3 t + 1.001) + 0.394 \sin(4 \times 10^3 t + 0.663) \\ + 0.233 \sin(8 \times 10^3 t + 0.372)\,\text{A}$$

Determine for the circuit (a) the average power supplied, and (b) the value of the two circuit components.

$$[\text{(a)} \; 42.3\,\text{W} \qquad \text{(b)} \; R = 80\,\Omega, C = 4\,\mu\text{F}]$$

26. A coil having inductance L and resistance R is supplied with a complex voltage given by

$$v = 240 \sin \omega t + V_3 \sin\left(3\omega t + \frac{\pi}{3}\right) + V_5 \sin\left(5\omega t - \frac{\pi}{12}\right) \text{ volts}$$

The resulting current is given by

$$i = 4.064 \sin(\omega t - 0.561) + 0.750 \sin(3\omega t - 0.036) + 0.182 \sin(5\omega t - 1.525) \text{ A}$$

The fundamental frequency is 500 Hz. Determine (a) the impedance of the circuit at the fundamental frequency, and hence the values of R and L, (b) the values of V_3 and V_5, (c) the rms voltage, (d) the rms current, (e) the circuit power, and (f) the power factor.

$$\left[\begin{matrix} \text{(a) } 59.05\,\Omega, \; R = 50\,\Omega, \; L = 10\,\text{mH} \qquad \text{(b) } 80\,\text{V, } 30\,\text{V} \\ \text{(c) } 180.1\,\text{V} \qquad \text{(d) } 2.93\,\text{A} \qquad \text{(e) } 429\,\text{W} \qquad \text{(f) } 0.81 \end{matrix}\right]$$

27. An alternating supply voltage represented by

$$v = (240 \sin 300t - 40 \sin 1500t + 60 \sin 2100t) \text{ volts}$$

is applied to the terminals of a circuit containing a $40\,\Omega$ resistor, a $200\,\text{mH}$ inductor and a $25\,\mu\text{F}$ capacitor in series. (a) Derive the expression for the current waveform and (b) calculate the power dissipated by the circuit.

$$\left[\begin{matrix} \text{(a) } i = 2.873 \sin(300t + 1.071) - 0.145 \sin(1500t - 1.425) \\ + 0.149 \sin(2100t - 1.471) \text{ A} \\ \text{(b) } 166\,\text{W} \end{matrix}\right]$$

28. A voltage v represented by

$$v = 120 \sin 314t + 25 \sin\left(942t + \frac{\pi}{6}\right) \text{ volts}$$

is applied to the circuit shown in Fig. 24. Determine (a) an expression for current i, (b) the percentage harmonic content of the supply current, (c) the total power dissipated, (d) an expression for the p.d. shown as v_1 and (e) expressions for the currents shown as i_R and i_C.

$$\left[\begin{matrix} \text{(a) } i = 0.134 \sin(314t + 0.464) + 0.047 \sin(942t + 0.987) \text{ A} \\ \text{(b) } 35.07\% \qquad \text{(c) } 7.72\,\text{W} \\ \text{(d) } v_1 = 53.6 \sin(314t + 0.464) + 18.8 \sin(942t + 0.987) \text{ V} \\ \text{(e) } i_R = 0.095 \sin(314t - 0.321 + 0.013 \sin(942t - 0.262) \text{ A} \\ \quad\; i_C = 0.095 \sin(314t + 1.249) + 0.045 \sin(942t + 1.309) \text{ A} \end{matrix}\right]$$

29. A circuit consisting of a resistor in series with an inductance has a current flowing in it given by

$$i = 5 + 20 \sin 2 \times 10^3 t + 30 \sin\left(10^4 t - \frac{\pi}{4}\right) \text{ A}$$

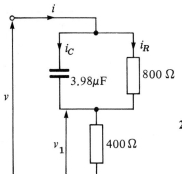

Figure 24

Calculate the rms value of current. Determine an expression to represent the terminal p.d. if the resistance is $10\,\Omega$ and the inductance is $5\,\text{mH}$. Also determine the rms terminal p.d. and the power dissipated in the circuit.

$$\left[\begin{matrix} 25.98\text{A}; \; v = 50 + 282.8 \sin(2 \times 10^3 t + (\pi/4)) + 1530 \sin(10^4 t + 0.588) \text{ V}; \\ 1101\,\text{V}; \; 13.25\,\text{kW} \end{matrix}\right]$$

Harmonic resonance

30. A voltage waveform having a fundamental of maximum value 250 V and a third harmonic of maximum value 20 V is applied to a series circuit comprising a $5\,\Omega$ resistor, a $400\,\text{mH}$ inductance and a $0.5\,\mu\text{F}$ capacitor. Determine (a) the frequency for resonance with the third harmonic and (b) the maximum values of the fundamental and third harmonic components of the current.

$$[\text{(a) } 118.6\,\text{Hz} \qquad \text{(b) } 0.105\,\text{A, } 4\,\text{A}]$$

31. A complex voltage waveform has a maximum value of 500 V at the fundamental frequency of 60 Hz and contains a 17th harmonic having an amplitude of 2% of the fundamental. The voltage is applied to a series circuit containing resistance $2\,\Omega$, inductance 732 mH and capacitance 33.26 nF. Determine (a) the maximum value of the 17th harmonic current, (b) the maximum value of the 17th harmonic p.d. across the capacitor, and (c) the amplitude of the fundamental current.

[(a) 5 A   (b) 23.46 kV   (c) 6.29 mA]

32 A complex voltage waveform v is given by the expression

$$v = 150\sin\omega t + 25\sin\left(3\omega t - \frac{\pi}{6}\right) + 10\sin\left(5\omega t + \frac{\pi}{3}\right)\text{volts}$$

where $\omega = 314\,\text{rad/s}$. The voltage is applied to a circuit consisting of a coil of resistance $10\,\Omega$ and inductance 50 mH in series with a variable capacitor.
(a) Calculate the value of the capacitance which will give resonance with the triple frequency component of the voltage. (b) Write down the corresponding equation for the current waveform. (c) Determine the rms value of current. (d) Find the power dissipated in the circuit.

$$\begin{bmatrix} \text{(a) } 22.54\,\mu\text{F} & \text{(b) } i = 1.191\sin(314t + 1.491) + 2.500\sin(942t - 0.524) \\ & \qquad\qquad\quad + 0.195\sin(1570t - 0.327)\,\text{A} \\ \text{(c) } 1.963\,\text{A} & \text{(d) } 38.53\,\text{W} \end{bmatrix}$$

33. (a) Explain what is meant by harmonic resonance in a.c. circuits.
  (b) A complex voltage of fundamental frequency 50 Hz is applied to a series circuit comprising resistance $20\,\Omega$, inductance $800\,\mu\text{H}$ and capacitance $74.94\,\mu\text{F}$. Resonance occurs at the nth harmonic. Determine the value of n.

[13]

34.   A complex voltage given by $v = 1200\sin\omega t + 300\sin 3\omega t + 100\sin 5\omega t$ volts is applied to a circuit containing a $25\,\Omega$ resistor, a $12\,\mu\text{F}$ capacitor and a 37 mH inductance connected in series. The fundamental frequency is 79.58 Hz. Determine (a) the rms value of the voltage, (b) an expression for the current waveform, (c) the rms value of current, (d) the amplitude of the third harmonic voltage across the capacitor, (e) the circuit power, and (f) the overall power factor.

$$\begin{bmatrix} \text{(a) } 877.5\,\text{V} & \text{(b) } i = 7.986\sin(\omega t + 1.404) + 12\sin 3\omega t \\ & \qquad\qquad\quad + 1.557\sin(5\omega t - 1.171)\,\text{A} \\ \text{(c) } 10.25\,\text{A} & \text{(d) } 666.7\,\text{V} \qquad \text{(e) } 2627\,\text{W} \qquad \text{(f) } 0.292 \end{bmatrix}$$

11 Magnetic and dielectric materials

1. Revision of terms and units used with magnetic circuits

(a) A **magnetic field** is the state of the space in the vicinity of a permanent magnet or an electric current throughout which the magnetic forces produced by the magnet or current are discernible.

(b) **Magnetic flux Φ** is the amount of magnetic field produced by a magnetic source. The unit of magnetic flux is the **weber, Wb**. If the flux linking one turn in a circuit changes by one weber in one second, a voltage of one volt will be induced in that turn.

(c) **Magnetic flux density B** is the amount of flux passing through a defined area that is perpendicular to the direction of the flux.

$$\text{Magnetic flux density} = \frac{\text{magnetic flux}}{\text{area}}$$

i.e., $B = \Phi/A$, where A is the area in square metres. The unit of magnetic flux density is the **tesla T**, where $1\,\text{T} = 1\,\text{Wb/m}^2$.

(d) **Magnetomotive force (mmf)** is the cause of the existence of a magnetic flux in a magnetic circuit.

$$\text{mmf, } F_m = NI \text{ amperes}$$

where N is the number of conductor (or turns) and I is the current in amperes. The unit of mmf is sometimes expressed as "ampere-turns". However since 'turns' have no dimension, the S.I. unit of mmf is the ampere.

(e) **Magnetic field strength** (or **magnetising force), $H = NI/l$ ampere per metre**, where l is the mean length of the flux path in metres. Thus **mmf $= NI = Hl$ amperes.**

(f) μ_0 is a constant called the **permeability of free space** (or the magnetic space constant). The value of μ_0 is $4\pi \times 10^{-7}\,\text{H/m}$. **For air, or any nonmagnetic medium**, the ratio $B/H = \mu_0$.

   (Although all nonmagnetic materials, including air, exhibit slight magnetic properties, these can effectively be neglected.)

(g) μ_r is the **relative permeability** and is defined as

$$\frac{\text{flux density in material}}{\text{flux density in a vacuum}}$$

μ_r varies with the type of magnetic material and, since it is a ratio of flux densities, it has no unit. From its definition, μ_r for a vacuum is 1. **For all media other than free space, $B/H = \mu_0\mu_r$.**

(h) Absolute permeability $\mu = \mu_0 \mu_r$.

(i) By plotting measured values of flux density B against magnetic field strength H a **magnetisation curve** (or B/H curve) is produced. For nonmagnetic materials this is a straight line having the approximate gradient of μ_0.

(j) From (g), $\mu_r = B/(\mu_0 H)$. Thus the relative permeability μ_r of a ferromagnetic material is proportional to the gradient of the B/H curve and varies with the magnetic field strength H.

(k) **Reluctance** S (or R_M) is the "magnetic resistance" of a magnetic circuit to the presence of magnetic flux.

$$\text{Reluctance } S = \frac{F_m}{\Phi} = \frac{NI}{\Phi} = \frac{Hl}{BA} = \frac{l}{(B/H)A} = \frac{l}{\mu_0 \mu_r A}$$

The unit of reluctance is $1/H$ (or H^{-1}) or A/Wb.

(l) **Permeance** is the magnetic flux per ampere of total magnetomotive force in the path of a magnetic field. It is the reciprocal of reluctance.

2. Magnetic properties of materials

The full theory of magnetism is one of the most complex of subjects. However the phenomenon may be satisfactorily explained by the use of a simple model. Bohr and Rutherford, who discovered atomic structure, suggested that electrons move around the nucleus confined to a plane, like planets around the sun. An even better model is to consider each electron as having a surface, which may be spherical or elliptical or something more complicated.

Magnetic effects in materials are due to the electrons contained in them, the electrons giving rise to magnetism in the following two ways:

(i) by revolving around the nucleus

(ii) by their angular momentum about their own axis, called spin.

In each of these cases the charge of the electron can be thought of as moving round in a closed loop and therefore acting as a current loop.

The main measurable quantity of an atomic model is the **magnetic moment**. When applied to a loop of wire carrying a current,

$$\text{magnetic moment} = \text{current} \times \text{area of the loop}$$

Electrons associated with atoms possess magnetic moment which gives rise to their magnetic properties.

Diamagnetism is a phenomenon exhibited by materials having a relative permeability less than unity. When electrons move more or less in a spherical orbit around the nucleus, the magnetic moment due to this orbital is zero, all the current due to moving electrons being considered as averaging to zero. If the net magnetic moment of the electron spins were also zero then there would be no tendency for the electron motion to line up in the presence of a magnetic field. However, as a field is being turned on, the flux through the electron orbitals increases. Thus, considering the orbital as a circuit, there will be, by Faraday's laws, an emf induced in it which will change the current in the circuit. The flux change will accelerate the electrons in its orbit, causing an induced magnetic moment. By Lenz's law the flux due to the induced magnetic moment will be such as to oppose the applied flux. As a result, the net flux through the material becomes less than in a vacuum. Since relative permeability is defined as

$$\frac{\text{flux density in material}}{\text{flux density in vacuum}}$$

with diamagnetic materials the relative permeability is less than one.

Paramagnetism is a phenomenon exhibited by materials where the relative permeability is greater than unity. Paramagnetism occurs in substances where atoms have a permanent magnetic moment. This may be caused by the orbitals not being spherical or by the spin of the electrons. Electron spins tend to pair up and cancel each other. However, there are many atoms with odd numbers of electrons, or in which pairing is incomplete. Such atoms have what is called a permanent dipole moment. When a field is applied to them they tend to line up with the field, like compass needles, and so strengthen the flux in that region. (Diamagnetic materials do not tend to line up with the field in this way.) When this effect is stronger than the diamagnetic effect, the overall effect is to make the relative permeability greater than one. Such materials are called paramagnetic.

Ferromagnetic materials

Ferromagnetisation is the phenomenon exhibited by materials having a relative permeability which is considerably greater than 1 and which varies with flux density. Iron, cobalt and nickel are the only elements that are ferromagnetic at ordinary working temperatures, but there are several alloys containing one or more of these metals as constituents, with widely varying ferromagnetic properties.

Consider the simple model of a single iron atom represented in Fig. 1. It consists of a small heavy central nucleus surrounded by a total of 26 electrons. Each electron has an orbital motion about the nucleus in a limited region, or shell, such shells being represented by circles K, L, M and N. The numbers in Fig. 1 represent the number of electrons in each shell.

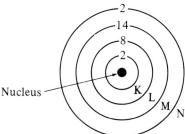

Figure 1 Single iron atom

The outer shell N contains two loosely held electrons, these electrons becoming the carriers of electric current, making iron electrically conductive. There are 14 electrons in the M shell and it is this group that is responsible for magnetism. An electron carries a negative charge and a charge in motion constitutes an electric current with which is associated a magnetic field. Magnetism would therefore result from the orbital motion of each electron in the atom. However, experimental evidence indicates that the resultant magnetic effect due to all the orbital motions in the metal solid is zero; thus the orbital currents may be disregarded.

In addition to the orbital motion, each electron spins on its own axis. A rotating charge is equivalent to a circular current and gives rise to a magnetic field. In any atom, all the axes about which the electrons spin are parallel, but rotation may be in either direction. In the single atom shown in Fig. 1, in each of the K, L and N shells equal numbers of electrons spin in the clockwise and anticlockwise directions respectively and therefore these shells are magnetically neutral. However, in shell M, nine of the electrons spin in one direction while five spin in the opposite direction. There is therefore a resultant effect due to four electrons.

The atom of cobalt has 15 electrons in the M shell, nine spinning in one direction and six in the other. Thus with cobalt there is a resultant effect due to 3 electrons. A nickel atom has a resultant effect due to 2 electrons. The atoms of the paramagnetic elements, such as manganese, chromium or aluminium, also have a resultant effect for the same reasons as that of iron, cobalt and nickel. However, in the diamagnetic materials there is an exact equality between the clockwise and anticlockwise spins.

The total magnetic field of the resultant effect due to the four electrons in the iron atom is large enough to influence other atoms. Thus the orientation of one atom tends to spread through the material, with atoms acting together in groups instead of behaving independently. These groups of atoms, called **domains** (which tend to remain permanently magnetised), act as units. Thus, when a field is applied to a piece of iron, these domains as a whole tend to line up and large flux densities can be produced. This means that the relative permeability of such materials is much greater than one. As the applied field is increased, more and more domains align and the induced flux increases.

The overall magnetic properties of iron alloys and materials containing iron, such as ferrite (ferrite is a mixture of iron oxide together with other oxides—lodestone is a ferrite), depend upon the structure and composition of the material. However, the presence of iron ensures marked magnetic properties of some kind in them. Ferromagnetic effects decrease with temperature, as do those due to paramagnetism. The loss of ferromagnetism with temperature is more sudden, however; the temperature at which it has all disappeared is called the **Curie temperature**. The ferromagnetic properties reappear on cooling, but any magnetism will have disappeared. Thus a permanent magnet will be demagnetised by heating above the Curie temperature (1040 K for iron) but can be remagnetised after cooling. Above the Curie temperature, ferromagnetics behave as paramagnetics.

3. Hysteresis and hysteresis loss

Hysteresis loop

Let a ferromagnetic material which is completely demagnetised, i.e., one in which $B = H = 0$ (either by heating the sample above its Curie temperature or by reversing the magnetising current a large number of times while at the

Figure 2

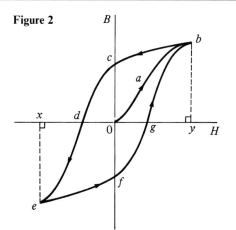

Figure 3

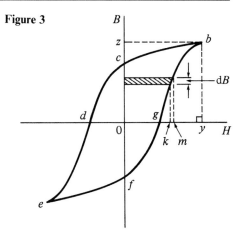

same time gradually reducing the current to zero) be subjected to increasing values of magnetic field strength H and the corresponding flux density B measured. The domains begin to align and the resulting relationship between B and H is shown by the curve $0ab$ in Fig. 2. At a particular value of H, shown as $0y$, most of the domains will be aligned and it becomes difficult to increase the flux density any further. The material is said to be saturated. Thus by is the **saturation flux density**.

If the value of H is now reduced it is found that the flux density follows curve bc, i.e., the domains will tend to stay aligned even when the field is removed. When H is reduced to zero, flux remains in the iron. This **remanent flux density** or **remanence** is shown as $0c$ in Fig. 2. When H is increased in the opposite direction, the domains begin to realign in the opposite direction and the flux density decreases until, at a value shown as $0d$, the flux density has been reduced to zero. The magnetic field strength $0d$ required to remove the residual magnetism, i.e., reduce B to zero, is called the **coercive force**.

Further increase of H in the reverse direction causes the flux density to increase in the reverse direction until saturation is reached, as shown by curve de. If the reversed magnetic field strength $0x$ is adjusted to the same value of $0y$ in the initial direction, then the final flux density xe is the same as yb. If H is varied backwards from $0x$ to $0y$, the flux density follows the curve $efgb$, similar to curve $bcde$.

It is seen from Fig. 2 that the flux density changes lag behind the changes in the magnetic field strength. This effect is called **hysteresis.** The closed figure $bcdefgb$ is called the **hysteresis loop** (or the B/H loop).

Hysteresis loss

A disturbance in the alignment of the domains of a ferromagnetic material causes energy to be expended in taking it through a cycle of magnetisation. This energy appears as heat in the specimen and is called the **hysteresis loss**.

Let the hysteresis loop shown in Fig. 3 be that obtained for an iron ring of mean circumference l and cross-sectional area a m^2 and let the number of turns on the magnetising coil be N.

Let the increase of flux density be dB when the magnetic field strength H is increased by a very small amount km (see Fig. 3) in time dt seconds, and let the current corresponding to $0k$ be i amperes. Thus since $H = NI/l$ then $0k = Ni/l$, from which

$$i = \frac{l(0k)}{N} \qquad (1)$$

The instantaneous emf e induced in the winding is given by

$$e = -N\frac{d\Phi}{dt} = -N\frac{d(Ba)}{dt} = -aN\frac{dB}{dt}$$

The applied voltage to neutralise this emf,

$$v = aN\frac{dB}{dt}$$

The instantaneous power supplied to a magnetic field,

$$p = vi = i\left(aN\frac{dB}{dt}\right) \text{watts}$$

Energy supplied to the magnetic field in time dt seconds

$$= \text{power} \times \text{time} = iaN\frac{dB}{dt}dt$$

$$= iaN\,dB \text{ joules} = \left(\frac{l(0k)}{N}\right)aN\,dB \text{ from equation (1)}$$

$$= (0k)\,dB(la)\,\text{joules} = (\text{area of shaded strip})(\text{volume of ring})$$

i.e., energy supplied in time dt seconds = (area of shaded strip) J/m³. Hence the energy supplied to the magnetic field when H is increased from zero to $0y = $ (area $fgbzf$) J/m³.

Similarly, the energy returned from the magnetic field when H is reduced from $0y$ to zero = (area $bzcb$) J/m³.

Hence

net energy absorbed by the magnetic field = (area $fgbcf$) J/m³

Thus the hysteresis loss for a complete cycle = **area of loop $efgbcde$ J/m³.**

If the hysteresis loop is plotted to a scale of 1 cm = α ampere/metre along the horizontal axis and 1 cm = β tesla along the vertical axis, and if A represents the area of the loop in square centimetres, then

$$\textbf{hysteresis loss/cycle} = A\,\alpha\beta \textbf{ joules per metre}^3 \qquad (2)$$

If hysteresis loops for a given ferromagnetic material are determined for different maximum values of H, they are found to lie within one another as shown in Fig. 4. The maximum sized hysteresis loop for a particular material is obtained at saturation. If, for example, the maximum flux density is reduced to half its value at saturation, the area of the resulting loop is considerably less than half the area of the loop at saturation. From

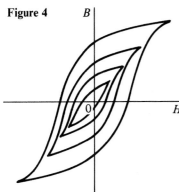

Figure 4

Figure 5

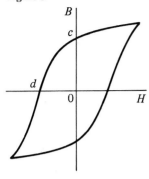

(a)

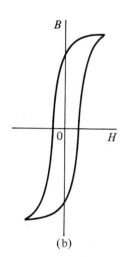

(b)

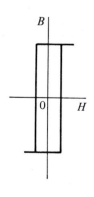

(c)

the areas of a number of such hysteresis loops, as shown in Fig. 4, the hysteresis loss per cycle was found by Steinmetz (the American electrical engineer) to be proportional to $(B_m)^n$, where n is called the **Steinmetz index** and can have a value between about 1.6 and 3.0, depending on the quality of the ferromagnetic material and the range of flux density over which the measurements are made.

From the above it is found that the hysteresis loss is proportional to the volume of the specimen and the number of cycles through which the magnetisation is taken. Thus

$$\text{hysteresis loss, } P_h = k_h v f (B_m)^n \text{ watts} \qquad (3)$$

where $v =$ volume in cubic metres, $f =$ frequency in hertz, and k_h is a constant for a given specimen and given range of B.

The magnitude of the hysteresis loss depends on the composition of the specimen and on the heat treatment and mechanical handling to which the specimen has been subjected.

Figure 5 shows typical hysteresis loops for (a) hard steel, which has a high remanence $0c$ and a large coercivity $0d$, (b) soft steel, which has a large remanence and small coercivity and (c) ferrite, this being a ceramic-like magnetic substance made from oxides of iron, nickel, cobalt, magnesium, aluminium and manganese. The hysteresis of ferrite is very small.

Worked problems on hysteresis loss

Problem 1. The area of a hysteresis loop obtained from a ferromagnetic specimen is 12.5 cm^2. The scales used were: horizontal axis $1 \text{ cm} = 500 \text{ A/m}$; vertical axis $1 \text{ cm} = 0.2 \text{ T}$. Determine (a) the hysteresis loss per m^3 per cycle, and (b) the hysteresis loss per m^3 at a frequency of 50 Hz.

(a) From equation (2),

$$\text{hysteresis loss per cycle} = A\alpha\beta = (12.5)(500)(0.2) = \mathbf{1250 \text{ J/m}^3}$$

(Note that, since $\alpha = 500 \text{ A/m}$ per centimetre and $\beta = 0.2 \text{ T}$ per centimetre, then 1 cm^2 of the loop represents

$$500 \frac{\text{A}}{\text{m}} \times 0.2 \text{ T} = 100 \frac{\text{A}}{\text{m}}\frac{\text{Wb}}{\text{m}^2} = 100 \frac{\text{AVs}}{\text{m}^3}$$

$$= 100 \frac{\text{Ws}}{\text{m}^3} = 100 \text{ J/m}^3$$

Hence 12.5 cm^2 represents $12.5 \times 100 = \mathbf{1250 \text{ J/m}^3}$.)

(b) At 50 Hz frequency,

$$\text{hysteresis loss} = \left(1250 \frac{\text{J}}{\text{m}^3}\right)\left(50 \frac{1}{\text{s}}\right) = \mathbf{62500 \text{ W/m}^3}$$

Problem 2. If, in problem 1, the maximum flux density is 1.5 T at a frequency of 50 Hz, determine the hysteresis loss per m^3 for a maximum flux density of 1.1 T and a frequency of 25 Hz. Assume the Steinmetz index to be 1.6.

From equation (3), hysteresis loss $P_h = k_h v f (B_m)^n$.

The loss at $f = 50\,\text{Hz}$ and $B_m = 1.5\,\text{T}$ is $62500\,\text{W/m}^3$. Thus $62500 = k_h(1)(50)(1.5)^{1.6}$, from which

$$\text{constant } k_h = \frac{62500}{(50)(1.5)^{1.6}} = 653.4$$

When $f = 25\,\text{Hz}$ and $B_m = 1.1\,\text{T}$,

$$\text{hysteresis loss, } P_h = k_h v f (B_m)^n$$
$$= (653.4)(1)(25)(1.1)^{1.6} = \mathbf{19026\,W/m^3}$$

Problem 3. A ferromagnetic ring has a uniform cross-sectional area of $2000\,\text{mm}^2$ and a mean circumference of $1000\,\text{mm}$. A hysteresis loop obtained for the specimen is plotted to scales of $10\,\text{mm} = 0.1\,\text{T}$ and $10\,\text{mm} = 400\,\text{A/m}$ and is found to have an area of $10^4\,\text{mm}^2$. Determine the hysteresis loss at a frequency of $80\,\text{Hz}$.

From equation (2),

hysteresis loss per cycle $= A\alpha\beta$

$$= (10^4 \times 10^{-6}\,\text{m}^2)\left(\frac{400\,\text{A/m}}{10 \times 10^{-3}\,\text{m}}\right)\left(\frac{0.1\,\text{T}}{10 \times 10^{-3}\,\text{m}}\right)$$

$$= 4000\,\text{J/m}^3$$

At a frequency of $80\,\text{Hz}$,

$$\text{hysteresis loss} = \left(4000\,\frac{\text{J}}{\text{m}^3}\right)\left(80\,\frac{1}{\text{s}}\right) = 320000\,\text{W/m}^3$$

$$\text{Volume of ring} = (\text{cross-sectional area})(\text{mean circumference})$$
$$= (2000 \times 10^{-6}\,\text{m}^2)(1000 \times 10^{-3}\,\text{m}) = 2 \times 10^{-3}\,\text{m}^3$$

Thus

$$\text{hysteresis loss, } P_h = \left(320000\,\frac{\text{W}}{\text{m}^3}\right)(2 \times 10^{-3}\,\text{m}^3) = \mathbf{640\,W}$$

Problem 4. The cross-sectional area of a transformer limb is $80\,\text{cm}^2$ and the volume of the transformer core is $5000\,\text{cm}^3$. The maximum value of the core flux is $10\,\text{mWb}$ at a frequency of $50\,\text{Hz}$. Taking the Steinmetz constant as 1.7, the hysteresis loss is found to be $100\,\text{W}$. Determine the value of the hysteresis loss when the maximum core flux is $8\,\text{mWb}$ and the frequency is $50\,\text{Hz}$.

When the maximum core flux is $10\,\text{mWb}$ and the cross-sectional area is $80\,\text{cm}^2$,

$$\text{maximum flux density, } B_{m1} = \frac{\Phi_1}{A} = \frac{10 \times 10^{-3}}{80 \times 10^{-4}} = 1.25\,\text{T}$$

From equation (3), hysteresis loss, $P_{h1} = k_h v f (B_{m1})^n$. Hence

$$100 = k_h(5000 \times 10^{-6})(50)(1.25)^{1.7}$$

from which

$$\text{constant } k_h = \frac{100}{(5000 \times 10^{-6})(50)(1.25)^{1.7}} = 273.7$$

When the maximum core flux is 8 m Wb,

$$B_{m2} = \frac{8 \times 10^{-3}}{80 \times 10^{-4}} = 1\,\text{T}$$

Hence

hysteresis loss, $P_{h_2} = k_h v f (B_{m2})^n$
$$= (273.7)(5000 \times 10^{-6})(50)(1)^{1.7} = \mathbf{68.4\,W}$$

Further problems on hysteresis loss may be found in section 11, problems 1 to 6, page 280.

4. Eddy current loss

If a coil is wound on a ferromagnetic core (such as in a transformer) and alternating current is passed through the coil, an alternating flux is set up in the core. The alternating flux induces an emf e in the coil given by $e = N(d\phi/dt)$. However, in addition to the desirable effect of inducing an emf in the coil, the alternating flux induces undesirable voltages in the iron core. These induced emfs set up circulating currents in the core, known as **eddy currents**. Since the core possesses resistance, the eddy currents heat the core: this represents wasted energy.

Eddy currents can be reduced by laminating the core, i.e., splitting it into thin sheets with very thin layers of insulating material inserted between each pair of the laminations (this may be achieved by simply varnishing one side of the lamination or by placing paper between each lamination). The insulation presents a high resistance and this reduces any induced circulating currents.

The eddy current loss may be determined as follows. Let Fig. 6 represent one strip of the core, having a thickness of t metres, and consider just a rectangular prism of the strip having dimensions $t\,\text{m} \times 1\,\text{m} \times 1\,\text{m}$ as shown. The area of the front face $ABCD$ is $(t \times 1)\,\text{m}^2$ and, since the flux enters this

Figure 6

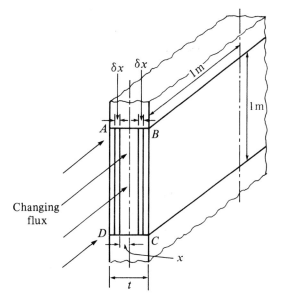

face at right angles, the eddy currents will flow along paths parallel to the long sides.

Consider two such current paths each of width δx and distance x m from the centre line of the front face. The area of the rectangle enclosed by the two paths, $A = (2x)(1) = 2x$ m^2. Hence the maximum flux entering the rectangle,

$$\Phi_m = (B_m)(A) = (B_m)(2x) \text{ weber} \tag{4}$$

Induced emf e is given by $e = N(d\phi/dt)$. Since the flux varies sinusoidally, $\phi = \Phi_m \sin \omega t$. Thus

$$\text{emf } e = N\frac{d}{dt}(\Phi_m \sin \omega t) = N\omega\Phi_m \cos \omega t$$

The maximum value of emf occurs when $\cos \omega t = 1$, i.e., $E_m = N\omega\Phi_m$. Rms value of emf,

$$E = \frac{E_m}{\sqrt{2}} = \frac{N\omega\Phi_m}{\sqrt{2}}$$

Now $\omega = 2\pi f$, hence

$$E = \left(\frac{2\pi}{\sqrt{2}}\right)f N\Phi_m = 4.44 f N\Phi_m$$

i.e.,

$$E = 4.44 f N(B_m)(A) \tag{5}$$

From equation (4), $\Phi_m = (B_m)(2x)$. Hence induced emf $E = 4.44 f N(B_m)(2x)$ and, since the number of turns $N = 1$,

$$E = 8.88 B_m f x \text{ volts} \tag{6}$$

Resistance R is given by $R = \rho l/a$, where ρ is the resistivity of the lamination material. Since the current set up is confined to the two loop sides (thus $l = 2$ m and $a = (\delta x \times l)$ m^2), the total resistance of the path is given by

$$R = \frac{\rho(2)}{\delta x} = \frac{2\rho}{\delta x} \tag{7}$$

The eddy current loss in the two strips is given by

$$\frac{E^2}{R} = \frac{8.88^2 B_m^2 f^2 x^2}{2\rho/\delta x} \qquad \text{from equations (6) and (7)}$$

$$= \frac{8.88^2 B_m^2 f^2 x^2 \delta x}{2\rho}$$

The total eddy current loss P_e in the rectangular prism considered is given by

$$P_e = \int_0^{t/2} \left(\frac{8.88^2 B_m^2 f^2}{2\rho}\right)x^2 \, dx = \left(\frac{8.88^2 B_m^2 f^2}{2\rho}\right)\left[\frac{x^3}{3}\right]_0^{t/2}$$

$$= \left(\frac{8.88^2 B_m^2 f^2}{2\rho}\right)\left(\frac{t^3}{24}\right) \text{watts}$$

The volume of the prism is $(t \times l \times l)\,\mathrm{m}^3$. Hence the eddy current loss per m^3 is given by

$$P_e = \frac{8.88^2\,B_m^2 f^2 t^2}{48\rho} \tag{8}$$

i.e.,

$$P_e = k_e (B_m)^2 f^2 t^2 \text{ watts per } \mathbf{m}^3 \tag{9}$$

where k_e is a constant.

From equation (9) it is seen that eddy current loss is proportional to the square of the thickness of the core strip. It is therefore desirable to make lamination strips as thin as possible. However, at high frequencies where it is not practicable to make very thin laminations, core losses may be reduced by using ferrite cores or dust cores. Ferrite is a ceramic material having magnetic properties similar to silicon steel, and dust cores consist of fine particles of carbonyl iron or permalloy (i.e. nickel and iron), each particle of which is insulated from its neighbour by a binding material. Such materials have a very high value of resistivity.

Worked problems on eddy current loss

Problem 1. The eddy current loss in a particular magnetic circuit is 10 W. If the frequency of operation is reduced from 50 Hz to 30 Hz with the flux density remaining unchanged, determine the new value of eddy current loss.

From equation (9), eddy current loss, $P_e = k_e (B_m)^2 f^2 t^2$ or $P_e = kf^2$, where $k = k_e (B_m)^2 t^2$, since B_m and t are constant.

When the eddy current loss is 10 W, frequency f is 50 Hz. Hence $10 = k(50)^2$, from which

$$\text{constant } k = \frac{10}{(50)^2}$$

When the frequency is 30 Hz, eddy current loss

$$P_e = k(30)^2 = \frac{10}{(50)^2}(30)^2 = \mathbf{3.6\ W}$$

Problem 2. The core of a transformer operating at 50 Hz has an eddy current loss of 100 W and the core laminations have a thickness of 0.50 mm. The core is redesigned so as to operate with the same eddy current loss but at a different voltage and at a frequency of 250 Hz. Assuming that at the new voltage the maximum flux density is one-third of its original value and the resistivity of the core remains unaltered, determine the necessary new thickness of the laminations.

From equation (9), $P_e = k_e (B_m)^2 f^2 t^2$.

Hence, at 50 Hz frequency, $100 = k_e(B_m)^2 (50)^2 (0.50 \times 10^{-3})^2$, from which

$$k_e = \frac{100}{(B_m)^2 (50)^2 (0.50 \times 10^{-3})^2}$$

At 250 Hz frequency,

$$100 = k_e \left(\frac{B_m}{3}\right)^2 (250)^2 (t)^2$$

i.e.,

$$100 = \left(\frac{100}{(B_m)^2(50)^2(0.50 \times 10^{-3})^2} \right)\left(\frac{B_m}{3} \right)^2 (250)^2(t)^2$$

$$= \frac{100(250)^2(t)^2}{(3)^2(50)^2(0.5 \times 10^{-3})^2}$$

from which

$$t^2 = \frac{(100)(3)^2(50 \times 10^{-3})^2(0.50)^2}{(100)(250)^2}$$

i.e.,

lamination thickness, $t =$ $\dfrac{(3)(50)(0.50 \times 10^{-3})}{(250)} = 0.3 \times 10^{-3}\,\text{m}$ or   **0.30 mm**

Problem 3. The core of an inductor has a hysteresis loss of 40 W and an eddy current loss of 20 W when operating at 50 Hz frequency. (a) Determine the values of the losses if the frequency is increased to 60 Hz. (b) What will be the total core loss if the frequency is 50 Hz and the laminations are made one-half of their original thickness? Assume that the flux density remains unchanged in each case.

(a) From equation (3), hysteresis loss, $P_h = k_h v f (B_m)^n = k_1 f$ (where $k_1 = k_h v (B_m)^n$), since the flux density and volume are constant. Thus when the hysteresis is 40 W and the frequency 50 Hz,

$$40 = k_1(50)$$

from which

$$k_1 = \tfrac{40}{50} = 0.8$$

If the frequency is increased to 60 Hz,

hysteresis loss, $P_h = k_1(60) = (0.8)(60) =$ 48 W

From equation (9),

$$\text{eddy current loss, } P_e = k_e(B_m)^2 f^2 t^2$$

$$= k_2 f^2 \text{(where } k_2 = k_e(B_m)^2 t^2),$$

since the flux density and lamination thickness are constant.
 When the eddy current loss is 20 W the frequency is 50 Hz. Thus

$$20 = k_2(50)^2$$

from which

$$k_2 = \tfrac{20}{(50)^2} = 0.008$$

If the frequency is increased to 60 Hz,

eddy current loss, $P_e = k_2(60)^2 = (0.008)(60)^2 =$ 28.8 W

(b) The hysteresis loss, $P_h = k_h v f (B_m)^n$, is independent of the thickness of the laminations. Thus, if the thickness of the laminations is halved, the hysteresis loss remains at **40 W.**
 Eddy current loss $P_e = k_e(B_m)^2 f^2 t^2$, i.e. $P_e = k_3 f^2 t^2$, where $k_3 = k_e(B_m)^2$. Thus

$$20 = k_3(50)^2 t^2$$

from which

$$k_3 = \frac{20}{(50)^2 t^2}$$

When the thickness is $t/2$,

$$P_e = k_3(50)^2 \left(\frac{t}{2}\right)^2$$

$$= \left(\frac{20}{(50)^2 t^2}\right)(50)^2 \left(\frac{t}{2}\right)^2 = \mathbf{5\,W}$$

Hence the total core loss when the thickness of the laminations is halved is given by hysteresis loss + eddy current loss $= 40 + 5 = \mathbf{45\,W}$.

Problem 4. When a transformer is connected to a 500 V, 50 Hz supply, the hysteresis and eddy current losses are 400 W and 150 W respectively. The applied voltage is increased to 1 kV and the frequency to 100 Hz. Assuming the Steinmetz index to be 1.6, determine the new total core loss.

From equation (9), the hysteresis loss, $P_h = k_h v f(B_m)^n$. From equation (5), emf, $E = 4.44\, f N(B_m)(A)$, from which $B_m \propto (E/f)$ since turns N and cross-sectional area A are constants. Hence

$$P_h = k_1 f \left(\frac{E}{f}\right)^{1.6} = k_1 f^{-0.6} E^{1.6}$$

At 500 V and 50 Hz, $400 = k_1(50)^{-0.6}(500)^{1.6}$, from which

$$k_1 = \frac{400}{(50)^{-0.6}(500)^{1.6}} = 0.20095$$

At 1000 V and 100 Hz,

$$\text{hysteresis loss, } P_h = k_1(100)^{-0.6}(1000)^{1.6}$$
$$= (0.20095)(100)^{-0.6}(1000)^{1.6} = 800\ \text{W}$$

From equation (9)

$$\text{eddy current loss, } P_e = k_e(B_m)^2 f^2 t^2 = k_2 \left(\frac{E}{f}\right)^2 f^2 = k_2 E^2$$

At 500 V, $150 = k_2(500)^2$, from which

$$k_2 = \frac{150}{(500)^2} = 6 \times 10^{-4}$$

At 1000 V,

$$\text{eddy current loss, } P_e = k_2(1000)^2 = (6 \times 10^{-4})(1000)^2 = 600\ \text{W}$$

Hence the new **total core loss** $= 800 + 600 = \mathbf{1400\,W}$.

Further problems on eddy current loss may be found in section 11, problems 7 to 12, page 281.

5. Separation of hysteresis and eddy current losses

From equation (3),

$$\text{hysteresis loss, } P_h = k_h v f(B_m)^n$$

From equation (9),

$$\text{eddy current loss, } P_e = k_e(B_m)^2 f^2 t^2$$

The total core loss P_c is given by $P_c = P_h + P_e$.

If, for a particular inductor or transformer, the core flux density is maintained constant, then $P_h = k_1 f$, where constant $k_1 = k_h v(B_m)^n$, and $P_e = k_2 f^2$, where constant $k_2 = k_e(B_m)^2 t^2$. Thus the total core loss $P_c = k_1 f + k_2 f^2$ or

$$\frac{P_c}{f} = k_1 + k_2 f$$

which is of the straight line form $y = mx + c$. Thus if P_c/f is plotted vertically against f horizontally, a straight line graph results having a gradient k_2 and a vertical-axis intercept k_1.

If the total core loss P_c is measured over a range of frequencies, then k_1 and k_2 may be determined from the graph of P_c/f against f. Hence the hysteresis loss $P_h (= k_1 f)$ and the eddy current loss $P_e (= k_2 f^2)$ at a given frequency may be determined.

The above method of separation of losses is an approximate one since the Steinmetz index n is not a constant value but tends to increase with increase of frequency. However, a reasonable indication of the relative magnitudes of the hysteresis and eddy current losses in an iron core may be determined.

Worked problems on the separation of hysteresis and eddy current losses

Problem 1. The total core loss of a ferromagnetic cored transformer winding is measured at different frequencies and the results obtained are:

Total core loss, P_c (watts)	45	105	190	305
Frequency, f (hertz)	30	50	70	90

Determine the separate values of the hysteresis and eddy current losses at frequencies of (a) 50 Hz and (b) 60 Hz.

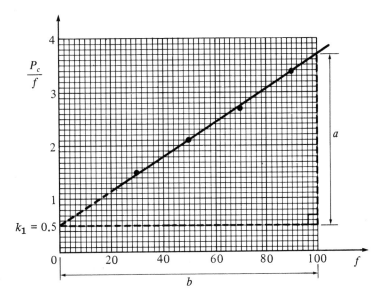

Figure 7

To obtain a straight line graph, values of P_c/f are plotted against f.

f (Hz)	30	50	70	90
P_c/f	1.5	2.1	2.7	3.4

A graph of P_c/f against f is shown in Fig. 7. The graph is a straight line of the form $P_c/f = k_1 + k_2 f$.

The vertical axis intercept at $f = 0$, $k_1 = 0.5$.

The gradient of the graph, $k_2 = \dfrac{a}{b} = \dfrac{3.7-0.5}{100} = 0.032$.

Since $P_c/f = k_1 + k_2 f$, then $P_c = k_1 f + k_2 f^2$, i.e.,

$$\text{total core losses} = \text{hysteresis loss} + \text{eddy current loss}.$$

(a) At a frequency of 50 Hz,

$$\text{hysteresis loss} = k_1 f = (0.5)(50) = \mathbf{25\ W}$$

$$\text{eddy current loss} = k_2 f^2 = (0.032)(50)^2 = \mathbf{80\ W}$$

(b) At a frequency of 60 Hz,

$$\text{hysteresis loss} = k_1 f = (0.5)(60) = \mathbf{30\ W}$$

$$\text{eddy current loss} = k_2 f^2 = (0.032)(60)^2 = \mathbf{115.2\ W}$$

Problem 2. The core of a synchrogenerator has total losses of 400 W at 50 Hz and 498 W at 60 Hz, the flux density being constant for the two tests. (a) Determine the hysteresis and eddy current losses at 50 Hz. (b) If the flux density is increased by 25% and the lamination thickness is increased by 40%, determine the hysteresis and eddy current losses at 50 Hz. Assume the Steinmetz index to be 1.7.

(a) From equation (3),

$$\text{hysteresis loss, } P_h = k_h v f (B_m)^n = k_1 f$$

(if volume v and the maximum flux density are constant)
    From equation (9),

$$\text{eddy current loss, } P_e = k_e (B_m)^2 f^2 t^2 = k_2 f^2$$

(if the maximum flux density and the lamination thickness are constant). Hence the total core loss $P_c = P_h + P_e$ i.e., $P_c = k_1 f + k_2 f^2$.

$$\text{At 50 Hz frequency,} \qquad 400 = k_1(50) + k_2(50)^2 \qquad (1)$$

$$\text{At 60 Hz frequency,} \qquad 498 = k_1(60) + k_2(60)^2 \qquad (2)$$

Solving equations (1) and (2) gives values of k_1 and k_2.

$$6 \times \text{equation (1) gives} \qquad 2400 = 300\,k_1 + 15000\,k_2 \qquad (3)$$

$$5 \times \text{equation (2) gives} \qquad 2490 = 300\,k_1 + 18000\,k_2 \qquad (4)$$

$$\text{Equation (4)} - \text{equation (3) gives} \qquad 90 = 3000\,k_2$$

from which

$$k_2 = \tfrac{90}{3000} = 0.03$$

Substituting $k_2 = 0.03$ in equation (1) gives $400 = 50\,k_1 + 75$, from which $k_1 = 6.5$. Thus, at 50 Hz frequency,

$$\text{hysteresis loss } P_h = k_1 f = (6.5)(50) = \mathbf{325\ W}$$

$$\text{eddy current loss } P_e = k_2 f^2 = (0.03)(50)^2 = \mathbf{75\ W}$$

(b) Hysteresis loss, $P_h = k_h v f (B_m)^n$. Since at 50 Hz the flux density is increased by 25%, the new hysteresis loss is $(1.25)^{1 \cdot 7}$ times greater than 325 W, i.e.,

$$P_h = (1.25)^{1 \cdot 7}(325) = \mathbf{474.9\ W}$$

Eddy current loss, $P_e = k_e (B_m)^2 f^2 t^2$. Since at 50 Hz the flux density is increased by 25%, and the lamination thickness is increased by 40%, the new eddy current loss is $(1.25)^2 (1.4)^2$ times greater than 75 W,

$$\text{i.e., } P_e = (1.25)^2 (1.4)^2 (75) = \mathbf{229.7\ W}$$

Further problems on the separation of hysteresis and eddy current losses may be found in section 11, problems 13 to 16, page 281.

6. Nonpermanent magnetic materials

General

Nonpermanent magnetic materials are those in which magnetism may be induced. With the magnetic circuits of electrical machines, transformers and heavy current apparatus a high value of flux density B is desirable so as to limit the cross-sectional area A ($\Phi = BA$) and therefore the weight and cost involved. At the same time the magnetic field strength H ($= NI/l$) should be as small as possible so as to limit the $I^2 R$ losses in the exciting coils. The relative permeability ($\mu_r = B/(\mu_0 H)$) and the saturation flux density should therefore be high. Also, when flux is continually varying, as in transformers, inductors and armature cores, low hysteresis and eddy current losses are essential.

Silicon–iron alloys

In the earliest electrical machines the magnetic circuit material used was iron with low content of carbon and other impurities. However, it was later discovered that the deliberate addition of silicon to the iron brought about a great improvement in magnetic properties. The laminations now used in electrical machines and in transformers at supply frequencies are made of silicon–steel in which the silicon in different grades of the material varies in amounts from about 0.5% to 4.5% by weight. The silicon added to iron increases the resistivity. This in turn increases the resistance ($R = (\rho l/A)$) and thus helps to reduce eddy current loss. The hysteresis loss is also reduced; however, the silicon reduces the saturation flux density.

A limit to the amount of silicon which may be added in practice is set by the mechanical properties of the material, since the addition of silicon causes a material to become brittle. Also the brittleness of a silicon–iron alloy depends on temperature. About 4.5% silicon is found to be the upper practical limit for silicon–iron sheets. Lohys is a typical example of a silicon–iron alloy and is used for the armatures of d.c. machines and for the rotors and stators of a.c. machines. Stalloy, which has a higher proportion of silicon and lower losses, is used for transformer cores.

Silicon steel sheets are often produced by a hot-rolling process. In these finished materials the constituent crystals are not arranged in any particular manner with respect, for example, to the direction of rolling or the plane of the sheet. If silicon steel is reduced in thickness by rolling in the cold state and the material is then annealed it is possible to obtain a finished sheet in which the crystals are nearly all approximately parallel to one another. The material has strongly directional magnetic properties, the rolling direction being the direction of highest permeability. This direction is also the direction of lowest hysteresis loss. This type of material is particularly suitable for use in transformers, since the axis of the core can be made to correspond with the rolling direction of the sheet and thus full use is made of the high permeability, low loss direction of the sheet.

With silicon–iron alloys a maximum magnetic flux density of about 2 T is possible. With cold-rolled silicon steel, used for large machine construction, a maximum flux density of 2.5 T is possible, whereas the maximum obtainable with the hot-rolling process is about 1.8 T. (In fact, with any material, only under the most abnormal of conditions will the value of flux density exceed 3 T.)

It should be noted that the term "iron-core" implies that the core is made of iron; it is, in fact, almost certainly made from steel, pure iron being extremely hard to come by. Equally, an iron alloy is generally a steel and so it is preferred to describe a core as being a steel rather than an iron core.

Nickel–iron alloys

Nickel and iron are both ferromagnetic elements and when they are alloyed together in different proportions a series of useful magnetic alloys is obtained. With about 25%–30% nickel content added to iron, the alloy tends to be very hard and almost nonmagnetic at room temperature. However, when the nickel content is increased to, say, 75%–80% (together with small amounts of molybdenum and copper), very high values of initial and maximum permeabilities and very low values of hysteresis loss are obtainable if the alloys are given suitable heat treatment. For example, Permalloy, having a content of 78% nickel, 3% molybdenum and the remainder iron, has an initial permeability of 20 000 and a maximum permeability of 100 000 compared with values of 250 and 5000 respectively for iron. The maximum flux density for Permalloy is about 0.8 T. Mumetal (76% nickel, 5% copper and 2% chromium) has similar characteristics.

Such materials are used for the cores of current and a.f. transformers, for magnetic amplifiers and also for magnetic screening. However, nickel–iron alloys are limited in that they have a low saturation value when compared with iron. Thus, in applications where it is necessary to work at a high flux density, nickel–iron alloys are inferior to both iron and silicon–iron. Also nickel–iron alloys tend to be more expensive than silicon–iron alloys.

Eddy current loss is proportional to the thickness of lamination squared, thus such losses can be reduced by using laminations as thin as possible. Nickel–iron alloy strip as thin as 0.004 mm, wound in a spiral, may be used.

Dust cores

In many circuits high permeability may be unnecessary or it may be more important to have a very high resistivity. Where this is so, metal powder or dust cores are widely used up to frequencies of 150 MHz. These consist of particles of nickel–iron–molybdenum for lower frequencies and iron for the higher frequencies. The particles, which are individually covered with an insulating film, are mixed with an insulating, resinous binder and pressed into shape.

Ferrites

Magnetite, or ferrous ferrite, is a compound of ferric oxide and ferrous oxide and possesses magnetic properties similar to those of iron. However, being a semiconductor, it has a very high resistivity. Manufactured ferrites are compounds of ferric oxide and an oxide of some other metal such as manganese, nickel or zinc. Ferrites are free from eddy current losses at all but the highest frequencies (i.e., $> 100\,\mathrm{MHz}$) but have a much lower initial permeability compared with nickel–iron alloys or silicon–iron alloys. Ferrites have typically a maximum flux density of about 0.4 T. Ferrite cores are used in audio-frequency transformers and inductors.

7. Permanent magnetic materials

A permanent magnet is one in which the material used exhibits magnetism without the need for excitation by a current-carrying coil. The silicon–iron and nickel–iron alloys discussed in section 6 are "soft" magnetic materials having high permeability and hence low hysteresis loss. The opposite characteristics are required in the "hard" materials used to make permanent magnets. In permanent magnets, high remanent flux density and high coercive force, after magnetisation to saturation, are desirable in order to resist demagnetisation. The hysteresis loop should embrace the maximum possible area. Possibly the best criterion of the merit of a permanent magnet is its maximum energy product $(BH)_m$, i.e., the maximum value of the product of the flux density B and the magnetic field strength H along the demagnetisation curve (shown as cd in Fig. 2). A rough criterion is the product of coercive force and remanent flux density, i.e. $(0d)/(0c)$ in Fig. 2. The earliest materials used for permanent magnets were tungsten and chromium steel, followed by a series of cobalt steels, to give both a high remanent flux density and a high value of $(BH)_m$.

Alni was the first of the aluminium–nickel–iron alloys to be discovered, and with the addition of cobalt, titanium and niobium, the Alnico series of magnets was developed, the properties of which vary according to composition. These materials are very hard and brittle. Many alloys with other compositions and trade names are commercially available.

A considerable advance was later made when it was found that directional magnetic properties could be induced in alloys of suitable composition if they were heated in a strong magnetic field. This discovery led to the powerful Alcomex and Hycomex series of magnets. By using special casting techniques to give a grain-orientated structure, even better properties are obtained if the field applied during heat treatment is parallel to the columnar crystals in the magnet. The values of coercivity, the remanent flux density and hence $(BH)_m$ are high for these alloys.

The most recent and most powerful permanent magnets discovered are made by powder metallurgy techniques and are based on an intermetallic compound of cobalt and samarium. These are very expensive and are only available in a limited range of small sizes.

8. Electric fields and capacitance

Figure 8 represents two parallel metal plates, *A* and *B*, charged to different potentials. If an electron that has a negative charge is placed between the plates, a force will act on the electron tending to push it away from the negative plate *B* towards the positive plate, *A*. Similarly, a positive charge would be acted on by a force tending to move it toward the negative plate. Any region such as that shown between the plates in Fig. 8, in which an electric charge experiences a force, is called an **electrostatic field**. The direction of the field is defined as that of the force acting on a positive charge placed in the field. In Fig. 8, for example, the direction of the force is from the positive plate to the negative plate.

Such a field may be represented in magnitude and direction by **lines of electric force** drawn between the charged surfaces. The closeness of the lines is an indication of the field strength. Whenever a p.d. is established between two points, an electric field will always exist. Figure 9(a) shows a typical field pattern for an isolated point charge, and Fig. 9(b) shows the field pattern for adjacent charges of opposite polarity. Electric lines of force (often called electric flux lines) are continuous and start and finish on point charges. Also, the lines cannot cross each other. When a charged body is placed close to an uncharged body, an induced charge of opposite sign appears on the surface of the uncharged body. This is because lines of force from the charged body terminate on its surface.

The concept of field lines or lines of force is used to illustrate the properties of an electric field. However, it should be remembered that they are only aids to the imagination.

Figure 10 shows two parallel conducting plates separated from each other by air. They are connected to opposite terminals of a battery of voltage *V* volts. There is therefore an electric field in the space between the plates. If the plates are close together, the electric lines of force will be straight and parallel and equally spaced, except near the edge where fringing will occur (see Fig. 8). Over the area in which there is negligible fringing,

$$\text{Electric field strength, } E = \frac{V}{d} \text{ volts/metre}$$

where *d* is the distance between the plates. Electric field strength is also called **potential gradient**.

Figure 8 Elecrostatic field

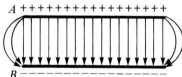

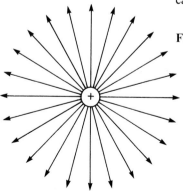

(a)

Figure 9 (a) Isolated point charge
  (b) Adjacent charges of opposite polarity

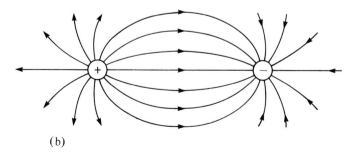

(b)

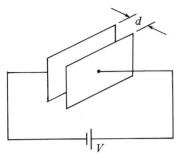

Figure 10

Static electric fields arise from electric charges, electric field lines beginning and ending on electric charges. Thus the presence of the field indicates the presence of equal positive and negative electric charges on the two plates of Fig. 10. Let the charge be $+Q$ coulombs on one plate and $-Q$ coulombs on the other. The property of this pair of plates which determines how much charge corresponds to a given p.d. between the plates is called their capacitance:

$$\text{capacitance } C = \frac{Q}{V} \text{ farads}$$

Unit flux is defined as emanating from a positive charge of 1 coulomb. Thus electric flux ψ is measured in coulombs, and for a charge of Q coulombs, the flux $\psi = Q$ coulombs.

Electric flux density D is the amount of flux passing through a defined area A that is perpendicular to the direction of the flux:

$$\text{electric flux density, } D = \frac{Q}{A} \text{ coulombs/metre}^3$$

Electric flux density is also called **charge density, σ.**

9. Dielectrics

Permittivity

At any point in an electric field, the electric field strength E maintains the electric flux and produces a particular value of electric flux density D at that point. For a field established in **vacuum** (or for practical purposes in air), the ratio D/E is a constant ε_0, i.e.,

$$\frac{D}{E} = \varepsilon_0$$

where ε_0 is called the **permittivity of free space** or the free space constant. The value of ε_0 is 8.85×10^{-12} F/m.

When an insulating medium, such as mica, paper, plastic or ceramic, is introduced into the region of an electric field the ratio of D/E is modified:

$$\frac{D}{E} = \varepsilon_0 \varepsilon_r$$

where ε_r, the **relative permittivity** of the insulating material, indicates its insulating power compared with that of vacuum:

$$\text{relative permittivity } \varepsilon_r = \frac{\text{flux density in material}}{\text{flux density in vacuum}}$$

ε_r has no unit. Typical values of ε_r include air, 1.00; polythene, 2.3; mica, 3–7; glass, 5–10; water, 80; ceramics, 6–1000.

The product $\varepsilon_0 \varepsilon_r$ is called the **absolute permittivity, ε,** i.e.,

$$\varepsilon = \varepsilon_0 \varepsilon_r$$

The insulating medium separating charged surfaces is called a **dielectric.**

Compared with conductors, dielectric materials have very high resistivities (and hence low conductance, since $\rho = (1/\sigma)$). They are therefore used to separate conductors at different potentials, such as capacitor plates or electric power lines.

Parallel-plate capacitor

For a parallel-plate capacitor experiments show that capacitance C is proportional to the area A of a plate, inversely proportional to the plate spacing d (i.e., the dielectric thickness) and depends on the nature of the dielectric and the number of plates n:

$$\textbf{capacitance } C = \frac{\varepsilon_0 \varepsilon_r A(n-1)}{d} \textbf{ farads}$$

Polarisation

When a dielectric is placed between charged plates, the capacitance of the system increases. The mechanism by which a dielectric increases capacitance is called **polarization**. In an electric field the electrons and atomic nuclei of the dielectric material experience forces in opposite directions. Since the electrons in an insulator cannot flow, each atom becomes a tiny dipole (i.e., an arrangement of two electric charges of opposite polarity) with positive and negative charges slightly separated, i.e., the material becomes polarised.

Within the material this produces no discernible effects. However, on the surfaces of the dielectric, layers of charge appear. Electrons are drawn toward the positive potential, producing a negative charge layer, and away from the side toward the negative potential, leaving positive surface charge behind. Therefore the dielectric becomes a volume of neutral insulator with surface charges of opposite polarity on opposite surfaces. The result of this is that the electric field inside the dielectric is less than the electric field causing the polarization, because these two charge layers give rise to a field which opposes the electric field causing it. Since electric field strength, $E = V/d$, the p.d. between the plates, $V = Ed$. Thus, if E decreases when the dielectric is inserted, then V falls too and this drop in p.d. occurs without change of charge on the plates. Thus, since capacitance $C = Q/V$, capacitance increases, this increase being by a factor equal to ε_r above that obtained with a vacuum dielectric.

There are two main ways in which polarization takes place:

(i) The electric field, as explained above, pulls the electrons and nucleii in opposite directions because they have opposite charges, which makes each atom into an electric dipole. The movement is only small and takes place very fast since the electrons are very light. Thus, if the applied electric field is varied periodically, the polarization, and hence the permittivity due to these induced dipoles, is independent of the frequency of the applied field.

(ii) Some atoms have a permanent electric dipole as a result of their structure and, when an electric field is applied, they turn and tend to align along the field. The response of the permanent dipoles is slower than the response of the induced dipoles and that part of the relative permittivity which arises from this type of polarization decreases with increase of frequency.

Most materials contain both induced and permanent dipoles, so the relative permittivity usually tends to decrease with increase of frequency.

Dielectric strength

The maximum amount of field strength that a dielectric can withstand is called the dielectric strength of the material. When an electric field is established across the faces of a material, molecular alignment and distortion of the electron orbits around the atoms of the dielectric occur. This produces a mechanical stress which in turn generates heat. The production of heat represents a dissipation of power, such a loss being present in all practical dielectrics, especially when used in high-frequency systems where the field polarity is continually and rapidly changing.

A dielectric whose conductivity is not zero between the plates of a capacitor provides a conducting path along which charges can flow and thus discharge the capacitor. The resistance R of the dielectric is given by $R = \rho l/a$, l being the thickness of the dielectric film (which may be as small as 0.001 mm) and a being the area of the capacitor plates. The resistance R of the dielectric may be represented as a leakage resistance across an ideal capacitor (see section 10 on dielectric loss). The required lower limit for acceptable resistance between the plates varies with the use to which the capacitor is put. High-quality capacitors have high shunt-resistance values. A measure of dielectric quality is the time taken for a capacitor to discharge a given amount through the resistance of the dielectric. This is related to the product CR.

$$\text{Capacitance, } C \propto \frac{\text{area}}{\text{thickness}} \quad \text{and} \quad \frac{1}{R} \propto \frac{\text{area}}{\text{thickness}}$$

thus CR is a characteristic of a given dielectric. In practice, circuit design is considerably simplified if the shunt conductance of a capacitor can be ignored (i.e. $R \to \infty$) and the capacitor therefore regarded as an open circuit for direct current.

Since capacitance C of a parallel plate capacitor is given by $C = \varepsilon_0 \varepsilon_r A/d$, reducing the thickness d of a dielectric film increases the capacitance, but decreases the resistance. It also reduces the voltage the capacitor can withstand without breakdown (since $V = Q/C$). Any material will eventually break down, usually destructively, when subjected to a sufficiently large electric field. A spark may occur at breakdown which produces a hole through the film. The metal film forming the metal plates may be welded together at the point of breakdown.

Breakdown depends on electric field strength E (where $E = V/d$), so thinner films will break down with smaller voltages across them. This is the main reason for limiting the voltage that may be applied to a capacitor. All practical capacitors have a safe working voltage stated on them, generally

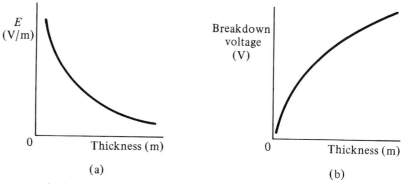

Figure 11

(a)   (b)

at a particular maximum temperature. Figure 11 shows the typical shapes of graphs expected for electric field strength E plotted against thickness and for breakdown voltage plotted against thickness. The shape of the curves depend on a number of factors, and these include:

(i) the type of dielectric material,
(ii) the shape and size of the conductors associated with it,
(iii) the atmospheric pressure,
(iv) the humidity/moisture content of the material,
(v) the operating temperature.

Dielectric strength is an important factor in the design of capacitors as well as transformers and high voltage insulators, and in motors and generators.

Dielectrics vary in their ability to withstand large fields. Some typical values of dielectric strength, together with resistivity and relative permittivity are shown in Table 1. The ceramics have very high relative permittivities and they tend to be 'ferroelectric', i.e., they do not lose their polarities when the electric field is removed. When ferroelectric effects are present, the charge on a capacitor is given by $Q = (CV) +$ (remanent polarization). These dielectrics often possess an appreciable negative temperature coefficient of permittivity. Despite this, a high permittivity is often very desirable and ceramic dielectrics are widely used.

Table 1. Dielectric properties of some common materials

Material	Resistivity, ρ (Ω m)	Relative permittivity, ε_r	Dielectric strength (V/m)
Air		1.0	3×10^6
Paper	10^{10}	3.7	1.6×10^7
Mica	5×10^{11}	5.4	$10^8 – 10^9$
Titaniumdioxide	10^{12}	100	6×10^6
Polythene	$> 10^{11}$	2.3	4×10^7
Polystyrene	$> 10^{13}$	2.5	2.5×10^7
Ceramic (type 1)	4×10^{11}	6–500	4.5×10^7
Ceramic (type 2)	$10^6 – 10^{13}$	500–1000	$2 \times 10^6 – 10^7$

Thermal effects

As the temperature of most dielectrics is increased, the insulation resistance falls rapidly. This causes the leakage current to increase, which generates further heat. Eventually a condition known as thermal avalanche or thermal runaway may develop, when the heat is generated faster than it can be dissipated to the surrounding environment. The dielectric will burn and thus fail.

Thermal effects may often seriously influence the choice and application of insulating materials. Some important factors to be considered include:

(i) the melting-point (for example, for waxes used in paper capacitors),
(ii) aging due to heat,
(iii) the maximum temperature that a material will withstand without serious deterioration of essential properties,
(iv) flash-point or ignitability,
(v) resistance to electric arcs,
(vi) the specific heat capacity of the material,
(vii) thermal resistivity,
(viii) the coefficient of expansion,
(ix) the freezing-point of the material.

Mechanical properties

Mechanical properties determine, to varying degree, the suitability of a solid material for use as an insulator: tensile strength, transverse strength, shearing strength and compressive strength are often specified. Most solid insulations have a degree of inelasticity and many are quite brittle, thus it is often necessary to consider features such as compressibility, deformation under bending stresses, impact strength and extensibility, tearing strength, machinability and the ability to fold without damage.

Types of practical capacitor

Practical types of capacitor are characterised by the material used for their dielectric.

An **air capacitor** usually consists of two sets of metal plates, one fixed and the other variable. As the moving plate is rotated on a spindle, the meshing, and thus the capacitance, is varied. Air capacitors are used in radio and electronic circuits where very low losses are required or where a variable capacitance is needed. The maximum value of this type of capacitor is usually between 500 pF and 1000 pF.

Paper capacitors usually consist of thin aluminium, lead or tin foil separated by a layer of waxed paper, the length corresponding to the capacitance required. The whole is usually wound in a roll so as to occupy as little space as possible. Fixed paper capacitors are made in various working voltages up to about 150 kV and are used where loss is not very important. The maximum value of this type of capacitor is between 500 pF and about 10 μF. The capacitance tends to change appreciably with temperature and the service life is shorter than with most other types of capacitor.

Mica capacitors are used where loss must be low. Mica is easily obtained in thin sheets and is a good insulator. However, mica is expensive and is not usually used in capacitors above about $0.2\,\mu F$. Such capacitors used to comprise a stack of lead or aluminium foil and mica tightly clamped together, the whole being impregnated with wax and then placed in a moulded case for protection. A modified form of mica capacitor now commonly used is the silvered mica type, where the mica is coated on both sides with a thin layer of silver which forms the plates. Since the silver is in intimate contact with the mica, the capacitance is more stable in value and less likely to change with age. These are made in fixed values up to about 1000 pF. They have a stable temperature coefficient, a high working voltage rating and a comparatively long service life and are used mainly in high-frequency circuits.

Plastic capacitors using such materials as polystyrene and Teflon for the dielectrics are increasingly employed. Their construction is similar to that of the paper capacitor, but using a plastic film instead of paper. Plastic capacitors operate satisfactorily under conditions of high temperature, provide a precise value of capacitance, an extremely long service life and high reliability.

Titanium oxide capacitors have a very high capacitance with a small physical size when used at low voltages.

Ceramic capacitors are made in various forms, each type of construction depending on the value of capacitance required. Ceramic material can have a very high permittivity which enables capacitors of high capacitance to be made which are of small physical size with a high working voltage rating. Such capacitors are available in the range 1 pF to $0.1\,\mu F$ and are used in high-frequency electronic circuits subject to a wide range of temperatures.

Electrolytic capacitors are similar in construction to paper capacitors, with aluminium foil used for the plates and with a relatively thick absorbent material, such as paper, impregnated with a chemical electrolyte (such as ammonium borate), separating the plates. Operation depends on the formation of a thin aluminium oxide layer on the positive plate by electrolytic action when a suitable direct potential is maintained between the plates. This oxide layer is very thin and forms the dielectric. This type of capacitor must always be used with a d.c. supply and must be connected with the correct polarity. Electrolytic capacitors are manufactured with working voltages from 6 V to about 600 V. Accuracy is generally not very high and there are large tolerances on marked capacitor values. However, electrolytic capacitors possess a much larger capacitance than other capacitors of similar dimensions due to the oxide film only being a few micrometres thick.

Liquid dielectrics

Liquid dielectrics used for insulation purposes are refined mineral oils, silicone fluids and synthetic oils such as chlorinated diphenyl. The principal uses of liquid dielectrics are as a filling and cooling medium for transformers, capacitors and rheostats, as an insulating and arc-quenching medium in switchgear such as circuit breakers, and as an impregnant of absorbent insulations—for example, wood, slate, paper and pressboard, used mainly in transformers, switchgear, capacitors and cables.

Gas insulation

Two gases used as insulation are nitrogen and sulphur hexafluoride. Nitrogen is used as an insulation medium in some sealed transformers and in power cables, and sulphur hexafluoride is finding increasing use in switchgear both as an insulant and an arc-extinguishing medium.

10. Dielectric loss and loss angle

In capacitors with solid dielectrics, losses can be attributed to two causes:

(i) dielectric hysteresis, a phenomonen by which energy is expended and heat produced as the result of the reversal of electrostatic stress in a dielectric subjected to alternating electric stress—this loss is analogous to hysteresis loss in magnetic materials;

(ii) leakage currents that may flow through the dielectric and along surface paths between the terminals.

The total dielectric loss may be represented as the loss in an additional resistance connected between the plates. This may be represented as either a small resistance in series with an ideal capacitor or as a large resistance in parallel with an ideal capacitor.

Series representation

The circuit and phasor diagrams for the series representation are shown in Fig. 12. The circuit phase angle is shown as angle ϕ. If resistance R_S were zero then current I would lead voltage V by 90°, this being the case of a perfect capacitor. The difference between 90° and the circuit phase angle ϕ is the angle shown as δ. This is known as the **loss angle** of the capacitor, i.e.,

$$\text{loss angle, } \delta = (90° - \phi)$$

For the equivalent series circuit,

$$\tan \delta = \frac{V_{R_S}}{V_{C_S}} = \frac{IR_S}{IX_{C_S}}$$

i.e.,

$$\tan \delta = \frac{R_S}{1/(\omega C_S)} = R_S \omega C_S \qquad (10)$$

Power factor of capacitor,

$$\cos \phi = \frac{V_{R_S}}{V} = \frac{IR_S}{IZ_S} = \frac{R_S}{Z_S} \approx \frac{R_S}{X_{C_S}}$$

since $X_{C_S} \approx Z_S$ when δ is small. Hence **power factor** $= \cos \phi \approx R_S \omega C_S$, i.e.,

$$\cos \phi \approx \tan \delta \qquad (11)$$

Figure 12 (a) Circuit diagram
(b) Phasor diagram

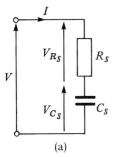

(a)

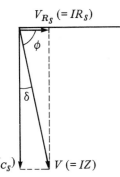

(b)

Parallel representation

The circuit and phasor diagrams for the parallel representation are shown in Fig. 13. From the phasor diagram,

$$\tan \delta = \frac{I_{R_p}}{I_{C_p}} = \frac{V/R_p}{V/X_{C_p}} = \frac{X_{C_p}}{R_p}$$

i.e.,

$$\tan \delta = \frac{1}{R_p \omega C_p} \tag{12}$$

Figure 13 (a) Circuit diagram
(b) Phasor diagram

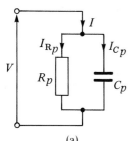

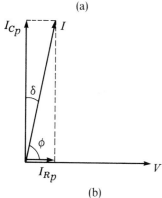

(a)

(b)

Power factor of capacitor,

$$\cos \phi = \frac{I_{R_p}}{I} = \frac{V/R_p}{V/Z_p} = \frac{Z_p}{R_p} \approx \frac{X_{C_p}}{R_p}$$

since $X_{C_p} \approx Z_p$ when δ is small. Hence

$$\text{power factor} = \cos \phi \approx \frac{1}{R_p \omega C_p}$$

i.e.,

$$\cos \phi \approx \tan \delta$$

(For equivalence between the series and the parallel circuit representations,

$$C_S \approx C_p = C \quad \text{and} \quad R_S \omega C_S \approx \frac{1}{R_p \omega C_p}$$

from which $R_S \approx 1/R_p \omega^2 C^2$.)

Power loss in the dielectric $= VI \cos \phi$. From the phasor diagram of Fig. 13,

$$\cos \delta = \frac{I_{C_p}}{I} = \frac{V/X_{C_p}}{I} = \frac{V \omega C}{I}$$

or

$$I = \frac{V \omega C}{\cos \delta}$$

Hence

$$\text{power loss} = VI \cos \phi = V \left(\frac{V \omega C}{\cos \delta} \right) \cos \phi$$

However, $\cos \phi = \sin \delta$ (complementary angles), thus

$$\text{power loss} = V \left(\frac{V \omega C}{\cos \delta} \right) \sin \delta = V^2 \omega C \tan \delta$$

(since $\sin \delta / \cos \delta = \tan \delta$).
Hence

$$\textbf{dielectric power loss} = V^2 \omega C \tan \delta \textbf{ watts} \tag{13}$$

Worked problems on dielectric loss and loss angle

Problem 1. The equivalent series circuit for a particular capacitor consists of a $1.5\,\Omega$ resistance in series with a $400\,\text{pF}$ capacitor. Determine, at a frequency of $8\,\text{MHz}$ (a) the loss angle of the capacitor, and (b) the power factor of the capacitor.

(a) From equation (10), for a series equivalent circuit,

$$\tan \delta = R_s \omega C_s$$
$$= (1.5)(2\pi 8 \times 10^6)(400 \times 10^{-12}) = 0.030159$$

Hence **loss angle, δ** = arctan $(0.030159) = \mathbf{1.727°}$ or **0.030 rad.**

(b) From equation (11), **power factor** = $\cos \phi \approx \tan \delta = \mathbf{0.030}$.

Problem 2. A capacitor has a loss angle of $0.025\,\text{rad}$, and when it is connected across a $5\,\text{kV}$, $50\,\text{Hz}$ supply, the power loss is $20\,\text{W}$. Determine the component values of the equivalent parallel circuit.

From equation (13),

$$\text{power loss} = V^2 \omega C \tan \delta$$

i.e.,

$$20 = (5000)^2 (2\pi 50) C \tan(0.025)$$

from which

$$\text{capacitance } C = \frac{20}{(5000)^2 (2\pi 50) \tan(0.025)} = \mathbf{0.102\,\mu F}$$

(Note tan(0.025) means "the tangent of $0.025\,\text{rad}$".)
From equation (12), for a parallel equivalent circuit,

$$\tan \delta = \frac{1}{R_p \omega C_p}$$

from which parallel resistance,

$$R_p = \frac{1}{\omega C_p \tan \delta}$$

$$= \frac{1}{(2\pi 50)(0.102 \times 10^{-6}) \tan 0.025}$$

i.e., $$R_p = \mathbf{1.248\,M\Omega.}$$

Problem 3. A $2000\,\text{pF}$ capacitor has an alternating voltage of $20\,\text{V}$ connected across it at a frequency of $10\,\text{kHz}$. If the power dissipated in the dielectric is $500\,\mu\text{W}$, determine (a) the loss angle, (b) the equivalent series loss resistance, and (c) the equivalent parallel loss resistance.

(a) From equation (13), power loss $= V^2 \omega C \tan \delta$, i.e.,

$$500 \times 10^{-6} = (20)^2 (2\pi 10 \times 10^3)(2000 \times 10^{-12}) \tan \delta$$

Hence

$$\tan \delta = \frac{500 \times 10^{-6}}{(20)^2 (2\pi 10 \times 10^3)(2000 \times 10^{-12})} = 9.947 \times 10^{-3}$$

from which **loss angle, $\delta = \mathbf{0.57°}$** or $\mathbf{9.95 \times 10^{-3}\,rad.}$

(b) From equation (10), for an equivalent series circuit, $\tan \delta = R_s \omega C_s$, from which equivalent series resistance,

$$R_s = \frac{\tan \delta}{\omega C_s} = \frac{9.947 \times 10^{-3}}{(2\pi\, 10 \times 10^3)(2000 \times 10^{-12})}$$

i.e.,

$$R_s = 79.16\,\Omega$$

(c) From equation (12), for an equivalent parallel circuit,

$$\tan \delta = \frac{1}{R_p \omega C_p}$$

from which equivalent parallel resistance,

$$R_p = \frac{1}{\tan \delta \omega C_p}$$

$$= \frac{1}{9.947 \times 10^{-3}(2\pi\, 10 \times 10^3)(2000 \times 10^{-12})}$$

i.e.,

$$R_p = 800\,\text{k}\Omega$$

Further problems on dielectric loss and loss angle may be found in the following section (11), problems 17 to 21, page 282.

11. Further problems

Hysteresis loss

1. The area of a hysteresis loop obtained from a specimen of steel is 2000 mm². The scales used are: horizontal axis 1 cm = 400 A/m; vertical axis 1 cm = 0.5 T. Determine (a) the hysteresis loss per m³ per cycle, (b) the hysteresis loss per m³ at a frequency of 60 Hz. (c) If the maximum flux density is 1.2 T at a frequency of 60 Hz, determine the hysteresis loss per m³ for a maximum flux density of 1 T and a frequency of 20 Hz, assuming the Steinmetz index to be 1.7.
[(a) 4 kJ/m³   (b) 240 kW/m³   (c) 58.68 kW/m³]

2. A steel ring has a uniform cross-sectional area of 1500 mm² and a mean circumference of 800 mm. A hysteresis loop obtained for the specimen is plotted to scales of 1 cm = 0.05 T and 1 cm = 100 A/m and it is found to have an area of 720 cm². Determine the hysteresis loss at a frequency of 50 Hz.
[216 W]

3. What is hysteresis? Explain how a hysteresis loop is produced for a ferromagnetic specimen and how its area is representative of the hysteresis loss.
   The area of a hysteresis loop plotted for a ferromagnetic material is 80 cm², the maximum flux density being 1.2 T. The scales of B and H are such that 1 cm = 0.15 T and 1 cm = 10 A/m. Determine the loss due to hysteresis if 1.25 kg of the material is subjected to an alternating magnetic field of maximum flux density 1.2 T at a frequency of 50 Hz. The density of the material is 7700 kg/m³.
[0.974 W]

4. The cross-sectional area of a transformer limb is 8000 mm² and the volume of the transformer core is 4×10^6 mm³. The maximum value of the core flux is 12 mWb and the frequency is 50 Hz. Assuming the Steinmetz constant is 1.6, the hysteresis loss is found to be 250 W. Determine the hysteresis loss when the maximum core flux is 9 mWb, the frequency remaining unchanged.
[157.8 W]

5. The hysteresis loss in a transformer is 200 W when the maximum flux density is 1 T and the frequency is 50 Hz. Determine the hysteresis loss if the maximum flux density is increased to 1.2 T and the frequency reduced to 32 Hz. Assume the hysteresis loss over this range to be proportional to $(B_m)^{1.6}$.

[171.4 W]

6. A hysteresis loop is plotted to scales of 1 cm = 0.004 T and 1 cm = 10 A/m and has an area of 200 cm². If the ferromagnetic circuit for the loop has a volume of 0.02 m³ and operates at 60 Hz frequency, determine the hysteresis loss for the ferromagnetic specimen.

[9.6 W]

Eddy current loss

7. In a magnetic circuit operating at 60 Hz, the eddy current loss is 25 W. If the frequency is reduced to 30 Hz with the flux density remaining unchanged, determine the new value of eddy current loss.

[6.25 W]

8. A transformer core operating at 50 Hz has an eddy current loss of 150 W and the core laminations are 0.4 mm thick. The core is redesigned so as to operate with the same eddy current loss but at a different voltage and at 200 Hz frequency. Assuming that at the new voltage the flux density is half of its original value and the resistivity of the core remains unchanged, determine the necessary new thickness of the laminations.

[0.20 mm]

9. An inductor core has an eddy current loss of 25 W and a hysteresis loss of 35 W when operating at 50 Hz frequency. Assuming that the flux density remains unchanged, determine (a) the value of the losses if the frequency is increased to 75 Hz, and (b) the total core loss if the frequency is 50 Hz and the laminations are $\frac{2}{5}$ of their original thickness.

[(a) $P_h = 52.5$ W, $P_e = 56.25$ W   (b) 39 W]

10. A transformer is connected to a 400 V, 50 Hz supply. The hysteresis loss is 250 W and the eddy current loss is 120 W. The supply voltage is increased to 1.2 kV and the frequency to 80 Hz. Determine the new total core loss if the Steinmetz index is assumed to be 1.6.

[2173.6 W]

11. The hysteresis and eddy current losses in a magnetic circuit are 5 W and 8 W respectively. If the frequency is reduced from 50 Hz to 30 Hz, the flux density remaining the same, determine the new values of hysteresis and eddy current loss.

[3 W; 2.88 W]

12. The core loss in a transformer connected to a 600 V, 50 Hz supply is 1.5 kW of which 60% is hysteresis loss and 40% eddy current loss. Determine the total core loss if the same winding is connected to a 750 V, 60 Hz supply. Assume the Steinmetz constant to be 1.6.

[2090 W]

Separation of hysteresis and eddy current loss

13. Tests to determine the total loss of the steel core of a coil at different frequencies gave the following results:

Frequency (Hz)	40	50	70	100
Total core loss (W)	40	57.5	101.5	190

Determine the hysteresis and eddy current losses at (a) 50 Hz and (b) 80 Hz.

[(a) 20 W; 37.5 W   (b) 32 W; 96 W]

14. Explain why, when steel is subjected to alternating magnetisation energy, losses occur due to both hysteresis and eddy currents.

    The core loss in a transformer core at normal flux density was measured at frequencies of 40 Hz and 50 Hz, the results being 40 W and 52.5 W respectively. Calculate, at a frequency of 50 Hz, (a) the hysteresis loss and (b) the eddy current loss.

[(a) 40 W   (b) 12.5 W]

15. Results of a test used to separate the hysteresis and eddy current losses in the core of a transformer winding gave the following results:

Total core loss (W)	48	96	160	240
Frequency (Hz)	40	60	80	100

If the flux density is held constant throughout the test, determine the values of the hysteresis and eddy current losses at 50 Hz.

[20 W; 50 W]

16. A transformer core has a total core loss of 275 W at 50 Hz and 600 W at 100 Hz, the flux density being constant for the two tests. (a) Determine the hysteresis and eddy current losses at 75 Hz. (b) If the flux density is increased by 40% and the lamination thickness is increased by 20% determine the hysteresis and eddy current losses at 75 Hz. Assume the Steinmetz index to be 1.6.

[(a) 375 W; 56.25 W   (b) 642.4 W; 158.8 W]

Dielectric loss and loss angle

17. The equivalent series circuit for a capacitor consists of a $3\,\Omega$ resistance in series with a 250 pF capacitor. Determine the loss angle of the capacitor at a frequency of 5 MHz, giving the answer in degrees and in radians. Find also the power factor of the capacitor.

[$1.35°$ or 0.024 rad; 0.024]

18. A capacitor has a loss angle of 0.008 rad and when it is connected across a 4 kV, 60 Hz supply the power loss is 15 W. Determine the component values of (a) the equivalent parallel circuit, and (b) the equivalent series circuit.

[(a) $0.311\,\mu$F, $1.066\,M\Omega$ (b) $0.311\,\mu$F, $68.23\,\Omega$]

19. A coaxial cable has a capacitance of $4\,\mu$F and a dielectric power loss of 12 kW when operated at 50 kV and frequency 50 Hz. Calculate (a) the value of the loss angle, and (b) the equivalent parallel resistance of the cable.

[(a) $0.219°$ or 3.82×10^{-3} rad   (b) $208.2\,k\Omega$]

20. What are the main reasons for power loss in capacitors with solid dielectrics? Explain the term "loss angle".

    A voltage of 10 V and frequency 20 kHz is connected across a 1 nF capacitor. If the power dissipated in the dielectric is 0.2 mW, determine (a) the loss angle, (b) the equivalent series loss resistance, and (c) the equivalent parallel loss resistance.

[(a) $0.912°$ or 0.0159 rad   (b) $126.5\,\Omega$ (c) $0.5\,M\Omega$]

21. The equivalent series circuit for a capacitor consists of a $0.5\,\Omega$ resistor in series with a capacitor of reactance $2\,k\Omega$. Determine for the capacitor (a) the loss angle, (b) the power factor, and (c) the equivalent parallel resistance.

[(a) $0.014°$ or 2.5×10^{-4} rad   (b) 2.5×10^{-4} (c) $8\,M\Omega$]

12 Field theory

1. Field plotting by curvilinear squares

Electric fields, magnetic fields and conduction fields (i.e., a region in which an electric current flows) are analogous, i.e., they all exhibit similar characteristics. Thus they may all be analysed by similar processes. In the following the electric field is analysed.

Figure 1 shows two parallel plates A and B. Let the potential on plate A be $+V$ volts and that on plate B be $-V$ volts. The force acting on a point charge of 1 coulomb placed between the plates is the electric field strength E. It is measured in the direction of the field and its magnitude depends on the p.d. between the plates and the distance between the plates. In Fig. 1, moving along a line of force from plate B to plate A means moving from $-V$ to $+V$ volts. The p.d. between the plates is therefore $2V$ volts and this potential changes linearly when moving from one plate to the other. Hence a potential gradient is followed which changes by equal amounts for each unit of distance moved.

Figure 1 Lines of force intersecting equipotential lines in an electric field

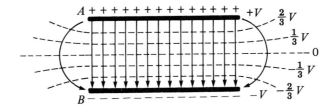

Lines may drawn connecting together all points within the field having equal potentials. These lines are called **equipotential lines** and these have been drawn in Fig. 1 for potentials of $\frac{2}{3}V, \frac{1}{3}V, 0, -\frac{1}{3}V$ and $-\frac{2}{3}V$. The zero equipotential line represents earth potential and the potentials on plates A and B are respectively above and below earth potential. Equipotential lines form part of an equipotential surface. Such surfaces are parallel to the plates shown in Fig. 1 and the plates themselves are equipotential surfaces. There can be no current flow between any given points on such a surface since all points on an equipotential surface have the same potential. Thus a line of force (or flux) must intersect an equipotential surface at right angles. A line of force in an electrostatic field is often termed a **streamline.**

An electric field distribution for a concentric cylinder capacitor is shown in Fig. 2. An electric field is set up in the insulating medium between two good conductors. Any volt drop within the conductors can usually be neglected compared with the p.d.'s across the insulation since the conductors have a high conductivity. All points in the conductors are thus at the

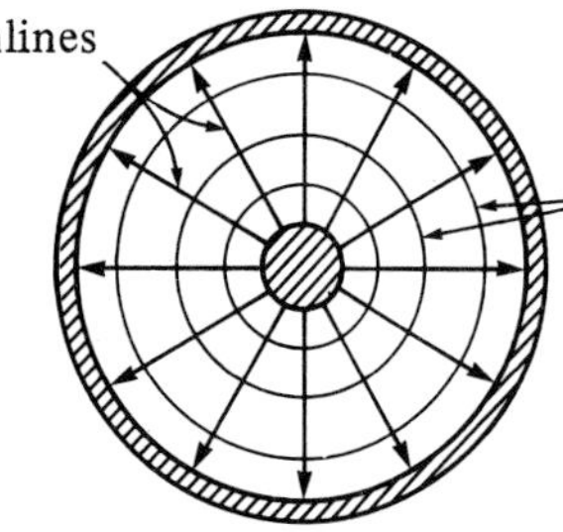

Figure 2 Electric field distribution for a concentric cylinder capacitor

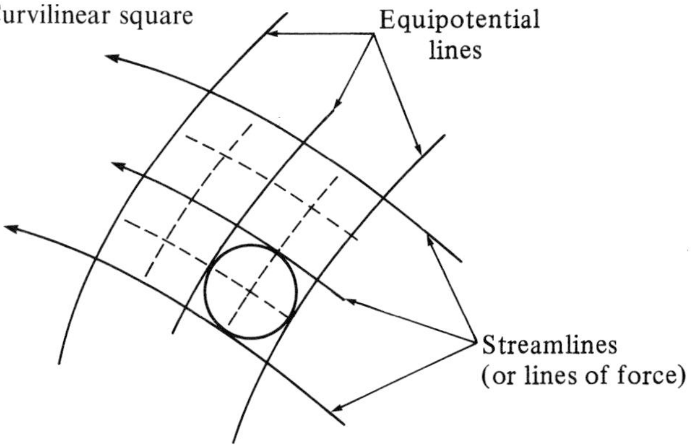

Streamlines

Equipotential lines

Figure 3 Curvilinear square

Equipotential lines

Streamlines (or lines of force)

same potential so that the conductors form the boundary equipotentials for the electrostatic field. Streamlines (or lines of force) which must cut all equipotentials at right angles leave one boundary at right angles, pass across the field, and enter the other boundary at right angles.

In a magnetic field, a streamline is a line so drawn that its direction is everywhere parallel to the direction of the magnetic flux. An equipotential surface in a magnetic field is the surface over which a magnetic pole may be moved without the expenditure of work or energy.

In a conduction field, a streamline is a line drawn with a direction which is everywhere parallel to the direction of the current flow.

A method of solving certain field problems by a form of graphical estimation is available which may only be applied, however, to plane linear fields; examples include the field existing between parallel plates or between two long parallel conductors. In general, the plane of a field may be divided into a number of squares formed between the line of force (i.e. streamline) and the equipotential. Figure 3 shows a typical pattern. In most cases true squares will not exist, since the streamlines and equipotentials are curved. However, since the streamlines and the equipotentials intersect at right angles, square-like figures are formed, and these are usually called "**curvilinear squares**". The square-like figure shown in Fig. 3 is a curvilinear square since, on successive subdivision by equal numbers of intermediate streamlines and equipotentials, the smaller figures are seen to approach a true square form.

When subdividing to give a field in detail, and in some cases for the initial equipotentials, "**Moores circle**" **technique** can be useful in that it tends to eliminate the trial and error process. If, say, two flux lines and an equipotential are given and it is required to draw a neighbouring equipotential, a circle tangential to the three given lines is constructed. The new equipotential is then approximately tangential to the circle, as shown in Fig. 3.

Consider the electric field established between two parallel metal plates, as shown in Fig. 4. The streamlines and the equipotential lines are shown sketched and are seen to form curvilinear squares. Consider a true square $abcd$ lying between equipotentials AB and CD. Let this square be the end of x metres depth of the field forming a flux tube between adjacent

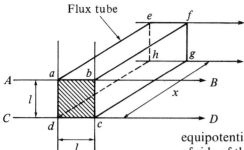

Figure 4

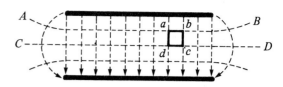

Figure 5

equipotential surfaces $abfe$ and $cdhg$ as shown in Fig. 5. Let l be the length of side of the squares. Then the capacitance C_t of the flux tube is given by

$$C_t = \frac{\varepsilon_0 \varepsilon_r (\text{area of plate})}{\text{plate separation}}$$

i.e.,

$$C_t = \frac{\varepsilon_0 \varepsilon_r (lx)}{l} = \varepsilon_0 \varepsilon_r x \qquad (1)$$

Thus the capacitance of the flux tube whose end is a true square is independent of the size of the square.

Let the distance between the plates of a capacitor be divided into an exact number of parts, say n (in Fig. 4, $n = 4$). Using the same scale, the breadth of the plate is divided into a number of parts (which is not always an integer value), say m (in Fig. 4, $m = 10$, neglecting fringing). Thus between equipotentials AB and CD in Fig. 4 there are m squares in parallel and so there are m capacitors in parallel. For m capacitors connected in parallel, the equivalent capacitance C_T is given by $C_T = C_1 + C_2 + C_3 + \cdots + C_m$. If the capacitors have the same value, i.e., $C_1 = C_2 = C_3 = \cdots = C_m = C_t$, then

$$C_T = mC_t \qquad (2)$$

Similarly, there are n squares in series in Fig. 4 and thus n capacitors in series. For n capacitors connected in series, the equivalent capacitance C_T is given by

$$\frac{1}{C_T} = \frac{1}{C_1} + \frac{1}{C_2} + \cdots + \frac{1}{C_n}$$

If $C_1 = C_2 = \cdots = C_n = C_t$ then $1/C_T = n/C_t$, from which

$$C_T = \frac{C_t}{n} \qquad (3)$$

Thus if m is the number of parallel squares measured along each equipotential and n is the number of series squares measured along each streamline (or line of force), then the total capacitance C of the field is given, from equations (1)–(3), by

$$C = \varepsilon_0 \varepsilon_r \times \frac{m}{n} \text{ farads} \qquad (4)$$

For example, let a parallel-plate capacitor have plates $8\,\text{mm} \times 5\,\text{mm}$ and spaced $4\,\text{mm}$ apart (see Fig. 6). Let the dielectric have a relative

Figure 6

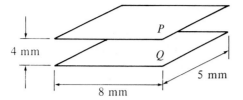

Figure 7

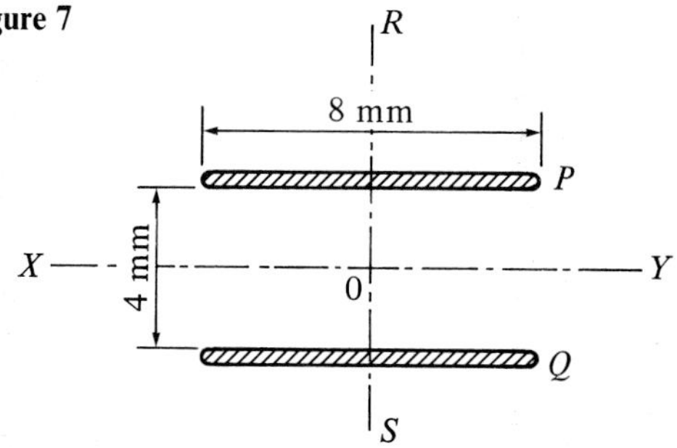

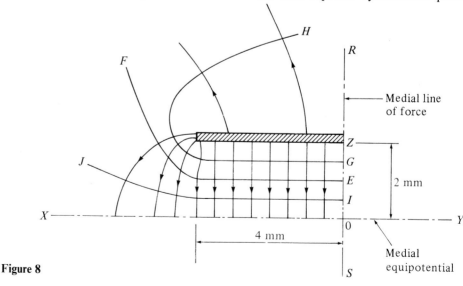

permittivity 3.5. If the distance between the plates is divided into, say, four equipotential lines, then each is 1 mm apart. Hence $n = 4$.

Using the same scale, the number of lines of force from plate P to plate Q must be 8, i.e., $m = 8$. This is, of course, neglecting any fringing. From equation (4), capacitance $C = \varepsilon_0\varepsilon_r x(m/n)$, where $x = 5$ mm or 0.005 m in this case. Hence

$$C = (8.85 \times 10^{-12})(3.5)(0.005)\tfrac{8}{4} = \textbf{0.31 pF}$$

(Using the normal equation for capacitance of a parallel-plate capacitor,

$$C = \frac{\varepsilon_0\varepsilon_r A}{d} = \frac{(8.85 \times 10^{-12})(3.5)(0.008 \times 0.005)}{0.004} = \textbf{0.31 pF}$$

The capacitance found by each method gives the same value; this is expected since the field is uniform between the plates, giving a field plot of true squares.)

The effect of fringing may be considered by estimating the capacitance by field plotting. This is described below.

In the side view of the plates shown in Fig. 7, RS is the medial line of force or medial streamline, by symmetry. Also XY is the medial equipotential. The field may thus be divided into four separate symmetrical parts.

Figure 8

Considering just the top left part of the field, the field plot is estimated as follows, with reference to Fig. 8:

(i) Estimate the position of the equipotential *EF* which has the mean potential between that of the plate and that of the medial equipotential *X*0. *F* is not taken too far since it is difficult to estimate. Point *E* will lie slightly closer to point *Z* than point 0.

(ii) Estimate the positions of intermediate equipotentials *GH* and *IJ*.

(iii) All the equipotential lines plotted are $\frac{2}{4}$, i.e., 0.5 mm apart. Thus a series of streamlines, cutting the equipotential at right angles, are drawn, the streamlines being spaced 0.5 mm apart, with the object of forming, as far as possible, curvilinear squares.

It may be necessary to erase the equipotentials and redraw them to fit the lines of force. The field between the plates is almost uniform, giving a field plot of true squares in this region. At the corner of the plates the squares are smaller, this indicating a great stress in this region. On the top of the plate the squares become very large, indicating that the main field exists between the plates.

From equation (4),

$$\text{total capacitance, } C = \varepsilon_0 \varepsilon_r x \frac{m}{n} \text{ farads}$$

The number of parallel squares measured along each equipotential is about 13 in this case and the number of series squares measured along each line of force is 4. Thus, for the plates shown in Fig. 7, $m = 2 \times 13 = 26$ and $n = 2 \times 4 = 8$. Since x is 5 mm,

$$\text{total capacitance} = \varepsilon_0 \varepsilon_r x \frac{m}{n} = (8.85 \times 10^{-12})(3.5)(0.005)\frac{26}{8}$$

$$= \mathbf{0.50\ pF}$$

Worked problems on field plotting by curvilinear squares

Problem 1. A field plot between two metal plates is shown in Fig. 9. The relative permeability of the dielectric is 2.8. Determine the capacitance per metre length of the system.

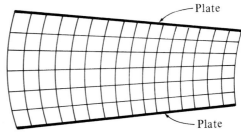

Figure 9

From equation (4), capacitance $C = \varepsilon_0 \varepsilon_r x (m/n)$. From Fig. 9, $m = 16$, i.e., the number of parallel squares measured along each equipotential, and $n = 6$, i.e., the number of series squares measured along each line of force. Hence capacitance for a 1 m length,

$$C = (8.85 \times 10^{-12})(2.8)(1)\frac{16}{6} = \mathbf{66.08\ pF}$$

Problem 2. A field plot for a cross-section of a concentric cable is shown in Fig. 10. If the relative permeability of the dielectric is 3.4, determine the capacitance of a 100 m length of the cable.

From equation (4), capacitance $C = \varepsilon_0 \varepsilon_r x (m/n)$. In this case, $m \approx 12.5$ and $n = 4$. Also $x = 100$ m. Thus

$$\text{capacitance } C = (8.85 \times 10^{-12})(3.4)(100)\frac{12.5}{4} = \textbf{9400 pF} \text{ or } \textbf{9.40 nF}$$

Further problems on field plotting by curvilinear squares may be found in section 9, problems 1 to 3, page 314.

Figure 10

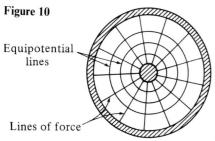

Equipotential lines

Lines of force

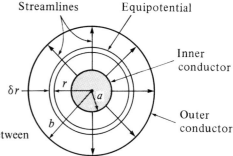

Streamlines   Equipotential

δr r a

b

Inner conductor

Outer conductor

Figure 11 Electric field between two concentric cylinders

2. Capacitance between concentric cylinders

A **concentric cable** is one which contains two or more separate conductors, arranged concentrically (i.e., having a common centre), with insulation between them. In a **coaxial cable**, the central conductor, which may be either solid or hollow, is surrounded by an outer tubular conductor, the space in between being occupied by a dielectric. If air is the dielectric then concentric insulating discs are used to prevent the conductors touching each other. The two kinds of cable serve different purposes. The main feature they have in common is a complete absence of external flux and therefore a complete absence of interference with and from other circuits.

The electric field between two concentric cylinders (i.e., a coaxial cable) is shown in the cross-section of Fig. 11. The conductors form the boundary equipotentials for the field, the boundary equipotentials in Fig. 11 being concentric cylinders of radii a and b. The streamlines, or lines of force, are radial lines cutting the equipotentials at right angles.

Let Q be the charge per unit length of the inner conductor. Then the total flux across the dielectric per unit length is Q coulombs/metre. This total flux will pass through the elemental cylinder of width δr at radius r (shown in Fig. 11) and a distance of 1 m into the plane of the paper. The surface area of a cylinder of length 1 m within the dielectric with radius r is $(2\pi r \times 1)$ m^2. Hence the electric flux density at radius r,

$$D = \frac{Q}{A} = \frac{Q}{2\pi r}$$

The electric field strength or electric stress, E, at radius r is given by

$$E = \frac{D}{\varepsilon_0 \varepsilon_r} = \frac{Q}{2\pi r \varepsilon_0 \varepsilon_r} \text{ volts/metre} \tag{5}$$

Let the p.d. across the element be δV volts. Since

$$E = \frac{\text{voltage}}{\text{thickness}}$$

voltage $= E \times$ thickness. Therefore

$$\delta V = E \delta r = \frac{Q}{2\pi r \varepsilon_0 \varepsilon_r} \delta r$$

The total p.d. between the boundaries,

$$V = \int_a^b \frac{Q}{2\pi r \varepsilon_0 \varepsilon_r} \, dr = \frac{Q}{2\pi \varepsilon_0 \varepsilon_r} \int_a^b \frac{1}{r} \, dr$$

$$= \frac{Q}{2\pi \varepsilon_0 \varepsilon_r} [\ln r]_a^b = \frac{Q}{2\pi \varepsilon_0 \varepsilon_r} [\ln b - \ln a]$$

i.e.,

$$V = \frac{Q}{2\pi \varepsilon_0 \varepsilon_r} \ln \frac{b}{a} \text{ volts} \qquad (6)$$

The capacitance per unit length,

$$C = \frac{\text{charge per unit length}}{\text{p.d.}}$$

Hence capacitance,

$$C = \frac{Q}{V} = \frac{Q}{(Q/(2\pi \varepsilon_0 \varepsilon_r)) \ln(b/a)}$$

i.e.,

$$C = \frac{2\pi \varepsilon_0 \varepsilon_r}{\ln(b/a)} \text{ farads/metre} \qquad (7)$$

Dielectric stress

Rearranging equation (6) gives

$$\frac{Q}{2\pi \varepsilon_0 \varepsilon_r} = \frac{V}{\ln(b/a)}$$

However, from equation (5),

$$E = \frac{Q}{2\pi r \varepsilon_0 \varepsilon_r}$$

Thus dielectric stress,

$$E = \frac{V}{r \ln(b/a)} \text{ volts/metre} \qquad (8)$$

From equation (8), the dielectric stress at any point is seen to be inversely proportional to r, i.e., $E \propto 1/r$.

The dielectric stress E will have a maximum value when r is at its minimum, i.e., when $r = a$. Thus

$$E_m = \frac{V}{a \ln(b/a)}$$ (9)

It follows that

$$E_{min} = \frac{V}{b \ln(b/a)}$$

Dimensions of most economical cable

It is important to obtain the most economical dimensions when designing a cable. A relationship between a and b may be obtained as follows. If E_m and V are both fixed values, then, from equation (9),

$$\frac{V}{E_m} = a \ln \frac{b}{a}$$

Letting $V/E_m = k$, a constant, gives

$$a \ln \frac{b}{a} = k$$

from which $\ln(b/a) = k/a$, $b/a = e^{k/a}$ and

$$b = ae^{k/a}$$ (10)

For the most economical cable, b will be a minimum value. Using the product rule of calculus,

$$\frac{db}{da} = (e^{k/a})(1) + (a)\left(-\frac{k}{a^2} e^{k/a} \right) = 0 \text{ for a minimum value.}$$

(Note, to differentiate $e^{k/a}$ with respect to a, an algebraic substitution may be used, letting $u = 1/a$).

$$e^{k/a} - \frac{k}{a} e^{k/a} = 0$$

Therefore

$$e^{k/a}\left(1 - \frac{k}{a} \right) = 0$$

from which $a = k$. Thus

$$a = \frac{V}{E_m}$$ (11)

From equation (10), internal sheath radius, $b = ae^{k/a} = ae^1 = ae$, i.e.,

$$b = 2.718\,a$$ (12)

Concentric cable field plotting

Figure 12 shows a cross-section of a concentric cable having a core radius r_1 and a sheath radius r_4. It was shown in section 1 that the capacitance of a true square is given by $C = \varepsilon_0 \varepsilon_r$ farads/metre.

Figure 12

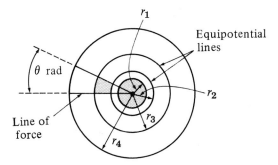

Now if $\theta = \ln(r_b/r_a)$ then $C = \varepsilon_0 \varepsilon_r$ F/m, the same as for a true square. If $\theta = \ln(r_b/r_a)$, then $e^\theta = (r_b/r_a)$. Thus if, say, two equipotential surfaces are chosen within the dielectric as shown in Fig. 12, then $e^\theta = r_2/r_1$, $e^\theta = r_3/r_2$ and $e^\theta = r_4/r_3$. Hence

A curvilinear square is shown shaded in Fig. 12. Such squares can be made to have the same capacitance as a true square by the correct choice of spacing between the lines of force and the equipotential surfaces in the field plot.

From equation (7), the capacitance between cylindrical equipotential lines at radii r_a and r_b is given by

$$C = \frac{2\pi\varepsilon_0\varepsilon_r}{\ln(r_b/r_a)}\ \text{farads/metre}$$

Thus for a sector of θ radians (see Fig. 12) the capacitance is given by

$$C = \frac{\theta}{2\pi}\left(\frac{2\pi\varepsilon_0\varepsilon_r}{\ln(r_b/r_a)}\right) = \frac{\theta\varepsilon_0\varepsilon_r}{\ln(r_b/r_a)}\ \text{farads/metre}$$

Now if $\theta = \ln(r_b/r_a)$ then $C = \varepsilon_0\varepsilon_r$ F/m, the same as for a true square. If $\theta = \ln(r_b/r_a)$, then $e^\theta = (r_b/r_a)$. Thus if, say, two equipotential surfaces are chosen within the dielectric as shown in Fig. 12, then $e^\theta = r_2/r_1$, $e^\theta = r_3/r_2$ and $e^\theta = r_4/r_3$. Hence

$$(e^\theta)^3 = \frac{r_2}{r_1} \times \frac{r_3}{r_2} \times \frac{r_4}{r_3}, \quad \text{i.e.,} \quad e^{3\theta} = \frac{r_4}{r_1} \tag{13}$$

It follows that $e^{2\theta} = r_3/r_1$.

Equation (13) is used to determine the value of θ and hence the number of sectors. Thus, for a concentric cable having a core radius 8 mm and inner sheath radius 32 mm, if two equipotential surfaces within the dielectric are chosen (and therefore form three capacitors in series in each sector),

$$e^{3\theta} = \frac{r_4}{r_1} = \frac{32}{8} = 4$$

Hence $3\theta = \ln 4$ and $\theta = \frac{1}{3}\ln 4 = 0.462$ rad (or $26.47°$). Thus there will be $2\pi/0.462 = 13.6$ sectors in the field plot. (Alternatively, $360°/26.47° = 13.6$ sectors.)

Figure 13

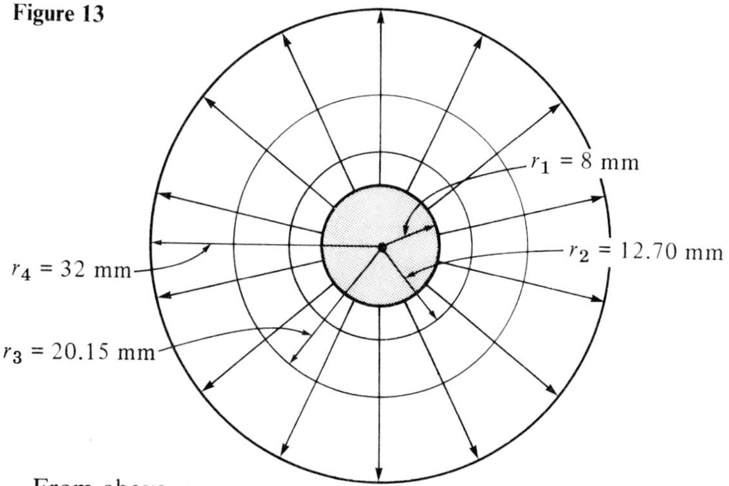

From above,

$$e^{2\theta} = r_3/r_1, \text{ i.e., } r_3 = r_1 e^{2\theta} = 8e^{2(0.462)} = 20.15 \, \text{mm}$$

$$e^\theta = \frac{r_2}{r_1}$$

from which

$$r_2 = r_1 e^\theta = 8e^{0.462} = 12.70 \, \text{mm}$$

The field plot is shown in Fig. 13. The number of parallel squares measured along each equipotential is 13.6 and the number of series squares measured along each line of force is 3. Hence in equation (4), where $C = \varepsilon_0 \varepsilon_r x (m/n)$, $m = 13.6$ and $n = 3$.

If the dielectric has a relative permittivity of, say, 2.5, then the capacitance per metre length,

$$C = (8.85 \times 10^{-12})(2.5)(1)\tfrac{13.6}{3} = \textbf{100 pF}$$

(From equation (7),

$$C = \frac{2\pi \varepsilon_0 \varepsilon_r}{\ln(r_4/r_1)} \, \text{F/m} = \frac{2\pi(8.85 \times 10^{-12})(2.5)}{\ln(32/8)} = \textbf{100 F/m})$$

Thus field plotting using curvilinear squares provides an alternative method of determining the capacitance between concentric cylinders.

Worked problems on the capacitance between concentric cylinders

Problem 1. A coaxial cable has an inner core radius of 0.5 mm and an outer conductor of internal radius 6.0 mm. Determine the capacitance per metre length of the cable if the dielectric has a relative permittivity of 2.7.

From equation (7),

$$\text{capacitance } C = \frac{2\pi \varepsilon_0 \varepsilon_r}{\ln(b/a)} = \frac{2\pi(8.85 \times 10^{-12})(2.7)}{\ln(6.0/0.5)}$$

$$= \textbf{60.4 pF}$$

Problem 2. A single-core concentric cable has a capacitance of 80 pF per metre length. The relative permittivity of the dielectric is 3.5 and the core diameter is 8.0 mm. Determine the internal diameter of the sheath.

From equation (7),

$$\text{capacitance } C = \frac{2\pi\varepsilon_0\varepsilon_r}{\ln(b/a)} \text{ F/m}$$

from which

$$\ln\frac{b}{a} = \frac{2\pi\varepsilon_0\varepsilon_r}{C} = \frac{2\pi(8.85 \times 10^{-12})(3.5)}{(80 \times 10^{-12})} = 2.433$$

Since the core radius, $a = 8.0/2 = 4.0$ mm, $\ln(b/4.0) = 2.433$ and $b/4.0 = e^{2.433}$. Thus the internal radius of the sheath, $b = 4.0e^{2.433} = 45.57$ mm. Hence the internal diameter of the sheath $= 2 \times 45.57 = $ **91.14 mm.**

Problem 3. A concentric cable has a core diameter of 32 mm and an inner sheath diameter of 80 mm. The core potential is 40 kV and the relative permittivity of the dielectric is 3.5. Determine (a) the capacitance per kilometre length of the cable, (b) the dielectric stress at a radius of 30 mm, and (c) the maximum and minimum values of dielectric stress.

(a) From equation (7), capacitance per metre length,

$$C = \frac{2\pi\varepsilon_0\varepsilon_r}{\ln(b/a)}$$

$$= \frac{2\pi(8.85 \times 10^{-12})(3.5)}{\ln(40/16)} = 212.4 \times 10^{-12} \text{ F/m}$$

$$= 212.4 \times 10^{-12} \times 10^3 \text{ F/km}$$

$$= \textbf{212 nF/km} \text{ or } \textbf{0.212 } \boldsymbol{\mu}\textbf{F/km}$$

(b) From equation (8), dielectric stress at radius r,

$$E = \frac{V}{r\ln(b/a)} = \frac{40 \times 10^3}{(30 \times 10^{-3})\ln(40/16)}$$

$$= \textbf{1.46} \times \textbf{10}^{\textbf{6}} \textbf{ V/m} \text{ or } \textbf{1.46 MV/m}$$

(c) Maximum dielectric stress,

$$E_m = \frac{V}{a\ln(b/a)} = 2.73 \text{ MV/m}$$

Minimum dielectric stress,

$$E_{\min} = \frac{V}{b\ln(b/a)} = 1.09 \text{ MV/m}$$

Problem 4. A single-core concentric cable is to be manufactured for a 60 kV, 50 Hz transmission system. The dielectric used is paper which has a maximum permissible safe dielectric stress of 10 MV/m rms and a relative permittivity of 3.5. Calculate (a) the core and inner sheath radii for the most economical cable, (b) the capacitance per metre length, and (c) the charging current per kilometre run.

(a) From equation (11),

$$\text{core radius, } a = \frac{V}{E_m} = \frac{60 \times 10^3 \text{ V}}{10^6 \text{ V/m}}$$

$$= 6 \times 10^{-3} \text{ m} = \textbf{6.0 mm}$$

From equation (12), internal sheath radius, $b = ae = 6.0e = \textbf{16.3 mm.}$

(b) From equation (7),

$$\text{capacitance } C = \frac{2\pi\varepsilon_0\varepsilon_r}{\ln(b/a)} \text{ F/m}$$

Since $b = ae$,

$$C = \frac{2\pi\varepsilon_0\varepsilon_r}{\ln e} = 2\pi\varepsilon_0\varepsilon_r = 2\pi(8.85 \times 10^{-12})(3.5)$$

$$= 195 \times 10^{-12} \text{ F/m or } \textbf{195 pF/m}$$

(c)

$$\text{Charging current} = \frac{V}{X_C} = \frac{V}{1/(\omega C)} = \omega CV = (2\pi 50)(195 \times 10^{-12})(60 \times 10^3)$$

$$= 3.68 \times 10^{-3} \text{ A/m}$$

Hence the charging current per kilometre = **3.68 A.**

Problem 5. A concentric cable has a core diameter of 25 mm and an inside sheath diameter of 80 mm. The relative permittivity of the dielectric is 2.5, the loss angle is 3.5×10^{-3} rad and the working voltage is 132 kV at 50 Hz frequency. Determine for a 1 km length of the cable (a) the capacitance, (b) the charging current and (c) the power loss.

(a) From equation (7),

$$\text{capacitance, } C = \frac{2\pi\varepsilon_0\varepsilon_r}{\ln(b/a)} \text{ F/m}$$

$$= \frac{2\pi(8.85 \times 10^{-12})(2.5)}{\ln(40/12.5)} \times 10^3 \text{ F/km}$$

$$= 0.120 \,\mu\text{F/km}$$

Thus the capacitance for a 1 km length of the cable is **0.120 μF.**

(b)

$$\text{Charging current } I = \frac{V}{X_C} = \frac{V}{1/(\omega C)} = \omega CV = (2\pi 50)(0.120 \times 10^{-6})(132 \times 10^3)$$

$$= \textbf{4.98 A/km}$$

(c) From equation (13), chapter 11,

$$\text{power loss} = V^2 \omega C \tan \delta$$

$$= (132 \times 10^3)^2 (2\pi 50)(0.120 \times 10^{-6}) \tan (3.5 \times 10^{-3})$$

$$= \textbf{2300 W}$$

Problem 6. A concentric cable has a core diameter of 20 mm and a sheath inside diameter of 60 mm. The permittivity of the dielectric is 3.2. Using three equipotential

Figure 14

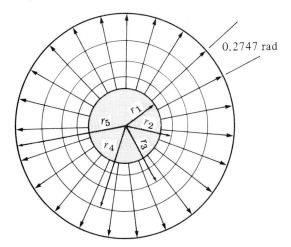

0.2747 rad

surfaces within the dielectric, determine the capacitance of the cable per metre length by the method of curvilinear squares. Draw the field plot for the cable.

The field plot consists of radial lines of force dividing the cable cross-section into a number of sectors, the lines of force cutting the equipotential surfaces at right angles. Since three equipotential surfaces are required in the dielectric, four capacitors in series are found in each sector of θ radians.

In Fig. 14, $r_1 = 20/2 = 10$ mm and $r_5 = 60/2 = 30$ mm. It follows from equation (13) that $e^{4\theta} = r_5/r_1 = 30/10 = 3$, from which $4\theta = \ln 3$ and $\theta = \frac{1}{4}\ln 3 = 0.2747$ rad.

Thus the number of sectors in the plot shown in Fig. 14 is $2\pi/0.2747 =$ **22.9**. The three equipotential lines are shown in Fig. 14 at radii of r_2, r_3 and r_4. From equation (13),

$$e^{3\theta} = \frac{r_4}{r_1}, \quad \text{from which } r_4 = r_1 e^{3\theta} = 10\, e^{3(0.2747)} = 22.80 \text{ mm}$$

$$e^{2\theta} = \frac{r_3}{r_1}, \quad \text{from which } r_3 = r_1 e^{2\theta} = 10\, e^{2(0.2747)} = 17.32 \text{ mm}$$

$$e^{\theta} = \frac{r_2}{r_1}, \quad \text{from which } r_2 = r_1 e^{\theta} = 10\, e^{0.2747} = 13.16 \text{ mm}$$

Thus the field plot for the cable is as shown in Fig. 14.

From equation (4), capacitance $C = \varepsilon_0\varepsilon_r x(m/n)$. The number of parallel squares along each equipotential, $m = 22.9$ and the number of series squares measured along each line of force, $n = 4$. Thus

$$\text{capacitance } C = (8.85 \times 10^{-12})(3.2)(1)\tfrac{22.9}{4} = \textbf{162 pF}$$

(Checking, from equation (7),

$$\text{capacitance } C = \frac{2\pi\varepsilon_0\varepsilon_r}{\ln(r_5/r_1)} = \frac{2\pi(8.85 \times 10^{-12})(3.2)}{\ln(30/10)}$$

$$= \textbf{162 pF})$$

Further problems on the capacitance between concentric cylinders may be found in section 9, problems 4 to 11, page 314.

3. Capacitance of an isolated twin line

The field distribution with two oppositely charged, long conductors, A and B, each of radius a is shown in Fig. 15. The distance D between the centres of the two conductors is such that D is much greater than a. Figure 16 shows the field of each conductor separately.

Initially, let conductor A carry a charge of $+Q$ coulombs per metre while conductor B is uncharged. Consider a cylindrical element of radius r about conductor A having a depth of 1 m and a thickness δr as shown in Fig. 16.

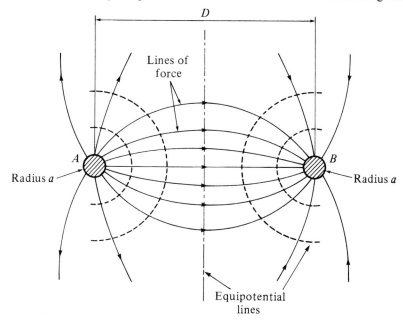

Figure 15

The electric flux density D at the element (i.e., at radius r) is given by

$$D = \frac{\text{charge}}{\text{area}} = \frac{Q}{(2\pi r \times 1)} \text{coulomb/metre}^2$$

Figure 16

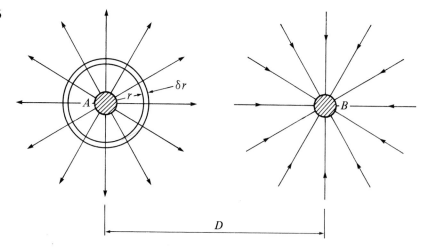

The electric field strength at the element,

$$E = \frac{D}{\varepsilon_0 \varepsilon_r} = \frac{Q/2\pi r}{\varepsilon_0 \varepsilon_r} = \frac{Q}{2\pi r \varepsilon_0 \varepsilon_r} \text{ volts/metre}$$

Since $E = V/d$, potential difference, $V = Ed$. Thus

$$\text{p.d. at the element} = E\delta r = \frac{Q\delta r}{2\pi r \varepsilon_0 \varepsilon_r} \text{ volts}$$

The potential may be considered as zero at a large distance from the conductor. Let this be at radius R. Then the potential of conductor A above zero, V_{A_1}, is given by

$$V_{A_1} = \int_a^R \frac{Q \, dr}{2\pi r \varepsilon_0 \varepsilon_r} = \frac{Q}{2\pi \varepsilon_0 \varepsilon_r} \int_a^R \frac{1}{r} \, dr = \frac{Q}{2\pi \varepsilon_0 \varepsilon_r} [\ln r]_a^R$$

$$= \frac{Q}{2\pi \varepsilon_0 \varepsilon_r} [\ln R - \ln a]$$

i.e.,

$$V_{A_1} = \frac{Q}{2\pi \varepsilon_0 \varepsilon_r} \ln \frac{R}{a}$$

Since conductor B lies in the field of conductor A, by reasoning similar to that above, the potential at conductor B above zero, V_{B_1}, is given by

$$V_{B_1} = \int_D^R \frac{Q \, dr}{2\pi r \varepsilon_0 \varepsilon_r} = \frac{Q}{2\pi \varepsilon_0 \varepsilon_r} [\ln r]_D^R = \frac{Q}{2\pi \varepsilon_0 \varepsilon_r} \ln \frac{R}{D}$$

Repeating the above procedure, this time assuming that conductor B carries a charge of $-Q$ coulombs per metre, while conductor A is uncharged, gives

$$\text{potential of conductor } B \text{ below zero, } V_{B_2} = \frac{-Q}{2\pi \varepsilon_0 \varepsilon_r} \ln \frac{R}{a}$$

and the potential of conductor A below zero, due to the charge on conductor B,

$$V_{A_2} = \frac{-Q}{2\pi \varepsilon_0 \varepsilon_r} \ln \frac{R}{D}$$

When both conductors carry equal and opposite charges, the total potential of A above zero is given by

$$V_{A_1} + V_{A_2} = \left(\frac{Q}{2\pi \varepsilon_0 \varepsilon_r} \ln \frac{R}{a} \right) + \left(\frac{-Q}{2\pi \varepsilon_0 \varepsilon_r} \ln \frac{R}{D} \right)$$

$$= \frac{Q}{2\pi \varepsilon_0 \varepsilon_r} \left(\ln \frac{R}{a} - \ln \frac{R}{D} \right)$$

$$= \frac{Q}{2\pi \varepsilon_0 \varepsilon_r} \left(\ln \frac{R/a}{R/D} \right) = \frac{Q}{2\pi \varepsilon_0 \varepsilon_r} \ln \frac{D}{a}$$

and the total potential of B below zero is given by

$$V_{B_1} + V_{B_2} = \frac{Q}{2\pi\varepsilon_0\varepsilon_r}\left(\ln\frac{R}{D} - \ln\frac{R}{a}\right)$$

$$= \frac{Q}{2\pi\varepsilon_0\varepsilon_r}\ln\frac{a}{D} = \frac{-Q}{2\pi\varepsilon_0\varepsilon_r}\ln\frac{D}{a}$$

Hence the p.d. between A and B is

$$2\left(\frac{Q}{2\pi\varepsilon_0\varepsilon_r}\ln\frac{D}{a}\right) \text{ volts/metre}$$

The capacitance between A and B per metre length,

$$C = \frac{\text{charge per metre}}{\text{p.d.}} = \frac{Q}{2(Q/(2\pi\varepsilon_0\varepsilon_r))\ln(D/a)}$$

i.e.,

$$C = \frac{1}{2}\frac{2\pi\varepsilon_0\varepsilon_r}{\ln(D/a)} \textbf{ farads/metre} \quad \text{or} \quad \frac{\pi\varepsilon_0\varepsilon_r}{\ln(D/a)} \textbf{ farads/metre} \qquad (14)$$

Worked problems on capacitance of an isolated twin line

Problem 1. Two parallel wires, each of diameter 5 mm, are uniformly spaced in air at a distance of 50 mm between centres. Determine the capacitance of the line if the total length is 200 m.

From equation (14), capacitance per metre length,

$$C = \frac{\pi\varepsilon_0\varepsilon_r}{\ln(D/a)} = \frac{\pi(8.85 \times 10^{-12})(1)}{\ln(50/(5/2))} \quad \text{since } \varepsilon_r = 1 \text{ for air,}$$

$$= \frac{\pi(8.85 \times 10^{-12})}{\ln 20} = 9.28 \times 10^{-12} \, \text{F}$$

Hence the capacitance of a 200 m length is $(9.28 \times 10^{-12} \times 200) \, \text{F} = \textbf{1860 pF}$ or **1.86 nF.**

Problem 2. A single-phase circuit is composed of two parallel conductors, each of radius 4 mm, spaced 1.2 m apart in air. The p.d. between the conductors at a frequency of 50 Hz is 15 kV. Determine, for a 1 km length of line, (a) the capacitance of the conductors, (b) the value of charge carried by each conductor, and (c) the charging current.

(a) From equation (14),

$$\text{capacitance } C = \frac{\pi\varepsilon_0\varepsilon_r}{\ln(D/a)} = \frac{\pi(8.85 \times 10^{-12})(1)}{\ln(1.2/4 \times 10^{-3})}$$

$$= \frac{\pi(8.85 \times 10^{-12})}{\ln 300} = 4.87 \, \text{pF/m}$$

Hence the capacitance per kilometre length is $(4.875 \times 10^{-12})(10^3) \, \text{F} = \textbf{4.87 nF.}$

(b) Charge $Q = CV = (4.87 \times 10^{-9})(15 \times 10^3) = \textbf{73.1} \, \boldsymbol{\mu}\textbf{C.}$

(c) Charging current $= \dfrac{V}{X_C} = \dfrac{V}{(1/\omega C)} = \omega C V$

$$= (2\pi 50)(4.87 \times 10^{-9})(15 \times 10^3) = \mathbf{0.023\,A} \text{ or } \mathbf{23\,mA}$$

Problem 3. The charging current for an 800 m run of isolated twin line is not to exceed 15 mA. The voltage between the lines is 10 kV at 50 Hz. If the line is air-insulated, determine (a) the maximum value required for the capacitance per metre length, and (b) the maximum diameter of each conductor if their distance between centres is 1.25 m.

(a)   Charging current $I = \dfrac{V}{X_C} = \dfrac{V}{(1/\omega C)} = \omega C V$

from which

$$\text{capacitance } C = \frac{1}{\omega V} = \frac{15 \times 10^{-3}}{(2\pi 50)(10 \times 10^3)} \text{farads per 800 metre run}$$

$$= 4.775\,\text{nF}$$

Hence the maximum required value of capacitance

$$= \frac{4.775 \times 10^{-9}}{800} \text{F/m} = \mathbf{5.97\,pF/m}$$

(b) From equation (14),

$$C = \frac{\pi \varepsilon_0 \varepsilon_r}{\ln(D/a)},$$

thus

$$5.97 \times 10^{-12} = \frac{\pi(8.85 \times 10^{-12})(1)}{\ln(1.25/a)}$$

from which

$$\ln\left(\frac{1.25}{a}\right) = \frac{\pi 8.85}{5.97} = 4.657$$

Hence

$$\frac{1.25}{a} = e^{4.657} = 105.3$$

and

$$\text{radius } a = \tfrac{1.25}{105.3}\,\text{m} = 0.01187\,\text{m or } 11.87\,\text{mm}$$

Thus **the maximum diameter of each conductor is** 2×11.87, i.e., **23.7 mm.**

Further problems on capacitance of an isolated twin line may be found in section 9, problems 12 to 17, page 315.

4. Energy stored in an electric field

Consider the p.d. across a parallel-plate capacitor of capacitance C farads being increased by dv volts in dt seconds. If the corresponding increase in charge is dq coulombs, then $dq = C\,dv$.

If the charging current at that instant is i amperes, then $dq = i\,dt$. Thus $i\,dt = C\,dv$, i.e.,

$$i = C\frac{dv}{dt}$$

(i.e., instantaneous current = capacitance × rate of change of p.d.)
The instantaneous value of power to the capacitor,

$$p = vi \text{ watts} = v\left(C\frac{dv}{dt}\right) \text{watts}$$

The energy supplied to the capacitor during time dt

$$= \text{power} \times \text{time} = \left(vC\frac{dv}{dt}\right)(dt) = Cv\,dv \text{ joules}$$

Thus the total energy supplied to the capacitor when the p.d. is increased from 0 to V volts is given by

$$W_f = \int_0^V Cv\,dv = C\left[\frac{v^2}{2}\right]_0^V$$

i.e.,

energy stored in the electric field, $W_f = \frac{1}{2}CV^2$ joules (15)

Consider a capacitor with dielectric of relative permittivity ε_r, thickness d metres and area A square metres. Capacitance $C = Q/V$, hence energy stored $= \frac{1}{2}(Q/V)V^2 = \frac{1}{2}QV$ joules. The electric flux density, $D = Q/A$, from which $Q = DA$. Hence the energy stored $= \frac{1}{2}(DA)V$ joules.

The electric field strength, $E = V/d$, from which $V = Ed$. Hence the energy stored $= \frac{1}{2}(DA)(Ed)$ joules. However Ad is the volume of the field. Hence

energy stored per unit volume, $\omega_f = \frac{1}{2}DE$ joules/cubic metre (16)

Since $D/E = \varepsilon_0\varepsilon_r$, then $D = \varepsilon_0\varepsilon_r E$. Hence, from equation (16), the energy stored per unit volume,

$$\omega_f = \frac{1}{2}(\varepsilon_0\varepsilon_r E)E = \frac{1}{2}\varepsilon_0\varepsilon_r E^2 \text{ joules/cubic metre} \quad (17)$$

Also, since $D/E = \varepsilon_0\varepsilon_r$ then $E = D/(\varepsilon_0\varepsilon_r)$. Hence, from equation (16), the energy stored per unit volume,

$$\omega_f = \frac{1}{2}D\left(\frac{D}{\varepsilon_0\varepsilon_r}\right) = \frac{D^2}{2\varepsilon_0\varepsilon_r} \text{ joules/cubic metre} \quad (18)$$

Summarising,

$$\text{energy stored in a capacitor} = \frac{1}{2}CV^2 \text{ joules}$$

and

$$\text{energy stored per unit volume of dielectric} = \frac{1}{2}DE = \frac{1}{2}\varepsilon_0\varepsilon_r E^2$$

$$= \frac{D^2}{2\varepsilon_0\varepsilon_r} \text{ joules/cubic metre}$$

Worked problems on energy stored in electric fields

Problem 1. Determine the energy stored in a 10 nF capacitor when charged to 1 kV, and the average power developed if this energy is dissipated in 10 μs.

From equation (15),

$$\text{energy stored, } W_f = \tfrac{1}{2}CV^2 = \tfrac{1}{2}(10 \times 10^{-9})(10^3)^2 = \textbf{5 mJ}$$

$$\text{average power developed} = \frac{\text{energy dissipated, } W}{\text{time, } t} = \frac{5 \times 10^{-3} \text{ J}}{10 \times 10^{-6} \text{ s}} = \textbf{500 W}$$

Problem 2. A capacitor is charged with 5 mC. If the energy stored is 625 mJ, determine (a) the voltage across the plates and (b) the capacitance of the capacitor.

(a) From equation (15),

$$\text{energy stored, } W_f = \tfrac{1}{2}CV^2 = \frac{1}{2}\left(\frac{Q}{V}\right)V^2 = \tfrac{1}{2}QV$$

from which voltage across the plates,

$$V = \frac{2 \times \text{energy stored}}{Q} = \frac{2 \times 0.625}{5 \times 10^{-3}}$$

$$= \textbf{250 V}$$

(b)

$$\text{Capacitance } C = \frac{Q}{V} = \frac{5 \times 10^{-3}}{250} \text{ F} = 20 \times 10^{-6} \text{ F} = \textbf{20 } \boldsymbol{\mu}\textbf{F}$$

Problem 3. A ceramic capacitor is to be constructed to have a capacitance of 0.01 μF and to have a steady working potential of 2.5 kV maximum. Allowing a safe value of field stress of 10 MV/m, determine (a) the required thickness of the ceramic dielectric, (b) the area of plate required if the relative permittivity of the ceramic is 10, and (c) the maximum energy stored by the capacitor.

(a) Field stress $E = V/d$, from which thickness of ceramic dielectric,

$$d = \frac{V}{E} = \frac{2.5 \times 10^3}{10 \times 10^6}$$

$$= 2.5 \times 10^{-4} \text{ m} = \textbf{0.25 mm}$$

(b) Capacitance $C = \varepsilon_0\varepsilon_r A/d$ for a two-plate parallel capacitor. Hence cross-sectional area of plate,

$$A = \frac{Cd}{\varepsilon_0\varepsilon_r} = \frac{(0.01 \times 10^{-6})(0.25 \times 10^{-3})}{(8.85 \times 10^{-12})(10)}$$

$$= \textbf{0.0282 m}^2 \text{ or } \textbf{282 cm}^2$$

(c) Maximum energy stored,

$$W_f = \tfrac{1}{2}CV^2 = \tfrac{1}{2}(0.01 \times 10^{-6})(2.5 \times 10^3)^2 = \textbf{0.0313 J} \text{ or } \textbf{31.3 mJ}$$

Problem 4. A 400 pF capacitor is charged to a p.d. of 100 V. The dielectric has a cross-sectional area of 200 cm^2 and a relative permittivity of 2.3. Calculate the energy stored per cubic metre of the dielectric.

From equation (18), energy stored per unit volume of dielectric,

$$\omega_f = \frac{D^2}{2\varepsilon_0\varepsilon_r}$$

Electric flux density

$$D = \frac{Q}{A} = \frac{CV}{A} = \frac{(400 \times 10^{-12})(100)}{200 \times 10^{-4}} = 2 \times 10^{-6} \, \text{C/m}^2$$

Hence energy stored,

$$\omega_f = \frac{D^2}{2\varepsilon_0 \varepsilon_r} = \frac{(2 \times 10^{-6})^2}{2(8.85 \times 10^{-12})(2.3)} = \textbf{0.098 J/m}^3 \text{ or } \textbf{98 mJ/m}^3$$

Problem 5. Two parallel plates each of dimensions 40 mm × 80 mm are spaced 10 mm apart in air. If a voltage of 25 kV is applied across the plates, determine the energy stored in the electric field.

Energy stored, $W_f = \frac{1}{2}CV^2$ joules.

$$\text{Capacitance } C = \frac{\varepsilon_0 \varepsilon_r A}{d} = \frac{(8.85 \times 10^{-12})(1)(40 \times 80 \times 10^{-6})}{10 \times 10^{-3}} \equiv 2.832 \, \text{pF}$$

Hence energy stored,

$$W_f = \frac{1}{2}CV^2 = \frac{1}{2}(2.832 \times 10^{-12})(25 \times 10^3)^2 \equiv \textbf{0.885 mJ} \text{ or } \textbf{885 } \mu\textbf{J}$$

(Alternatively, electric field strength,

$$E = \frac{V}{d} = \frac{25 \times 10^3}{10 \times 10^{-3}} = 2.5 \times 10^6 \, \text{V/m} \equiv 2.5 \, \text{MV/m}$$

From equation (17), energy stored,

$$\omega_f = \frac{1}{2}\varepsilon_0 \varepsilon_r E^2 = \frac{1}{2}(8.85 \times 10^{-12})(1)(2.5 \times 10^6)^2 \, \text{J/m}^3$$
$$= 27.66 \, \text{J/m}^3$$

Volume of electric field $= Ad = (40 \times 80 \times 10^{-6})(10 \times 10^{-3}) = 32 \times 10^{-6} \, \text{m}^3$. Hence energy stored in field,

$$W_f = (27.66 \, \text{J/m}^3)(32 \times 10^{-6} \, \text{m}^3) = \textbf{0.885 mJ} \text{ or } \textbf{885 } \mu\textbf{J})$$

Further problems on energy stored in electric fields may be found in section 9, problems 18 to 26, page 316.

5. Induced emf and inductance

A current flowing in a coil of wire is accompanied by a magnetic flux linking with the coil. If the current changes, the flux linkage (i.e., the product of flux and the number of turns) changes and an emf is induced in the coil. The induced emf e in a coil of N turns is given by

$$e = N\frac{\text{d}\phi}{\text{d}t} \text{ volts}$$

where $\text{d}\phi/\text{d}t$ is the rate of change of flux.

Inductance is the name given to the property of a circuit whereby there is an emf induced into the circuit by the change of flux linkages produced by a current change. The unit of inductance is the **henry, H.** A circuit has an inductance of 1H when an emf of 1 V is induced in it by a current changing uniformly at the rate of 1 A/s.

The emf induced in a coil of inductance L henry is given by

$$e = L \frac{di}{dt} \text{ volts}$$

where di/dt is the rate of change of current.

If a current changing uniformly from zero to I amperes produces a uniform flux change from zero to Φ webers in t seconds then (from above) average induced emf, $E_{av} = N\Phi/t = LI/t$, from which

$$\text{inductance of coil, } L = \frac{N\Phi}{I} \text{ henry}$$

Flux linkage means the product of flux, in webers, and the number of turns with which the flux is linked. Hence flux linkage $= N\Phi$. Thus since $L = N\Phi/I$, **inductance = flux linkages per ampere.**

6. Inductance of a concentric cylinder (or coaxial cable)

Skin effect

When a direct current flows in a uniform conductor the current will tend to distribute itself uniformly over the cross-section of the conductor. However, with alternating current, particularly if the frequency is high, the current carried by the conductor is not uniformly distributed over the available cross-section, but tends to be concentrated at the conductor surface. This is called **skin effect**. When current is flowing through a conductor, the magnetic flux that results is in the form of concentric circles. Some of this flux exists within the conductor and links with the current more strongly near the centre. The result is that the inductance of the central part of the conductor is greater than the inductance of the conductor near the surface. This is because of the greater number of flux linkages existing in the central region. At high frequencies the reactance $(X_L = 2\pi f L)$ of the extra inductance is sufficiently large to seriously affect the flow of current, most of which flows along the surface of the conductor where the impedance is low rather than near the centre where the impedance is high.

Inductance due to internal linkages at low frequency

When a conductor is used at high frequency the depth of penetration of the current is small compared with the conductor cross-section. Thus the internal linkages may be considered as negligible and the circuit inductance is that due to the fields in the surrounding space. However, at very low frequency the current distribution is considered uniform over the conductor cross-section and the inductance due to flux linkages has its maximum value.

Consider a conductor of radius R, as shown in Fig. 17, carrying a current I amperes uniformly distributed over the cross-section. At all points on the conductor cross-section

$$\text{current density, } J = \frac{\text{current}}{\text{area}} = \left(\frac{I}{\pi R^2}\right) \text{amperes/metre}^2$$

Figure 17

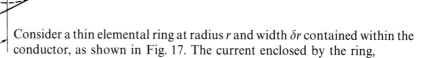

Consider a thin elemental ring at radius r and width δr contained within the conductor, as shown in Fig. 17. The current enclosed by the ring,

$$i = \text{current density} \times \text{area enclosed by the ring}$$

$$= \left(\frac{I}{\pi R^2}\right)(\pi r^2)$$

i.e.,

$$i = \frac{Ir^2}{R^2} \text{ amperes}$$

Magnetic field strength, $H = Ni/l$ amperes/metre.

At radius r, the mean length of the flux path, $l = 2\pi r$ (i.e., the circumference of the elemental ring) and $N = 1$ turn. Hence at radius r_1,

$$H_r = \frac{Ni}{l} = \frac{(1)(Ir^2/R^2)}{2\pi r} = \frac{Ir}{2\pi R^2} \text{ ampere/metre}$$

and the flux density,

$$B_r = \mu_0\mu_r H_r = \mu_0\mu_r \left(\frac{Ir}{2\pi R^2}\right) \text{ tesla}$$

Flux $\Phi = BA$ webers. For a 1 m length of the conductor, the cross-sectional area A of the element is $(\delta r \times 1)\,\text{m}^2$ (see Fig. 17). Thus the flux within the element of thickness δr,

$$\Phi = \left(\frac{\mu_0\mu_r Ir}{2\pi R^2}\right)(\delta r) \text{ webers}$$

The flux in the element links the portion $\pi r^2/\pi R^2$, i.e., r^2/R^2 of the total conductor. Hence

$$\text{linkages due to the flux within radius } r = \left(\frac{\mu_0\mu_r Ir}{2\pi R^2}\,\delta r\right)\frac{r^2}{R^2}$$

$$= \frac{\mu_0\mu_r Ir^3}{2\pi R^4}\,\delta r \text{ weber turns}$$

total linkages per metre due to the flux in the conductor

$$= \int_0^R \frac{\mu_0\mu_r Ir^3}{2\pi R^4}\,dr = \frac{\mu_0\mu_r I}{2\pi R^4}\int_0^R r^3\,dr$$

$$= \frac{\mu_0 \mu_r I}{2\pi R^4} \left[\frac{r^4}{4} \right]_0^R = \frac{\mu_0 \mu_r I}{2\pi R^4} \left[\frac{R^4}{4} \right]$$

$$= \frac{1}{4} \left(\frac{\mu_0 \mu_r I}{2\pi} \right) \text{weber turns}$$

inductance per metre due to the internal flux
= internal flux linkages per ampere

$$= \frac{1}{4} \left(\frac{\mu_0 \mu_r}{2\pi} \right) \quad \text{or} \quad \frac{\mu}{8\pi} \text{ henry/metre}$$

It is seen that the inductance is independent of the conductor radius R.

Inductance of a pair of concentric cylinders

The cross-section of a concentric (or coaxial) cable is shown in Fig. 18. Let a current of I amperes flow in one direction in the core and a current of I amperes flow in the opposite direction in the outer sheath conductor.

Consider an element of width δr at radius r, and let the radii of the inner and outer conductor be a and b respectively as shown. The magnetic field strength at radius r,

$$H_r = \frac{Ni}{l} = \frac{(1)(I)}{2\pi r} = \frac{I}{2\pi r}$$

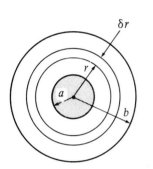

Figure 18 Cross-section of a concentric cable

The flux density at radius r,

$$B_r = \mu_0 \mu_r H_r = \frac{\mu_0 \mu_r I}{2\pi r}$$

For a 1 m length of the cable, the flux Φ within the element of width δr is given by

$$\Phi = B_r A = \left(\frac{\mu_0 \mu_r I}{2\pi r} \right) (\delta r \times 1) = \frac{\mu_0 \mu_r I}{2\pi r} \delta r \text{ webers}$$

This flux links the loop of the cable formed by the core and the outer sheath. Thus the flux linkage per metre length of the cable is $(\mu_0 \mu_r I/(2\pi r))\delta r$ weber turns, and

$$\text{total flux linkages per metre} = \int_a^b \frac{\mu_0 \mu_r I}{2\pi r} \mathrm{d}r = \frac{\mu_0 \mu_r I}{2\pi} \int_a^b \frac{1}{r} \mathrm{d}r$$

$$= \frac{\mu_0 \mu_r I}{2\pi} [\ln r]_a^b = \frac{\mu_0 \mu_r I}{2\pi} [\ln b - \ln a]$$

$$= \frac{\mu_0 \mu_r I}{2\pi} \ln \frac{b}{a} \text{ weber turns}$$

Thus

$$\text{inductance per metre length} = \text{flux linkages per ampere}$$

$$= \frac{\mu_0 \mu_r}{2\pi} \ln \frac{b}{a} \text{ henry/metre} \qquad (19)$$

At low frequencies the inductance due to the internal linkages is added to this result. Hence the total inductance per metre at low frequency is given by

$$L = \frac{1}{4}\left(\frac{\mu_0 \mu_r}{2\pi}\right) + \frac{\mu_0 \mu_r}{2\pi}\ln\frac{b}{a} \text{ henry/metre} \tag{20}$$

or

$$L = \frac{\mu}{2\pi}\left(\frac{1}{4} + \ln\frac{b}{a}\right) \text{ henry/metre} \tag{21}$$

Worked problems on the inductance of concentric cables

Problem 1. A coaxial cable has an inner core of radius 1.0 mm and an outer sheath of internal radius 4.0 mm. Determine the inductance of the cable per metre length. Assume that the relative permeability is unity.

From equation (21),

$$\text{inductance } L = \frac{\mu}{2\pi}\left(\frac{1}{4} + \ln\frac{b}{a}\right) \text{H/m}$$

$$= \frac{\mu_0 \mu_r}{2\pi}\left(\frac{1}{4} + \ln\frac{4.0}{1.0}\right) = \frac{(4\pi \times 10^{-7})(1)}{2\pi}(0.25 + \ln 4)$$

$$= 3.27 \times 10^{-7} \text{ H/m} \quad \text{or} \quad 0.327\,\mu\text{H/m}$$

Problem 2. A concentric cable has a core diameter of 10 mm. The inductance of the cable is 4×10^{-7} H/m. Ignoring inductance due to internal linkages, determine the diameter of the sheath. Assume that the relative permeability is 1.

From equation (19),

$$\text{inductance per metre length} = \frac{\mu_0 \mu_r}{2\pi}\ln\frac{b}{a},$$

where b = sheath radius and a = core radius. Hence

$$4 \times 10^{-7} = \frac{(4\pi \times 10^{-7})(1)}{2\pi}\ln\left(\frac{b}{5}\right)$$

from which

$$2 = \ln\left(\frac{b}{5}\right)$$

$$e^2 = \frac{b}{5}$$

and

$$\text{radius } b = 5e^2 = 36.95 \text{ mm}$$

Thus the diameter of the sheath is $2 \times 36.95 = \mathbf{73.9\,mm.}$

Problem 3. A coaxial cable 7.5 km long has a core 10 mm diameter and a sheath 25 mm diameter, the sheath having negligible thickness. Determine for the cable (a) the inductance, assuming nonmagnetic materials, and (b) the capacitance, assuming a dielectric of relative permittivity 3.

(a) From equation (21),

$$\text{inductance per metre length} = \frac{\mu}{2\pi}\left(\frac{1}{4} + \ln\frac{b}{a}\right) = \frac{\mu_0\mu_r}{2\pi}\left[\frac{1}{4} + \ln\left(\frac{12.5}{5}\right)\right]$$

$$= \frac{(4\pi \times 10^{-7})(1)}{2\pi}(0.25 + \ln 2.5)$$

$$= 2.33 \times 10^{-7}\,\text{H/m}$$

Since the cable is 7500 m long, the inductance $= 7500 \times 2.33 \times 10^{-7} = $ **1.75 mH.**

(b) From equation (7),

$$\text{capacitance, } C = \frac{2\pi\varepsilon_0\varepsilon_r}{\ln(b/a)} = \frac{2\pi(8.85 \times 10^{-12})(3)}{\ln(12.5/5)}$$

$$= 182.06\,\text{pF/m}$$

Since the cable is 7500 m long, the capacitance $= 7500 \times 182.06 \times 10^{-12}$

$$= \textbf{1.365}\,\boldsymbol{\mu}\textbf{F.}$$

Further problems on the inductance of concentric cables may be found in section 9, problems 27 to 32, page 317.

7. Inductance of an isolated twin line

Consider two isolated, long, parallel, straight conductors A and B, each of radius a metres, spaced D metres apart. Let the current in each be I amperes but flowing in opposite directions. Distance D is assumed to be much greater than radius a. The magnetic field associated with the conductors is as shown in Fig. 19. There is a force of repulsion between conductors A and B.

Figure 19

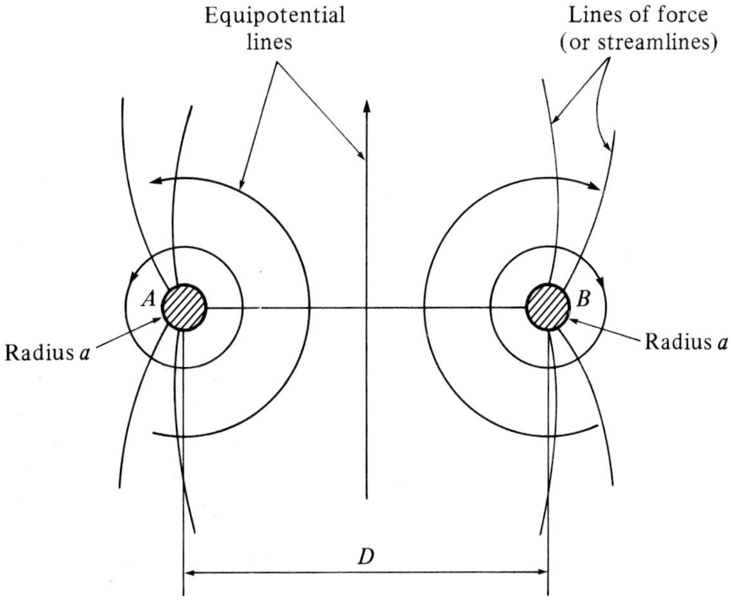

Equipotential lines

Lines of force (or streamlines)

Radius a

A

B

Radius a

D

Figure 20

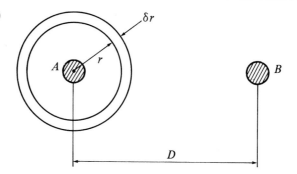

It is easier to analyse the field by initially considering each conductor alone (as in section 3). At any radius r from conductor A (see Fig. 20),

$$\text{magnetic field strength, } H_r = \frac{Ni}{l} = \frac{I}{2\pi r} \text{ ampere/metre}$$

and

$$\text{flux density, } B_r = \mu_0\mu_r H_r = \frac{\mu_0\mu_r I}{2\pi r} \text{ tesla}$$

The total flux in 1 m of the conductor,

$$\Phi = B_r A = \left(\frac{\mu_0\mu_r I}{2\pi r}\right)(\delta r \times 1) = \frac{\mu_0\mu_r I}{2\pi r}\,\delta r \text{ webers}$$

Since this flux links conductor A once, the linkages with conductor A due to this flux $= (\mu_0\mu_r I/(2\pi r))\delta r$ weber turns.

There is, in fact, no limit to the distance from conductor A at which a magnetic field may be experienced. However, let R be a very large radius at which the magnetic field strength may be regarded as zero. Then the total linkages with conductor A due to current in conductor A is given by

$$\int_a^R \frac{\mu_0\mu_r I}{2\pi r}\,dr = \frac{\mu_0\mu_r I}{2\pi}\int_a^R \frac{dr}{r}$$

$$= \frac{\mu_0\mu_r I}{2\pi}[\ln r]_a^R = \frac{\mu_0\mu_r I}{2\pi}[\ln R - \ln a]$$

$$= \frac{\mu_0\mu_r I}{2\pi}\ln\left(\frac{R}{a}\right)$$

Similarly, the total linkages with conductor B due to the current in A

$$= \int_D^R \frac{\mu_0\mu_r I}{2\pi r}\,dr = \frac{\mu_0\mu_r I}{2\pi}\ln\frac{R}{D}$$

Now consider conductor B alone, carrying a current of $-I$ amperes. By similar reasoning to above,

$$\text{total linkages with conductor } B \text{ due to the current in } B$$
$$= \frac{-\mu_0\mu_r I}{2\pi}\ln\left(\frac{R}{a}\right)$$

and

total linkages with conductor A due to the current in $B = \dfrac{-\mu_0\mu_r I}{2\pi}\ln\dfrac{R}{D}$

Hence

total linkages with conductor $A = \left(\dfrac{\mu_0\mu_r I}{2\pi}\ln\dfrac{R}{a}\right) + \left(\dfrac{-\mu_0\mu_r I}{2\pi}\ln\dfrac{R}{D}\right)$

$$= \dfrac{\mu_0\mu_r I}{2\pi}\left[\ln\dfrac{R}{a} - \ln\dfrac{R}{D}\right]$$

$$= \dfrac{\mu_0\mu_r I}{2\pi}\left[\ln\dfrac{R/a}{R/D}\right]$$

$$= \dfrac{\mu_0\mu_r I}{2\pi}\ln\dfrac{D}{a}\,\text{weber-turns/metre}$$

Similarly,

total linkages with conductor $B = \dfrac{\mu_0\mu_r I}{2\pi}\ln\dfrac{D}{a}\,\text{weber-turns/metre}$

For a 1 m length of the two conductors,

$$\text{total inductance} = \text{flux linkages per ampere}$$

$$= 2\left(\dfrac{\mu_0\mu_r}{2\pi}\ln\dfrac{D}{a}\right)\text{henry/metre}$$

i.e.,

$$\textbf{total inductance} = \dfrac{\boldsymbol{\mu_0\mu_r}}{\boldsymbol{\pi}}\,\textbf{ln}\,\dfrac{\boldsymbol{D}}{\boldsymbol{a}}\,\textbf{henry/metre} \qquad (22)$$

Equation (22) does not take into consideration the internal linkages of each line. From section (6),

inductance per metre due to internal linkages $= \dfrac{1}{4}\left(\dfrac{\mu_0\mu_r}{2\pi}\right)\text{henry/metre}$

Thus

inductance per metre due to internal linkages of two conductors

$$= 2\left(\dfrac{1}{4}\left(\dfrac{\mu_0\mu_r}{2\pi}\right)\right) = \dfrac{\mu_0\mu_r}{4\pi}\,\text{henry/metre}$$

Therefore, at low frequency,

total inductance per metre of the two conductors

$$= \dfrac{\mu_0\mu_r}{4\pi} + \dfrac{\mu_0\mu_r}{\pi}\ln\dfrac{D}{a}$$

$$= \dfrac{\boldsymbol{\mu_0\mu_r}}{\boldsymbol{\pi}}\left(\dfrac{1}{4} + \textbf{ln}\,\dfrac{\boldsymbol{D}}{\boldsymbol{a}}\right)\textbf{henry/metre} \qquad (23)$$

(This is often referred to as the "loop inductance".)

In most practical lines the relative permeability, $\mu_r = 1$.

Worked problems on the inductance of an isolated twin line

Problem 1. A single-phase power line comprises two conductors each with a radius 8.0 mm and spaced 1.2 m apart in air. Determine the inductance of the line per metre length ignoring internal linkages. Assume the relative permeability, $\mu_r = 1$.

From equation (22),

$$\text{inductance} = \frac{\mu_0 \mu_r}{\pi} \ln \frac{D}{a}$$

$$= \frac{(4\pi \times 10^{-7})(1)}{\pi} \ln \left(\frac{1.2}{8.0 \times 10^{-3}} \right) = 4 \times 10^{-7} \ln 150$$

$$= \mathbf{20.0 \times 10^{-7} \, H/m \text{ or } 2.0 \, \mu H/m}$$

Problem 2. Determine (a) the loop inductance, and (b) the capacitance of a 1 km length of a single-phase twin line having conductors of diameter 10 mm and spaced 800 mm apart in air.

(a) From equation (23),

$$\text{total inductance per loop metre} = \frac{\mu_0 \mu_r}{\pi} \left(\frac{1}{4} + \ln \frac{D}{a} \right)$$

$$= \frac{(4\pi \times 10^{-7})(1)}{\pi} \left(\frac{1}{4} + \ln \frac{800}{10/2} \right)$$

$$= (4 \times 10^{-7})(0.25 + \ln 160)$$

$$= 21.3 \times 10^{-7} \, H/m$$

Hence

loop inductance of a 1 km length of line $= 21.3 \times 10^{-7} \, H/m \times 10^3 \, m$

$$= \mathbf{21.3 \times 10^{-4} \, H \quad or \quad 2.13 \, mH}$$

(b) From equation (14),

$$\text{capacitance per metre length} = \frac{\pi \varepsilon_0 \varepsilon_r}{\ln(D/a)}$$

$$= \frac{\pi (8.85 \times 10^{-12})(1)}{\ln(800/5)}$$

$$= 5.478 \times 10^{-12} \, F/m$$

Hence

capacitance of a 1 km length of line $= 5.478 \times 10^{-12} \, F/m \times 10^3 \, m$

$$= \mathbf{5.478 \, nF}$$

Problem 3. The total loop inductance of an isolated twin power line is 2.185 μH/m. The diameter of each conductor is 12 mm. Determine the distance between their centres.

From equation (23),

$$\text{total loop inductance} = \frac{\mu_0 \mu_r}{\pi} \left(\frac{1}{4} + \ln \frac{D}{a} \right)$$

Hence

$$2.185 \times 10^{-6} = \frac{(4\pi \times 10^{-7})(1)}{\pi}\left(\frac{1}{4} + \ln\frac{D}{6}\right)$$

where D is the distance between centres in millimetres.

$$\frac{2.185 \times 10^{-6}}{4 \times 10^{-7}} = 0.25 + \ln\frac{D}{6}$$

$$\ln\frac{D}{6} = 5.4625 - 0.25 = 5.2125$$

$$\frac{D}{6} = e^{5.2125}$$

from which, distance $D = 6\,e^{5.2125} = \mathbf{1100\,mm}$ or $\mathbf{1.10\,m}$

Further problems on the inductance of an isolated twin line may be found in section 9, problems 33 to 38, page 317.

8. Energy stored in an electromagnetic field

Magnetic energy in a nonmagnetic medium

For a nonmagnetic medium the relative permeability, $\mu_r = 1$ and $B = \mu_0 H$. Thus the magnetic field strength H is proportional to the flux density B and a graph of B against H is a straight line, as shown in Fig. 21.

It was shown in section 3 of chapter 11 that, when the flux density is increased by an amount dB due to an increase dH in the magnetic field strength, then

energy supplied to the magnetic circuit = area of shaded strip

(in joules per cubic metre)

Thus, for a maximum flux density $0Y$ in Fig. 21,

total energy stored in the magnetic field

= area of triangle $0YX$

= $\frac{1}{2}$ × base × height

= $\frac{1}{2}(0Z)(0Y)$

If $0Y = B$ teslas and $0Z = H$ ampere/metre, then the total energy stored in a non-magnetic medium,

$$\omega_f = \tfrac{1}{2}HB \text{ joules/metre}^3 \qquad (24)$$

Since $B = \mu_0 H$ for a non-magnetic medium, the energy stored,

$$\omega_f = \tfrac{1}{2}H(\mu_0 H) = \tfrac{1}{2}\mu_0 H^2 \text{ joules/metre}^3 \qquad (25)$$

Alternatively, $H = B/\mu_0$, thus the energy stored,

$$\omega_f = \tfrac{1}{2}HB = \frac{1}{2}\left(\frac{B}{\mu_0}\right)B$$

$$= \frac{B^2}{2\mu_0} \text{ joules/metre}^3 \qquad (26)$$

Figure 21

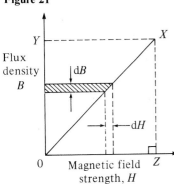

Flux density B (vertical axis), Magnetic field strength, H (horizontal axis)

Magnetic energy stored in an inductor

Establishing a magnetic field requires energy to be expended. However, once the field is established, the only energy expended is that supplied to maintain the flow of current in opposition to the circuit resistance, i.e., the I^2R loss, which is dissipated as heat.

For an inductive circuit containing resistance R and inductance L (see Fig. 22) the applied voltage V at any instant is given by $V = v_R + v_L$ i.e.,

$$V = iR + L\frac{di}{dt}$$

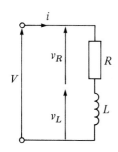

Figure 22

Multiplying throughout by current i gives the power equation:

$$Vi = i^2R + Li\frac{di}{dt}$$

Multiplying throughout by time dt seconds gives the energy equation:

$$Vi\,dt = i^2R\,dt + Li\,di$$

$Vi\,dt$ is the energy supplied by the source in time dt, $i^2R\,dt$ is the energy dissipated in the resistance and $Li\,di$ is the energy supplied in establishing the magnetic field or the energy absorbed by the magnetic field in time dt seconds.

Hence the total energy stored in the field when the current increases from 0 to I amperes is given by

$$\text{energy stored, } W_f = \int_0^I Li\,di = L\left[\frac{i^2}{2}\right]_0^I$$

i.e.,

$$\textbf{total energy stored, } W_f = \tfrac{1}{2}LI^2 \textbf{ joules} \qquad (27)$$

From section 5, inductance $L = N\Phi/I$, hence

$$\text{total energy stored} = \frac{1}{2}\left(\frac{N\Phi}{I}\right)I^2 = \tfrac{1}{2}N\Phi I \text{ joules}$$

Also $H = NI/l$, from which, $N = Hl/I$, and $\Phi = BA$. Thus the total energy stored,

$$W_f = \tfrac{1}{2}N\Phi I = \frac{1}{2}\left(\frac{Hl}{I}\right)(BA)I$$

$$= \tfrac{1}{2}HBlA \text{ joules}$$

$$= \tfrac{1}{2}HB \text{ joules/metre}^3$$

since lA is the volume of the magnetic field. This latter expression has already been derived in equation (24).

Summarising, the energy stored in a nonmagnetic medium,

$$\omega_f = \tfrac{1}{2}BH = \tfrac{1}{2}\mu_0 H^2 = \frac{B^2}{2\mu_0} \textbf{ joules/metre}^3$$

and the energy stored in an inductor, $W_f = \tfrac{1}{2}LI^2 \textbf{ joules.}$

Worked problems on energy stored in an electromagnetic field

Problem 1. Calculate the value of the energy stored when a current of 50 mA is flowing in a coil of inductance 200 mH. What value of current would double the energy stored?

From equation (27), energy stored in inductor,

$$W_f = \tfrac{1}{2}LI^2 = \tfrac{1}{2}(200 \times 10^{-3})(50 \times 10^{-3})^2$$

$$= 2.5 \times 10^{-4}\,\text{J} \quad \text{or} \quad \textbf{0.25 mJ} \quad \text{or} \quad \textbf{250}\,\boldsymbol{\mu}\textbf{J}$$

If the energy stored is doubled, then $(2)(2.5 \times 10^{-4}) = \tfrac{1}{2}(200 \times 10^{-3})I^2$, from which

$$\text{current } I = \sqrt{\frac{(4)(2.5 \times 10^{-4})}{(200 \times 10^{-3})}} = \textbf{70.71 mA}$$

Problem 2. The airgap of a moving coil instrument is 2.0 mm long and has a cross-sectional area of 500 mm². If the flux density is 50 mT, determine the total energy stored in the magnetic field of the airgap.

From equation (26), energy stored,

$$\omega_f = \frac{B^2}{2\mu_0} = \frac{(50 \times 10^{-3})^2}{2(4\pi \times 10^{-7})} = 9.95 \times 10^2\,\text{J/m}^3$$

Volume of airgap $= Al = (500 \times 2.0)\,\text{mm}^3 = 500 \times 2.0 \times 10^{-9}\,\text{m}^3$. Hence the energy stored in the airgap,

$$W_f = 9.95 \times 10^2\,\text{J/m}^3 \times 500 \times 2.0 \times 10^{-9}\,\text{m}^3$$

$$= 9.95 \times 10^{-4}\,\text{J} \equiv \textbf{0.995 mJ} \equiv \textbf{995}\,\boldsymbol{\mu}\textbf{J}$$

Problem 3. A flux of 20 mWb links with a 1000 turn coil when a current of 8 A passes through the coil. Calculate (a) the inductance of the coil, (b) the energy stored, and (c) the average emf induced if the current falls to zero in 80 ms.

(a) From section 5, inductance of coil

$$L = \frac{N\Phi}{I} = \frac{(1000)(20 \times 10^{-3})}{8} = \textbf{2.5 H}$$

(b) From equation (27), energy stored,

$$W_f = \tfrac{1}{2}LI^2 = \tfrac{1}{2}(2.5)(8)^2 = \textbf{80 J}$$

(c) From section 5, induced emf

$$e = N\frac{d\phi}{dt} = (1000)\frac{20 \times 10^{-3}}{80 \times 10^{-3}} = \textbf{250 V}$$

or

$$e = L\frac{di}{dt} = (2.5)\left(\frac{8}{80 \times 10^{-3}}\right) = \textbf{250 V}$$

Problem 4. Determine the strength of a uniform electric field if it is to have the same energy as that established by a magnetic field of flux density 0.8 T. Assume that the relative permeability of the magnetic field and the relative permittivity of the electric field are both unity.

Figure 23

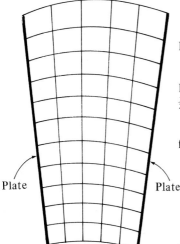

Plate   Plate

From equation (26), energy stored in magnetic field,

$$\omega_f = \frac{B^2}{2\mu_0} = \frac{(0.8)^2}{2(4\pi \times 10^{-7})}$$

$$= 2.546 \times 10^5 \, \text{J/m}^3$$

From equation (17), energy stored in electric field,

$$\omega_f = \tfrac{1}{2}\varepsilon_0\varepsilon_r E^2$$

Hence, if the energy stored in the magnetic and electric fields is to be the same, then $\tfrac{1}{2}\varepsilon_0\varepsilon_r E^2 = 2.546 \times 10^5$ i.e.,

$$\tfrac{1}{2}(8.85 \times 10^{-12})(1)E^2 = 2.546 \times 10^5$$

from which electric field strength,

$$E = \sqrt{\frac{(2)(2.546 \times 10^5)}{(8.85 \times 10^{-12})}} = \sqrt{(5.75 \times 10^{16})}$$

$$= \mathbf{2.40 \times 10^8 \, V/m \ or \ 240 \, MV/m}$$

Further problems on energy stored in an electromagnetic field may be found in the following section (9), problems 39 to 44, page 318.

9. Further problems

Field plotting by curvilinear squares

1. (a) Explain the meaning of the terms (i) streamline (ii) equipotential, with reference to an electric field.
   (b) A field plot between two metal plates is shown in Fig. 23. If the relative permittivity of the dielectric is 2.4, determine the capacitance of a 50 cm length of the system.

   [23.4 pF]

2. A field plot for a concentric cable is shown in Fig. 24. The relative permittivity of the dielectric is 5. Determine the capacitance of a 10 m length of the cable.

   [1.66 nF]

3. The plates of a capacitor are 10 mm long and 6 mm wide and are separated by a dielectric 3 mm thick and of relative permittivity 2.5. Determine the capacitance of the capacitor (a) when neglecting any fringing at the edges, (b) by producing a field plot taking fringing into consideration.

   $\begin{bmatrix}\text{(a) } 0.44 \, \text{pF} \quad \text{(b) } 0.60 \, \text{pF}-0.70 \, \text{pF, depending on the accuracy of the} \\ \text{plot}\end{bmatrix}$

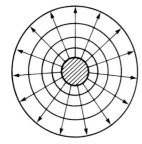

Figure 24

Capacitance between concentric cylinders

4. A coaxial cable has an inner conductor of radius 0.4 mm and an outer conductor of internal radius 4 mm. Determine the capacitance per metre length of the cable if the dielectric has a relative permittivity of 2.

   [48.30 pF]

5. A concentric cable has a core diameter of 40 mm and an inner sheath diameter of 100 mm. The relative permittivity of the dielectric is 2.5 and the core potential is 50 kV. Determine (a) the capacitance per kilometre length of the cable and (b) the dielectric stress at radii of 30 mm and 40 mm.

   [(a) 0.1517 μF   (b) 1.819 MV/m, 1.364 MV/m]

6. A coaxial cable has a capacitance of 100 pF per metre length. The relative permittivity of the dielectric is 3.2 and the core diameter is 1.0 mm. Determine the required inside diameter of the sheath.

[5.9 mm]

7. A single-core concentric cable is to be manufactured for a 100 kV, 50 Hz transmission system. The dielectric used is paper which has a maximum safe dielectric stress of 10 MV/m and a relative permittivity of 3.2. Calculate (a) the core and inner sheath radii for the most economical cable, (b) the capacitance per metre length, and (c) the charging current per kilometre run.

[(a) 10 mm; 27.2 mm   (b) 177.9 pF   (c) 5.59 A]

8. A single-core concentric cable used on a 20 kV, 50 Hz transmission system has a dielectric of relative permittivity 3 and of a safe dielectric stress 1.25 MV/m rms. Determine (a) the dimensions for the most economical cable, (b) the capacitance of the cable per 500 m run, and (c) the charging current per kilometre length of cable.

$$\begin{bmatrix} \text{(a) core radius 16 mm, inner sheath radius 43.5 mm} \\ \text{(b) 83.41 nF} \quad \text{(c) 1.048 A} \end{bmatrix}$$

9. A concentric cable has a core diameter of 30 mm and an inside sheath diameter of 75 mm. The relative permittivity is 2.6, the loss angle is 2.5×10^{-3} rad and the working voltage is 100 kV at 50 Hz frequency. Determine for a 1 km length of cable (a) the capacitance, (b) the charging current, and (c) the power loss.

[(a) 0.1578 μF   (b) 4.957 A   (c) 1239 W]

10. A concentric cable operates at 200 kV and 50 Hz. The maximum electric field strength within the cable is not to exceed 5 MV/m. Determine (a) the radius of the core and the inner radius of the sheath for ideal operation, and (b) the stress on the dielectric at the surface of the core and at the inner surface of the sheath.

[(a) 40 mm, 108.7 mm   (b) 5 MV/m, 1.84 MV/m]

11. A concentric cable has a core radius of 20 mm and a sheath inner radius of 40 mm. The permittivity of the dielectric is 2.5. Using two equipotential surfaces within the dielectric, determine the capacitance of the cable per metre length by the method of curvilinear squares. Draw the field plot for the cable.

[200.6 pF]

Capacitance of an isolated twin line

12. Two parallel wires, each of diameter 5.0 mm, are uniformly spaced in air at a distance of 40 mm between centres. Determine the capacitance of a 500 m run of the line.

[5.014 nF]

13. A single-phase circuit is comprised of two parallel conductors each of radius 5.0 mm and spaced 1.5 m apart in air. The p.d. between the conductors is 20 kV at 50 Hz. Determine (a) the capacitance per metre length of the conductors, and (b) the charging current per kilometre run.

[(a) 4.875 pF   (b) 30.63 mA]

14. The capacitance of a 300 m length of an isolated twin line is 1522 pF. The line comprises two air conductors which are spaced 1200 mm between centres. Determine the diameter of each conductor.

[10 mm]

15. (a) Derive an expression for the capacitance per metre length of an isolated twin line in terms of the absolute permittivity, the distance D between centres of the lines and the diameter d of each conductor.

(b) Two parallel infinitely long straight conductors of diameter 2.0 mm are spaced 120 mm apart in air and relatively far from conducting objects.

Determine the capacitance per kilometre run of the line.

$$\left[\text{(a)} \ C = \frac{\pi \varepsilon}{\ln\left(2D/d\right)} \qquad \text{(b)} \ 5.807\,\text{nF} \right]$$

16. An isolated twin line is comprised of two air-insulated conductors, each of radius 8.0 mm, which are spaced 1.60 m apart. The voltage between the lines is 7 kV at a frequency of 50 Hz. Determine for a 1 km length (a) the line capacitance, (b) the value of charge carried by each wire, and (c) the charging current.
[(a) 5.248 nF   (b) 36.74 μC   (c) 11.54 mA]

17. The charging current for a 1 km run of isolated twin line is not to exceed 30 mA. The p.d. between the lines is 20 kV at 50 Hz. If the line is air insulated and the conductors are spaced 1 m apart, determine (a) the maximum value required for the capacitance per metre length, and (b) the maximum diameter of each conductor.
[(a) 4.775 pF   (b) 5.92 mm]

Energy stored in an electric field

18. Determine the energy stored in a 5000 pF capacitor when charged to 800 V, and the average power developed if this energy is dissipated in 20 μs.
[1.6 mJ; 80 W]

19. A 0.25 μF capacitor is required to store 2 J of energy. Determine the p.d. to which the capacitor must be charged.
[4 kV]

20. A capacitor is charged with 6 mC. If the energy stored is 1.5 J determine (a) the voltage across the plates, and (b) the capacitance of the capacitor.
[(a) 500 V   (b) 12 μF]

21. After a capacitor is connected across a 250 V d.c. supply the charge is 5 μC. Determine (a) the capacitance, and (b) the energy stored.
[(a) 20 nF   (b) 0.625 mJ]

22. A capacitor consisting of two metal plates each of area 100 cm^2 and spaced 0.1 mm apart in air is connected across a 200 V supply. Determine (a) the electric flux density, (b) the potential gradient and (c) the energy stored in the capacitor.
[(a) 17.7 μC/m^2 (b) 2 MV/m   (c) 17.7 μJ]

23. A mica capacitor is to be constructed to have a capacitance of 0.05 μF and to have a steady working potential of 2 kV maximum. Allowing a safe value of field stress of 20 MV/m, determine (a) the required thickness of the mica dielectric, (b) the area of plate required if the relative permittivity of the mica is 5, (c) the maximum energy stored by the capacitor, and (d) the average power developed if this energy is dissipated in 25 μs.
[(a) 0.1 mm   (b) 0.113 m^2 (c) 0.1 J   (d) 4 kW]

24. A 500 pF capacitor is charged to a p.d. of 100 V. The dielectric has a cross-sectional area of 200 cm^2 and a relative permittivity of 2.4. Determine the energy stored per cubic metre in the dielectric.
[0.147 J/m^3]

25. A parallel-plate capacitor with the plates 20 mm apart is immersed in oil having a relative permittivity of 3. The plates are charged to a p.d. of 8 kV. Determine the energy stored in joules per cubic metre in the dielectric.
[2.124 J/m^3]

26. Two parallel plates each having dimensions 30 mm by 50 mm are spaced 8 mm apart in air. If a voltage of 40 kV is applied across the plates determine the energy stored in the electric field.
[1.328 mJ]

Inductance of a concentric cable

27. A coaxial cable has an inner core of radius 0.8 mm and an outer sheath of internal radius 4.8 mm. Determine the inductance of 25 m of the cable. Assume that the relative permeability of the material used is 1.

$$[10.2 \, \mu H]$$

28. A concentric cable has a core 12 mm diameter and a sheath 40 mm diameter, the sheath having negligible thickness. Determine the inductance and the capacitance of the cable per metre assuming nonmagnetic materials and a dielectric of relative permittivity 3.2.

$$[0.291 \, \mu H/m; \, 147.8 \, pF/m]$$

29. Show that the inductance L of 2π metres of a pair of concentric cylinders is given by $L = \mu(0.25 + \ln(D/d))$ henry, where μ is the absolute permeability, d is the diameter of the core and D is the diameter of outer cylinder.

30. A concentric cable has an inner sheath radius of 4.0 cm. The inductance of the cable is $0.5 \, \mu H/m$. Ignoring inductance due to internal linkages, determine the radius of the core. Assume that the relative permeability of the material is unity.

$$[3.28 \, mm]$$

31. Derive an expression for the inductance of a wire of length x metres and diameter d due to the internal flux linkages of the wire alone.

$$\left[L = \frac{\mu_0 \mu_r x}{8\pi} \right]$$

32. The inductance of a concentric cable of core radius 8 mm and inner sheath radius of 35 mm is measured as 2.0 mH. Determine (a) the length of the cable, and (b) the capacitance of the cable. Assume that nonmagnetic materials are used and the relative permittivity of the dielectric is 2.5.

$$[(a) \, 5.794 \, km \qquad (b) \, 0.546 \, \mu F]$$

Inductance of an isolated twin line

33. A single-phase power line comprises two conductors each with a radius of 15 mm and spaced 1.8 m apart in air. Determine the inductance per metre length, ignoring internal linkages and assuming the relative permeability, $\mu_r = 1$.

$$[1.915 \, \mu H/m]$$

34. In an isolated twin line the conductors have a diameter $d \times 10^{-3}$ metres and are spaced x metres apart in air. Show that the inductance L of the line, ignoring internal linkages, is given by

$$L = \frac{\mu_0}{\pi} \ln\left(\frac{2000 \, x}{d}\right) \text{ henry/metres}$$

35. Determine (a) the loop inductance, and (b) the capacitance of a 500 m length of single-phase twin line having conductors of diameter 8 mm and spaced 60 mm apart in air.

$$[(a) \, 0.592 \, mH \qquad (b) \, 5.133 \, nF]$$

36. An isolated twin power line has conductors 7.5 mm radius. Determine the distance between centres if the total loop inductance of 1 km of the line is 1.95 mH.

$$[765 \, mm]$$

37. An isolated twin line has conductors of diameter $d \times 10^{-3}$ metres and spaced D

millimetres apart in air. Derive an expression for the total loop inductance L of the line per metre length.

$$\left[L = \frac{\mu_0}{\pi}\left(\frac{1}{4} + \ln\frac{2D}{d}\right)\right]$$

38. A single-phase power line comprises two conductors spaced 2 m apart in air. The loop inductance of 2 km of the line is measured as 3.65 mH. Determine the diameter of the conductors.

[53.6 mm]

Energy stored in an electromagnetic field

39. Determine the value of the energy stored when a current of 120 mA flows in a coil of 500 mH. What value of current is required to double the energy stored?

[3.6 mJ; 169.7 mA]

40. A moving-coil instrument has two airgaps each 2.5 mm long and having a cross-sectional area of 8.0 cm^2. Determine the total energy stored in the magnetic field of the airgap if the flux density is 100 mT.

[15.92 mJ]

41. When a current of 5 A is passed through an 800 turn coil a flux of 500 μWb links with it. Determine (a) the inductance of the coil, (b) the energy stored in the magnetic field, and (c) the average emf induced if the current is reduced to zero in 160 ms.

[(a) 80 mH   (b) 1 J   (c) 2.5 V]

42. Determine the flux density of a uniform magnetic field if it is to have the same energy as that established by a uniform electric field of strength 45 MV/m. Assume the relative permeability of the magnetic field and the relative permittivity of the electric field are both unity.

[0.15 T]

43. A long single core concentric cable has inner and outer conductors of diameters D_1 and D_2 respectively. The conductors each carry a current of I amperes but in opposite directions. If the relative permeability of the material is unity and the inductance due to internal linkages is negligible, show that the magnetic energy stored in a 4 m length of the cable is given by

$$\frac{\mu_0 I^2}{\pi}\ln\left(\frac{D_2}{D_1}\right)\text{joules.}$$

44. 1 mJ of energy is stored in a uniform magnetic field having dimensions 20 mm by 10 mm by 1.0 mm. Determine for the field (a) the flux density, and (b) the magnetic field strength.

[(a) 0.112 T   (b) 89200 A/m]

13 Attenuators

1. Introduction

An **attenuator** is a device for introducing a specified loss between a signal source and a matched load without upsetting the impedance relationship necessary for matching. The loss introduced is constant irrespective of frequency; since reactive elements (L or C) vary with frequency, it follows that ideal attenuators are networks containing pure resistances. A fixed attenuator section is usually known as a "pad".

Attenuation is a reduction in the magnitude of a voltage or current due to its transmission over a line or through an attenuator. Any degree of attenuation may be achieved with an attenuator by suitable choice of resistance values but the input and output impedances of the pad must be such that the impedance conditions existing in the circuit into which it is connected are not disturbed. Thus an attenuator must provide the correct input and output impedances as well as providing the required attenuation.

Attenuator sections are made up of resistances connected as T or π arrangements (as introduced in chapter 9, section 2).

Two-port networks

Networks in which electrical energy is fed in at one pair of terminals and taken out at a second pair of terminals are called two-port networks. Thus an attenuator is a two-port network, as are transmission lines, transformers and electronic amplifiers. The network between the input port and the output port is a transmission network for which a known relationship exists between the input and output currents and voltages. If a network contains only passive circuit elements, such as in an attenuator, the network is said to be **passive**; if a network contains a source of emf, such as in an electronic amplifier, the network is said to be **active**.

Figure 1(a) shows a T-network, which is termed **symmetrical** if $Z_A = Z_B$ and Fig. 1(b) shows a π-network which is symmetrical if $Z_E = Z_F$. If $Z_A \neq Z_B$ in Fig. 1(a) and $Z_E \neq Z_F$ in Fig. 1(b), the sections are termed **asymmetrical**. Both networks shown have one common terminal, which may be earthed, and are therefore said to be **unbalanced**. The **balanced** form of the T-network is shown in Fig. 2(a) and the balanced form of the π-network is shown in Fig. 2(b).

Symmetrical T- and π-attenuators are discussed in section 4 and asymmetrical attenuators are discussed in sections 6 and 7. Before this it is

Figure 1 (a) T-network
(b) π-network

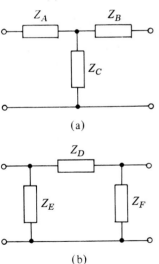

(a)

(b)

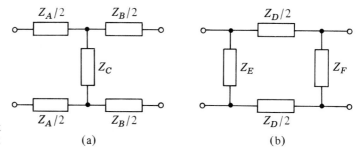

Figure 2 (a) Balanced T-network
(b) Balanced π-network

(a)   (b)

important to understand the concept of characteristic impedance, which is explained generally in section 2 (characteristic impedances will be used again in chapter 14), and logarithmic units, discussed in section 3. Another important aspect of attenuators, that of insertion loss, is discussed in section 5. To obtain greater attenuation, sections may be connected in cascade, and this is discussed in section 8.

2. Characteristic impedance

The input impedance of a network is the ratio of voltage to current (in complex form) at the input terminals. With a two-port network the input impedance often varies according to the load impedance across the output terminals. For any passive two-port network it is found that a particular value of load impedance can always be found which will produce an input impedance having the same value as the load impedance. This is called the **iterative impedance** for an asymmetrical network and its value depends on which pair of terminals is taken to be the input and which the output (there are thus two values of iterative impedance, one for each direction).

For a symmetrical network there is only one value for the iterative impedance and this is called the **characteristic impedance** of the symmetrical two-port network. Let the characteristic impedance be denoted by Z_0. Figure 3 shows a **symmetrical T-network** terminated in an impedance Z_0. Let the impedance "looking-in" at the input port also be Z_0. Then $V_1/I_1 = Z_0 = V_2/I_2$ in Fig. 3. From circuit theory,

$$\frac{V_1}{I_1} = Z_A + \frac{Z_B(Z_A + Z_0)}{Z_B + Z_A + Z_0}, \text{ since } (Z_A + Z_0) \text{ is in parallel with } Z_B,$$

$$= \frac{Z_A^2 + Z_A Z_B + Z_A Z_0 + Z_A Z_B + Z_B Z_0}{Z_A + Z_B + Z_0}$$

i.e.,

$$Z_0 = \frac{Z_A^2 + 2Z_A Z_B + Z_A Z_0 + Z_B Z_0}{Z_A + Z_B + Z_0}$$

Thus

$$Z_0(Z_A + Z_B + Z_0) = Z_A^2 + 2Z_A Z_B + Z_A Z_0 + Z_B Z_0$$

$$Z_0 Z_A + Z_0 Z_B + Z_0^2 = Z_A^2 + 2Z_A Z_B + Z_A Z_0 + Z_B Z_0$$

i.e., $Z_0^2 = Z_A^2 + 2Z_A Z_B$, from which

characteristic impedance, $\mathbf{Z_0 = \sqrt{(Z_A^2 + 2Z_A Z_B)}}$ (1)

If the output terminals of Fig. 3 are open-circuited, then the open-circuit impedance, $Z_{OC} = Z_A + Z_B$. If the output terminals of Fig. 3 are short-circuited, then the short-circuit impedance,

$$Z_{SC} = Z_A + \frac{Z_A Z_B}{Z_A + Z_B} = \frac{Z_A^2 + 2Z_A Z_B}{Z_A + Z_B}$$

Thus

$$Z_{OC} Z_{SC} = (Z_A + Z_B)\left(\frac{Z_A^2 + 2Z_A Z_B}{Z_A + Z_B}\right) = Z_A^2 + 2Z_A Z_B$$

Comparing this with equation (1) gives

$$Z_0 = \sqrt{(Z_{OC} Z_{SC})} \tag{2}$$

Figure 4

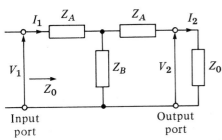

Figure 3

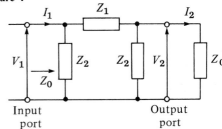

Figure 4 shows a symmetrical π-network terminated in an impedance Z_0. If the impedance "looking in" at the input port is also Z_0, then

$$\frac{V_1}{I_1} = Z_0 = (Z_2) \text{ in parallel with } [Z_1 \text{ in series with } (Z_0 \text{ and } Z_2) \text{ in parallel}]$$

$$= (Z_2) \text{ in parallel with } \left[Z_1 + \frac{Z_0 Z_2}{Z_0 + Z_2}\right]$$

$$= (Z_2) \text{ in parallel with } \left[\frac{Z_1 Z_0 + Z_1 Z_2 + Z_0 Z_2}{Z_0 + Z_2}\right]$$

i.e.,

$$Z_0 = \frac{(Z_2)((Z_1 Z_0 + Z_1 Z_2 + Z_0 Z_2)/(Z_0 + Z_2))}{Z_2 + ((Z_1 Z_0 + Z_1 Z_2 + Z_0 Z_2)/(Z_0 + Z_2))}$$

$$= \frac{(Z_1 Z_2 Z_0 + Z_1 Z_2^2 + Z_0 Z_2^2)/(Z_0 + Z_2)}{(Z_2 Z_0 + Z_2^2 + Z_1 Z_0 + Z_1 Z_2 + Z_0 Z_2)/(Z_0 + Z_2)}$$

$$= \frac{Z_1 Z_2 Z_0 + Z_1 Z_2^2 + Z_0 Z_2^2}{Z_2^2 + 2Z_2 Z_0 + Z_1 Z_0 + Z_1 Z_2}$$

Thus

$$Z_0(Z_2^2 + 2Z_2 Z_0 + Z_1 Z_0 + Z_1 Z_2) = Z_1 Z_2 Z_0 + Z_1 Z_2^2 + Z_0 Z_2^2$$

$$2Z_2 Z_0^2 + Z_1 Z_0^2 = Z_1 Z_2^2$$

from which

$$\text{characteristic impedance, } Z_0 = \sqrt{\left(\frac{Z_1 Z_2^2}{Z_1 + 2Z_2}\right)} \qquad (3)$$

If the output terminals of Fig. 4 are open-circuited, then the open-circuit impedance,

$$Z_{OC} = \frac{Z_2(Z_1 + Z_2)}{Z_2 + Z_1 + Z_2} = \frac{Z_2(Z_1 + Z_2)}{Z_1 + 2Z_2}$$

If the output terminals of Fig. 4 are short-circuited, then the short-circuit impedance,

$$Z_{SC} = \frac{Z_2 Z_1}{Z_1 + Z_2}$$

Thus

$$Z_{OC} Z_{SC} = \frac{Z_2(Z_1 + Z_2)}{(Z_1 + 2Z_2)} \left(\frac{Z_2 Z_1}{Z_1 + Z_2}\right) = \frac{Z_1 Z_2^2}{Z_1 + 2Z_2}$$

Comparing this expression with equation (3) gives

$$Z_0 = \sqrt{(Z_{OC} Z_{SC})},$$

which is the same as equation (2).

Thus the characteristic impedance Z_0 is given by $Z_0 = \sqrt{(Z_{OC} Z_{SC})}$ whether the network is a symmetrical T or a symmetrical π.

Equations (1) to (3) are used later in this chapter.

3. Logarithmic ratios

The ratio of two powers P_1 and P_2 may be expressed in logarithmic form. Let P_1 be the input power to a system and P_2 the output power.

If logarithms to base 10 are used, then the ratio is said to be in **bels**, i.e., power ratio in bels $= \lg(P_2/P_1)$. The bel is a large unit and the **decibel (dB)** is more often used, where 10 decibels = 1 bel, i.e.,

$$\text{power ratio in decibels} = 10 \lg \frac{P_2}{P_1} \qquad (4)$$

For example:

P_2/P_1	Power ratio (dB)
1	$10 \lg 1 = 0$
100	$10 \lg 100 = +20 \text{(power gain)}$
$\frac{1}{10}$	$10 \lg \frac{1}{10} = -10 \text{ (power loss or attenuation)}$

If **logarithms to base e** (i.e., natural or Naperian logarithms) are used, then the ratio of two powers is said to be in **nepers (Np)**, i.e.,

$$\text{power ratio in nepers} = \tfrac{1}{2} \ln \frac{P_2}{P_1} \qquad (5)$$

Thus when the power ratio $P_2/P_1 = 5$, the power ratio in nepers $= \frac{1}{2}\ln 5 = 0.805\,\text{Np}$, and when the power ratio $P_2/P_1 = 0.1$, the power ratio in nepers $= \frac{1}{2}\ln 0.1 = -1.15\,\text{Np}$. The attenuation along a transmission line is of an exponential form and it is in such applications that the unit of the neper is used (see chapter 14).

If the powers P_1 and P_2 refer to power developed in two equal resistors, R, then $P_1 = V_1^2/R$ and $P_2 = V_2^2/R$. Thus the ratio (from equation (4)) can be expressed, by the laws of logarithms, as

$$\textbf{ratio in decibels} = 10\lg\frac{P_2}{P_1} = 10\lg\left(\frac{V_2^2/R}{V_1^2/R}\right) = 10\lg\frac{V_2^2}{V_1^2}$$

$$= 10\lg\left(\frac{V_2}{V_1}\right)^2 = \textbf{20}\lg\frac{V_2}{V_1} \qquad (6)$$

Although this is really a power ratio, it is called the **logarithmic voltage ratio**. Alternatively (from equation (5)),

$$\text{ratio in nepers} = \frac{1}{2}\ln\frac{P_2}{P_1} = \frac{1}{2}\ln\left(\frac{V_2^2/R}{V_1^2/R}\right) = \frac{1}{2}\ln\left(\frac{V_2}{V_1}\right)^2$$

i.e.,

$$\textbf{ratio in nepers} = \textbf{ln}\,\frac{V_2}{V_1} \qquad (7)$$

Similarly, if currents I_1 and I_2 in two equal resistors R give powers P_1 and P_2 then (from equation (4))

$$\text{ratio in decibels} = 10\lg\frac{P_2}{P_1} = 10\lg\left(\frac{I_2^2 R}{I_1^2 R}\right) = 10\lg\left(\frac{I_2}{I_1}\right)^2$$

i.e.,

$$\textbf{ratio in decibels} = \textbf{20}\lg\frac{I_2}{I_1} \qquad (8)$$

Alternatively (from equation (5)),

$$\text{ratio in nepers} = \frac{1}{2}\ln\frac{P_2}{P_1} = \frac{1}{2}\ln\left(\frac{I_2^2 R}{I_1^2 R}\right) = \frac{1}{2}\ln\left(\frac{I_2}{I_1}\right)^2$$

i.e.,

$$\textbf{ratio in nepers} = \textbf{ln}\,\frac{I_2}{I_1} \qquad (9)$$

In equations (4) to (9) the output-to-input ratio has been used. However, the input-to-output ratio may also be used. For example, in equation (6), the output-to-input voltage ratio is expressed as $20\lg(V_2/V_1)\,\text{dB}$. Alternatively, the input-to-output voltage ratio may be expressed as $20\lg(V_1/V_2)\,\text{dB}$, the only difference in the values obtained being a difference in sign. If $20\lg(V_2/V_1) = 10\,\text{dB}$, say, then $20\lg(V_1/V_2) = -10\,\text{dB}$. Thus if an attenuator has a voltage input V_1 of $50\,\text{mV}$ and a voltage output V_2 of $5\,\text{mV}$, the voltage ratio V_2/V_1 is $\frac{5}{50}$ or $\frac{1}{10}$. Alternatively, this may be expressed as **"an attenuation of 10"**, i.e., $V_1/V_2 = 10$.

Worked problems on logarithmic ratios

Problem 1. The ratio of output power to input power in a system is

(a) 2   (b) 25   (c) 1000 and (d) $\frac{1}{100}$.

Determine the power ratio in each case (i) in decibels and (ii) in nepers.

(i) From equation (4), power ratio in decibels $= 10 \lg (P_2/P_1)$.
   (a) When $P_2/P_1 = 2$, power ratio $= 10 \lg 2 = $ **3 dB.**
   (b) When $P_2/P_1 = 25$, power ratio $= 10 \lg 25 = $ **14 dB.**
   (c) When $P_2/P_1 = 1000$, power ratio $= 10 \lg 1000 = $ **30 dB.**
   (d) When $P_2/P_1 = \frac{1}{100}$, power ratio $= 10 \lg \frac{1}{100} = $ **− 20 dB.**
(ii) From equation (5), power ratio in nepers $= \frac{1}{2} \ln (P_2/P_1)$.
   (a) When $P_2/P_1 = 2$, power ratio $= \frac{1}{2} \ln 2 = $ **0.347 Np.**
   (b) When $P_2/P_1 = 25$, power ratio $= \frac{1}{2} \ln 25 = $ **1.609 Np.**
   (c) When $P_2/P_1 = 1000$, power ratio $= \frac{1}{2} \ln 1000 = $ **3.454 Np.**
   (d) When $P_2/P_1 = \frac{1}{100}$, power ratio $= \frac{1}{2} \ln \frac{1}{100} = $ **− 2.303 Np.**

The power ratios in (a), (b) and (c) represent power gains, since the ratios are positive values; the power ratio in (d) represents a power loss or attenuation, since the ratio is a negative value.

Problem 2. 5% of the power supplied to a cable appears at the output terminals. Determine the attenuation in decibels.

If $P_1 = $ input power and $P_2 = $ output power, then

$$\frac{P_2}{P_1} = \frac{5}{100} = 0.05$$

From equation (4), power ratio in decibels $= 10 \lg (P_2/P_1) = 10 \lg 0.05 = -13 \, \text{dB}$.
Hence the attenuation (i.e., power loss) is 13 dB.

Problem 3. An amplifier has a gain of 15 dB. If the input power is 12 mW, determine the output power.

From equation (4), decibel power ratio $= 10 \lg (P_2/P_1)$. Hence $15 = 10 \lg (P_2/12)$, where P_2 is the output power in milliwatts.

$$1.5 = \lg \left(\frac{P_2}{12} \right)$$
$$\frac{P_2}{12} = 10^{1.5}$$

from the definition of a logarithm. Thus the output power,

$$P_2 = 12(10)^{1.5} = \textbf{379.5 mW}$$

Problem 4. The current output of an attenuator is 50 mA. If the current ratio of the attenuator is -1.32 Np, determine (a) the current input and (b) the current ratio expressed in decibels. Assume that the input and load resistances of the attenuator are equal.

(a) From equation (9), current ratio in nepers $= \ln (I_2/I_1)$. Hence $-1.32 = \ln (50/I_1)$, where I_1 is the input current in mA.

$$e^{-1.32} = \frac{50}{I_1}$$

from which

$$\textbf{current input, } I_1 = \frac{50}{e^{-1.32}} = 50e^{1.32} = \textbf{187.2 mA}$$

(b) From equation (8),

$$\text{current ratio in decibels} = 20\lg\frac{I_2}{I_1}$$

$$= 20\lg\left(\frac{50}{187.2}\right) = -\textbf{11.47 dB}$$

Further problems on logarithmic ratios may be found in section 9, problems 1 to 6, page 345.

4. Symmetrical T- and π-attenuators

(a) Symmetrical T-attenuator

As mentioned in section 1, the ideal attenuator is made up of pure resistances. A symmetrical T-pad attenuator is shown in Fig. 5 with a termination R_0 connected as shown. From equation (1),

$$R_0 = \sqrt{(R_1^2 + 2R_1R_2)} \qquad (10)$$

and from equation (2),

$$R_0 = \sqrt{(R_{oc}R_{sc})} \qquad (11)$$

With resistance R_0 as the termination, the input resistance of the pad will also be equal to R_0. If the terminating resistance R_0 is transferred to port A then the input resistance looking into port B will again be R_0.

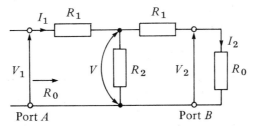

Figure 5 Symmetrical T-pad attenuator

The pad is therefore symmetrical in impedance in both directions of connection and may thus be inserted into a network whose impedance is also R_0. The value of R_0 is the characteristic impedance of the section.

As stated in section 3, attenuation may be expressed as a voltage ratio V_1/V_2 (see Fig. 5) or quoted in decibels as $20\lg(V_1/V_2)$ or, alternatively, as a power ratio as $10\lg(P_1/P_2)$. If a T-section is symmetrical, i.e., the terminals of the section are matched to equal impedances, then

$$10\lg\frac{P_1}{P_2} = 20\lg\frac{V_1}{V_2} = 20\lg\frac{I_1}{I_2}$$

since $R_{IN} = R_{LOAD} = R_0$, i.e.,

$$10 \lg \frac{P_1}{P_2} = 10 \lg \left(\frac{V_1}{V_2}\right)^2 = 10 \lg \left(\frac{I_1}{I_2}\right)^2$$

from which

$$\frac{P_1}{P_2} = \left(\frac{V_1}{V_2}\right)^2 = \left(\frac{I_1}{I_2}\right)^2$$

or

$$\sqrt{\left(\frac{P_1}{P_2}\right)} = \left(\frac{V_1}{V_2}\right) = \left(\frac{I_1}{I_2}\right)$$

Let $N = V_1/V_2$ or I_1/I_2 or $\sqrt{(P_1/P_2)}$, where N is the attenuation. In section 5, page 332, it is shown that, for a matched network, i.e., one terminated in its characteristic impedance, N is in fact the insertion loss ratio. (Note that in an asymmetrical network, only the expression $N = \sqrt{(P_1/P_2)}$ may be used—see section 7.) From Fig. 5,

$$\text{current } I_1 = \frac{V_1}{R_0}$$

$$\text{Voltage } V = V_1 - I_1 R_1 = V_1 - \left(\frac{V_1}{R_0}\right) R_1$$

i.e.,

$$V = V_1 \left(1 - \frac{R_1}{R_0}\right)$$

$$\text{Voltage } V_2 = \left(\frac{R_0}{R_1 + R_0}\right) V \qquad \text{by voltage division}$$

i.e.,

$$V_2 = \left(\frac{R_0}{R_1 + R_0}\right) V_1 \left(1 - \frac{R_1}{R_0}\right) = V_1 \left(\frac{R_0}{R_1 + R_0}\right)\left(\frac{R_0 - R_1}{R_0}\right)$$

Hence

$$\frac{V_2}{V_1} = \frac{R_0 - R_1}{R_0 + R_1} \quad \text{or} \quad \frac{V_1}{V_2} = N = \frac{R_0 + R_1}{R_0 - R_1} \tag{12}$$

From equation (12) and also equation (10), it is possible to derive expressions for R_1 and R_2 in terms of N and R_0, thus enabling an attenuator to be designed to give a specified attenuation and to be matched symmetrically into the network. From equation (12),

$$N(R_0 - R_1) = R_0 + R_1$$
$$NR_0 - NR_1 = R_0 + R_1$$
$$R_0(N - 1) = R_1(1 + N)$$

from which

$$R_1 = R_0 \frac{(N-1)}{(N+1)} \tag{13}$$

From equation (10), $R_0 = \sqrt{(R_1^2 + 2R_1 R_2)}$ i.e., $R_0^2 = R_1^2 + 2R_1 R_2$, from which

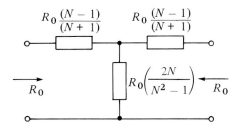

Figure 6

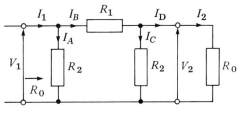

Figure 7 Symmetrical π-attenuator

$$R_2 = \frac{R_0^2 - R_1^2}{2R_1}$$

Substituting for R_1 from equation (13) gives

$$R_2 = \frac{R_0^2 - [R_0(N-1)/(N+1)]^2}{2[R_0(N-1)/(N+1)]} = \frac{[R_0^2(N+1)^2 - R_0^2(N-1)^2]/(N+1)^2}{2R_0(N-1)/(N+1)}$$

i.e.,

$$R_2 = \frac{R_0^2[(N+1)^2 - (N-1)^2]}{2R_0(N-1)(N+1)} = \frac{R_0[(N^2+2N+1)-(N^2-2N+1)]}{2(N^2-1)}$$

$$= \frac{R_0(4N)}{2(N^2-1)}$$

Hence

$$\boldsymbol{R_2 = R_0}\left(\frac{2N}{N^2-1}\right) \qquad (14)$$

Thus if the characteristic impedance R_0 and the attenuation $N(=V_1/V_2)$ are known for a symmetrical T-network then values of R_1 and R_2 may be calculated. Figure 6 shows a T-pad attenuator having input and output impedances of R_0 with resistances R_1 and R_2 expressed in terms of R_0 and N.

(b) Symmetrical π-attenuator

A symmetrical π-attenuator is shown in Fig. 7 terminated in R_0. From equation (3),

$$\text{characteristic impedance } \boldsymbol{R_0} = \sqrt{\left(\frac{R_1 R_2^2}{R_1 + 2R_2}\right)} \qquad (15)$$

and from equation (2),

$$\boldsymbol{R_0} = \sqrt{(R_{oc}R_{sc})} \qquad (16)$$

Given the attenuation factor

$$N = \frac{V_1}{V_2}\left(=\frac{I_1}{I_2}\right)$$

and the characteristic impedance R_0, it is possible to derive expressions for R_1 and R_2, in a similar way to the T-pad attenuator, to enable a π-attenuator to be effectively designed. Since $N = V_1/V_2$ then $V_2 = V_1/N$. From Fig. 7, current $I_1 = I_A + I_B$ and current $I_B = I_C + I_D$. Thus

$$\text{current } I_1 = \frac{V_1}{R_0} = I_A + I_C + I_D$$

$$= \frac{V_1}{R_2} + \frac{V_2}{R_2} + \frac{V_2}{R_0} = \frac{V_1}{R_2} + \frac{V_1}{NR_2} + \frac{V_1}{NR_0}$$

since $V_2 = V_1/N$, i.e.,

$$\frac{V_1}{R_0} = V_1\left(\frac{1}{R_2} + \frac{1}{NR_2} + \frac{1}{NR_0}\right)$$

Hence

$$\frac{1}{R_0} = \frac{1}{R_2} + \frac{1}{NR_2} + \frac{1}{NR_0}$$

$$\frac{1}{R_0} - \frac{1}{NR_0} = \frac{1}{R_2} + \frac{1}{NR_2}$$

$$\frac{1}{R_0}\left(1 - \frac{1}{N}\right) = \frac{1}{R_2}\left(1 + \frac{1}{N}\right)$$

$$\frac{1}{R_0}\left(\frac{N-1}{N}\right) = \frac{1}{R_2}\left(\frac{N+1}{N}\right)$$

Thus

$$R_2 = R_0\frac{(N+1)}{(N-1)} \tag{17}$$

From Fig. 7, current $I_1 = I_A + I_B$, and since the p.d. across R_1 is $(V_1 - V_2)$,

$$\frac{V_1}{R_0} = \frac{V_1}{R_2} + \frac{V_1 - V_2}{R_1}$$

$$\frac{V_1}{R_0} = \frac{V_1}{R_2} + \frac{V_1}{R_1} - \frac{V_2}{R_1}$$

$$\frac{V_1}{R_0} = \frac{V_1}{R_2} + \frac{V_1}{R_1} - \frac{V_1}{NR_1} \quad \text{since } V_2 = V_1/N$$

$$\frac{1}{R_0} = \frac{1}{R_2} + \frac{1}{R_1} - \frac{1}{NR_1}$$

$$\frac{1}{R_0} - \frac{1}{R_2} = \frac{1}{R_1}\left(1 - \frac{1}{N}\right)$$

$$\frac{1}{R_0} - \frac{(N-1)}{R_0(N+1)} = \frac{1}{R_1}\left(\frac{N-1}{N}\right) \quad \text{from equation (17),}$$

$$\frac{1}{R_0}\left(1 - \frac{N-1}{N+1}\right) = \frac{1}{R_1}\left(\frac{N-1}{N}\right)$$

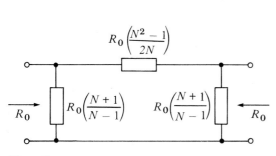

Figure 8

$$\frac{1}{R_0}\left(\frac{(N+1)-(N-1)}{(N+1)}\right)=\frac{1}{R_1}\left(\frac{N-1}{N}\right)$$

$$\frac{1}{R_0}\left(\frac{2}{N+1}\right)=\frac{1}{R_1}\left(\frac{N-1}{N}\right)$$

$$R_1=R_0\left(\frac{N-1}{N}\right)\left(\frac{N+1}{2}\right)$$

Hence

$$\boldsymbol{R_1=R_0\left(\frac{N^2-1}{2N}\right)} \tag{18}$$

Figure 8 shows a π-attenuator having input and output impedances of R_0 with resistance R_1 and R_2 expressed in terms of R_0 and N.

There is no difference in the functions of the T- and π-attenuator pads and either may be used in a particular situation.

Worked problems on symmetric T- and π-attenuators

Problem 1. Determine the characteristic impedance of each of the attenuator sections shown in Fig. 9.

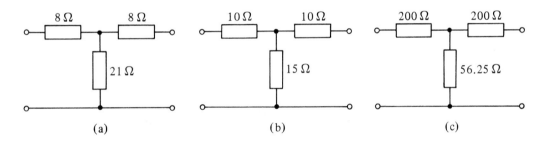

Figure 9   (a)   (b)   (c)

From equation (10), for a T-section attenuator the characteristic impedance, $R_0=\sqrt{(R_1^2+2R_1R_2)}$.

(a) $R_0=\sqrt{(8^2+(2)(8)(21))}=\sqrt{400}=\textbf{20 }\boldsymbol{\Omega}$.
(b) $R_0=\sqrt{(10^2+(2)(10)(15))}=\sqrt{400}=\textbf{20 }\boldsymbol{\Omega}$.
(c) $R_0=\sqrt{(200^2+(2)(200)(56.25))}=\sqrt{62500}=\textbf{250 }\boldsymbol{\Omega}$.

It is seen that the characteristic impedance of parts (a) and (b) is the same. In fact, there are numerous combinations of resistances R_1 and R_2 which would give the same value for the characteristic impedance.

Problem 2. A symmetrical π-attenuator pad has a series arm of 500 Ω resistance and each shunt arm of 1 kΩ resistance. Determine (a) the characteristic impedance, and (b) the attenuation (dB) produced by the pad.

The π-attenuator section is shown in Fig. 10 terminated in its characteristic impedance, R_0.

(a) From equation (15), for a symmetrical π-attenuator section,

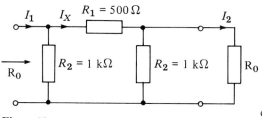

Figure 10

characteristic impedance, $R_0 = \sqrt{\left(\dfrac{R_1 R_2^2}{R_1 + 2R_2}\right)}$

Hence

$$R_0 = \sqrt{\left[\dfrac{(500)(1000)^2}{500 + 2(1000)}\right]} = 447\,\Omega.$$

(b) Attenuation $= 20\lg(I_1/I_2)\,\text{dB}$. From Fig. 10,

$$\text{current } I_X = \left(\dfrac{R_2}{R_2 + R_1 + (R_2 R_0/(R_2 + R_0))}\right)(I_1), \qquad \text{by current division}$$

i.e.,

$$I_X = \left(\dfrac{1000}{1000 + 500 + ((1000)(447)/(1000 + 447))}\right)I_1 = 0.553I_1$$

and

$$\text{current } I_2 = \left(\dfrac{R_2}{R_2 + R_0}\right)I_X = \left(\dfrac{1000}{1000 + 447}\right)I_X = 0.691I_X$$

Hence $I_2 = 0.691(0.553I_1) = 0.382I_1$ and $I_1/I_2 = 1/0.382 = 2.617$. Thus

$$\textbf{attenuation} = 20\lg 2.617 = \textbf{8.36 dB}$$

(Alternatively, since $I_1/I_2 = N$, then the formula

$$R_2 = R_0\left(\dfrac{N+1}{N-1}\right)$$

may be transposed for N, from which **attenuation** $= \textbf{20 lg } N.$)

Problem 3. For each of the attenuator networks shown in Fig. 11, determine (a) the input resistance when the output port is open-circuited, (b) the input resistance when the output port is short-circuited, and (c) the characteristic impedance.

(i) For the T-network shown in Fig. 11(i):
   (a) $R_{OC} = 15 + 10 = \textbf{25}\,\Omega$.

   (b) $R_{SC} = 15 + \dfrac{10 \times 15}{10 + 15} = 15 + 6 = \textbf{21}\,\Omega$.

Figure 11

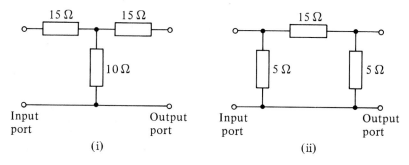

Input port   Output port   Input port   Output port

(i)   (ii)

(c) From equation (11), $R_0 = \sqrt{(R_{OC}R_{SC})} = \sqrt{[(25)(21)]} = \mathbf{22.9\,\Omega}$.
(Alternatively, from equation (10), $R_0 = \sqrt{(R_1^2 + 2R_1R_2)} = \sqrt{(15^2 + (2)(15)(10))}$
$= \mathbf{22.9\,\Omega}$.)

(ii) For the π-network shown in Fig. 11(ii):

(a) $R_{OC} = \dfrac{5 \times (15 + 5)}{5 + (15 + 5)} = \dfrac{100}{25} = \mathbf{4\,\Omega}$

(b) $R_{SC} = \dfrac{5 \times 15}{5 + 15} = \dfrac{75}{20} = \mathbf{3.75\,\Omega}$

(c) From equation (16),

$$R_0 = \sqrt{(R_{OC}R_{SC})} \qquad \text{as for a T-network}$$
$$= \sqrt{[(4)(3.75)]} = \sqrt{15} = \mathbf{3.87\,\Omega}$$

(Alternatively, from equation (15),

$$R_0 = \sqrt{\left(\frac{R_1 R_2^2}{R_1 + 2R_2}\right)} = \sqrt{\left(\frac{15(5)^2}{15 + 2(5)}\right)}$$
$$= \mathbf{3.87\,\Omega})$$

Problem 4. Design a T-section symmetrical attenuator pad to provide a voltage attenuation of 20 dB and having a characteristic impedance of 600 Ω.

Voltage attenuation in decibels $= 20\lg(V_1/V_2)$.
Attenuation, $N = V_1/V_2$, hence $20 = 20\lg N$, from which $N = 10$.
Characteristic impedance, $R_0 = 600\,\Omega$.

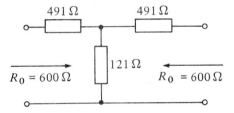

Figure 12

From equation (13),

$$\text{resistance } R_1 = \frac{R_0(N-1)}{(N+1)} = \frac{600(10-1)}{(10+1)} = \mathbf{491\,\Omega}$$

From equation (14),

$$\text{resistance } R_2 = R_0\left(\frac{2N}{N^2 - 1}\right) = 600\left(\frac{(2)(10)}{10^2 - 1}\right) = \mathbf{121\,\Omega}$$

Thus the T-section attenuator shown in Fig. 12 has a voltage attenuation of 20 dB and a characteristic impedance of 600 Ω.
(Check: From equation (10),

$$R_0 = \sqrt{(R_1^2 + 2R_1R_2)} = \sqrt{[491^2 + 2(491)(121)]} = 600\,\Omega)$$

Problem 5. Design a π-section symmetrical attenuator pad to provide a voltage attenuation of 20 dB and having a characteristic impedance of 600 Ω.

From problem 4, $N = 10$ and $R_0 = 600\,\Omega$.

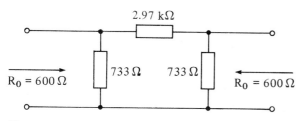

Figure 13

From equation (18),

$$\text{resistance } R_1 = R_0\left(\frac{N^2-1}{2N}\right) = 600\left(\frac{10^2-1}{(2)(10)}\right) = 2970\,\Omega \text{ or } 2.97\,k\Omega$$

From equation (17),

$$R_2 = R_0\left(\frac{N+1}{N-1}\right) = 600\left(\frac{10+1}{10-1}\right) = 733\,\Omega$$

Thus the π-section attenuator shown in Fig. 13 has a voltage attenuation of 20 dB and a characteristic impedance of $600\,\Omega$.

$\Bigg($ Check: From equation (15),

$$R_0 = \sqrt{\left(\frac{R_1 R_2^2}{R_1 + 2R_2}\right)} = \sqrt{\left(\frac{(2970)(733)^2}{2970 + (2)(733)}\right)} = 600\,\Omega \Bigg)$$

Further problems on symmetric T- and π-attenuators may be found in section 9, problems 7 to 18, page 345.

5. Insertion loss

Figure 14(a) shows a generator E connected directly to a load Z_L. Let the current flowing be I_L and the p.d. across the load V_L. z is the internal impedance of the source.

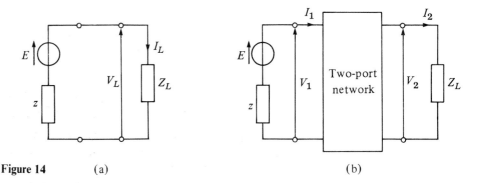

Figure 14   (a)   (b)

Figure 14(b) shows a two-port network connected between the generator E and load Z_L. The current through the load, shown as I_2, and the p.d. across the load, shown as V_2, will generally be less than current I_L and voltage V_L of Fig. 14(a), as a result of the insertion of the two-port network between generator and load.

The **insertion loss ratio, A_L,** is defined as

$$A_L = \frac{\text{voltage across load when connected directly to the generator}}{\text{voltage across load when the two-port network is connected}}$$

i.e.,

$$A_L = V_L/V_2 = I_L/I_2 \tag{19}$$

since $V_L = I_L Z_L$ and $V_2 = I_2 Z_L$. Since both V_L and V_2 refer to p.d.'s across the same impedance Z_L, the insertion loss ratio may also be expressed (from section 3) as

$$\textbf{insertion loss ratio} = 20\lg\left(\frac{V_L}{V_2}\right)\textbf{dB or } 20\lg\left(\frac{I_L}{I_2}\right)\textbf{dB} \tag{20}$$

When the two-port network is terminated in its characteristic impedance Z_0 the network is said to be **matched.** In such circumstances the input impedance is also Z_0, thus the insertion loss is simply the ratio of input to output voltage (i.e., V_1/V_2). Thus, for a network terminated in its characteristic impedance,

$$\textbf{insertion loss} = 20\lg\left(\frac{V_1}{V_2}\right)\textbf{dB or } 20\lg\left(\frac{I_1}{I_2}\right)\textbf{dB} \tag{21}$$

Worked problems on insertion loss

Problem 1. The attenuator shown in Fig. 15 feeds a matched load. Determine (a) the characteristic impedance R_0, and (b) the insertion loss in decibels.

(a) From equation (10), the characteristic impedance of a symmetric T-pad attenuator is given by $R_0 = \sqrt{(R_1^2 + 2R_1R_2)} = \sqrt{[300^2 + 2(300)(450)]} = \textbf{600 }\Omega$.
(b) Since the T-network is terminated in its characteristic impedance, then from equation (21), insertion loss $= 20\lg(V_1/V_2)\text{dB or } 20\lg(I_1/I_2)\text{dB}$.
   By current division in Fig. 15,

$$I_2 = \left(\frac{R_2}{R_2 + R_1 + R_0}\right)(I_1)$$

Hence

Figure 15

I_1 $R_1 = 300\,\Omega$ $R_1 = 300\,\Omega$

V_1 $R_2 = 450\,\Omega$ V_2 R_0 I_2

$$\textbf{insertion loss} = 20\lg\frac{I_1}{I_2} = 20\lg\left(\frac{I_1}{(R_2/(R_2 + R_1 + R_0))I_1}\right)$$

$$= 20\lg\left(\frac{R_2 + R_1 + R_0}{R_2}\right)$$

$$= 20\lg\left(\frac{450 + 300 + 600}{450}\right) = 20\lg 3$$

$$= \textbf{9.54 dB}$$

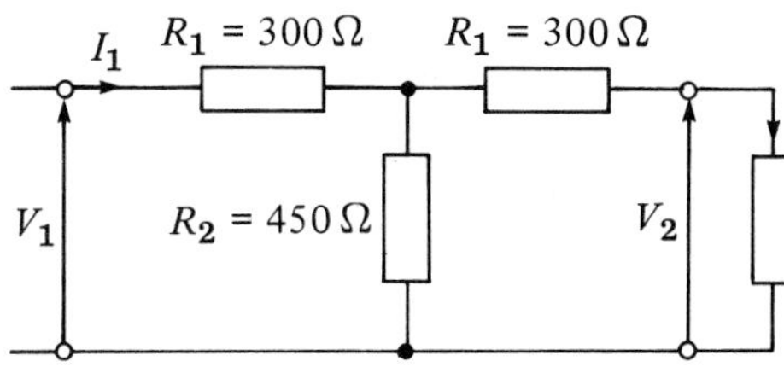

Problem 2. A 0–$3\,\text{k}\Omega$ rheostat is connected across the output of a signal generator of internal resistance $500\,\Omega$. If a load of $2\,\text{k}\Omega$ is connected across the rheostat, determine the insertion loss at a tapping of (a) $2\,\text{k}\Omega$, (b) $1\,\text{k}\Omega$.

The circuit diagram is shown in Fig. 16. Without the rheostat in the circuit the voltage across the 2 kΩ load, V_L (see Fig. 17), is given by

$$V_L = \left(\frac{2000}{2000 + 500}\right)E = 0.8\,E$$

(a) With the 2 kΩ tapping, the network of Fig. 16 may be redrawn as shown in Fig. 18, which in turn is simplified as shown in Fig. 19. From Fig. 19,

$$\text{voltage } V_2 = \left(\frac{1000}{1000 + 1000 + 500}\right)E = 0.4\,E$$

Hence, from equation (19), insertion loss ratio,

$$A_L = \frac{V_L}{V_2} = \frac{0.8E}{0.4E} = \mathbf{2},$$

or, from equation (20),

$$\text{insertion loss} = 20\lg(V_L/V_2) = 20\lg 2 = \mathbf{6.02\,dB}$$

(b) With the 1 kΩ tapping, voltage V_2 is given by

$$V_2 = \left(\frac{(1000 \times 2000)/(1000 + 2000)}{((1000 \times 2000)/(1000 + 2000)) + 2000 + 500}\right)E$$

$$= \left(\frac{666.7}{666.7 + 2000 + 500}\right)E = 0.211\,E$$

Hence, from equation (19),

$$\text{insertion loss ratio } A_L = \frac{V_L}{V_2} = \frac{0.8E}{0.211E} = \mathbf{3.79}$$

or, from equation (20),

$$\text{insertion loss in decibels} = 20\lg\left(\frac{V_L}{V_2}\right) = 20\lg 3.79$$

$$= \mathbf{11.57\,dB}$$

(Note that the insertion loss is not doubled by halving the tapping.)

Problem 3. A symmetrical π-attenuator pad has a series arm of resistance 1000 Ω and shunt arms each of 500 Ω. Determine (a) its characteristic impedance, and (b) the insertion loss (in decibels) when feeding a matched load.

The π-attenuator pad is shown in Fig. 20, terminated in its characteristic impedance, R_0.

Figure 16

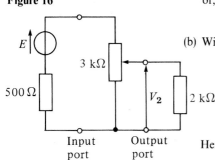

Input port   Output port

Figure 17

Figure 18

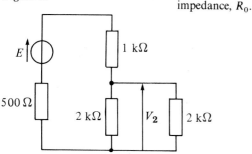

Figure 19

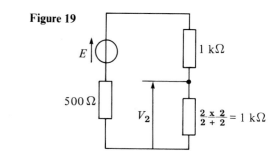

(a) From equation (15), the characteristic impedance of a symmetrical attenuator is given by

$$R_0 = \sqrt{\left(\frac{R_1 R_2^2}{R_1 + 2R_2}\right)} = \sqrt{\left(\frac{(1000)(500)^2}{1000 + 2(500)}\right)} = \mathbf{354\,\Omega}$$

(b) Since the attenuator network is feeding a matched load, from equation (21),

$$\text{insertion loss} = 20\lg\left(\frac{V_1}{V_2}\right)\text{dB} = 20\lg\left(\frac{I_1}{I_2}\right)\text{dB}.$$

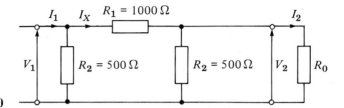

Figure 20

From Fig. 20, by current division,

$$\text{current } I_X = \left\{\frac{R_2}{R_2 + R_1 + (R_2 R_0/(R_2 + R_0))}\right\}(I_1),$$

and

$$\text{current } I_2 = \left(\frac{R_2}{R_2 + R_0}\right)I_X = \left(\frac{R_2}{R_2 + R_0}\right)\left(\frac{R_2}{R_2 + R_1 + (R_2 R_0/(R_2 + R_0))}\right)I_1$$

i.e.,

$$I_2 = \left(\frac{500}{500 + 354}\right)\left(\frac{500}{500 + 1000 + ((500)(354)/(500 + 354))}\right)I_1$$

$$= (0.5855)(0.2929)I_1 = 0.1715 I_1$$

Hence $I_1/I_2 = 1/0.1715 = 5.83$.

Thus the insertion loss in decibels $= 20\lg(I_1/I_2) = 20\lg 5.83 = \mathbf{15.3\,dB}$.

Further problems on insertion loss may be found in section 9, problems 19 to 21, page 347.

6. Asymmetrical T- and π-sections

Figure 21(a) shows an asymmetrical T-pad section where resistance $R_1 \neq R_3$. Figure 21(b) shows an asymmetrical π-section where $R_2 \neq R_3$. When viewed from port A, in each of the sections, the output impedance is

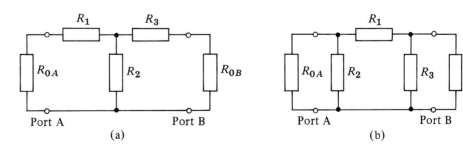

Figure 21
(a) Asymmetrical
    T-pad section
(b) Asymmetrical
    π-section

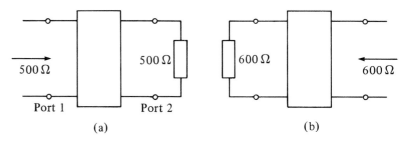

Figure 22   (a)   (b)

R_{OB}; when viewed from port B, the input impedance is R_{OA}. Since the sections are asymmetrical R_{OA} does not have the same value as R_{OB}.

Iterative impedance is the term used for the impedance measured at one port of a two-part network when the other port is terminated with an impedance of the same value. For example, the impedance looking into port 1 of Fig. 22(a) is, say, $500\,\Omega$ when port 2 is terminated in $500\,\Omega$ and the impedance looking into port 2 of Fig. 22(b) is, say, $600\,\Omega$ when port 1 is terminated in $600\,\Omega$. (In symmetric T- and π-sections the two iterative impedances are equal, this value being the characteristic impedance of the section.)

An **image impedance** is defined as the impedance which, when connected to the terminals of a network, equals the impedance presented to it at the opposite terminals. For example, the impedance looking into port 1 of Fig. 23(a) is, say, $400\,\Omega$ when port 2 is terminated in, say $750\,\Omega$, and the impedance seen looking into port 2 (Fig. 23(b)) is $750\,\Omega$ when port 1 is terminated in $400\,\Omega$. An asymmetrical network is correctly terminated when it is terminated in its image impedance. (If the image impedances are equal, the value is the characteristic impedance.)

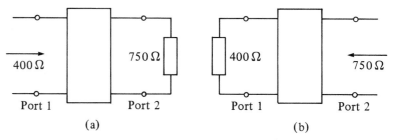

Figure 23   (a)   (b)

The following worked problems show how the iterative and image impedances are determined for asymmetrical T- and π-sections.

Worked problems on asymmetrical T- and π-sections

Problem 1. An asymmetrical T-section attenuator is shown in Fig. 24. Determine for the section (a) the image impedances, and (b) the iterative impedances.

(a) The image impedance R_{OA} seen at port 1 in Fig. 24 is given by equation (11): $R_{OA} = \sqrt{(R_{OC})(R_{SC})}$, where R_{OC} and R_{SC} refer to port 2 being respectively open-circuited and short-circuited.

$$R_{OC} = 200 + 100 = 300\,\Omega$$

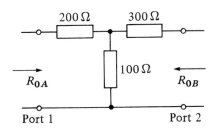

Figure 24

R_{OA}

Port 1   Port 2

Figure 25   Port 1   Port 2

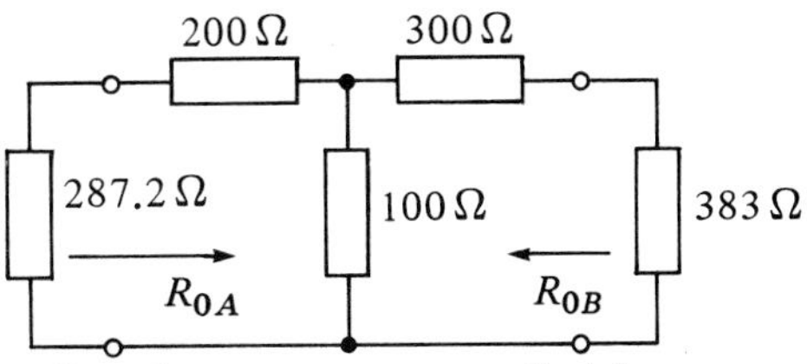

and
$$R_{SC} = 200 + \frac{(100)(300)}{100 + 300} = 275\,\Omega$$

Hence $R_{OA} = \sqrt{[(300)(275)]} = \mathbf{287.2\,\Omega.}$
  Similarly, $R_{OB} = \sqrt{(R_{OC})(R_{SC})}$, where R_{OC} and R_{SC} refer to port 1 being respectively open-circuited and short-circuited.
$$R_{OC} = 300 + 100 = 400\,\Omega$$
and
$$R_{SC} = 300 + \frac{(200)(100)}{200 + 100} = 366.7\,\Omega$$

Hence $R_{OB} = \sqrt{[(400)(366.7)]} = \mathbf{383\,\Omega.}$ **Thus the image impedances are 287.2 Ω and 383 Ω and are shown in the circuit of Fig. 25.**

$\Bigg($ Checking:

$$R_{OA} = 200 + \frac{(100)(300 + 383)}{100 + 300 + 383} = 287.2\,\Omega$$
and
$$R_{OB} = 300 + \frac{(100)(200 + 287.2)}{100 + 200 + 287.2} = 383\,\Omega \Bigg)$$

(b) The iterative impedance at port 1 in Fig. 26 is shown as R_1. Hence
$$R_1 = 200 + \frac{(100)(300 + R_1)}{100 + 300 + R_1} = 200 + \frac{30000 + 100R_1}{400 + R_1}$$

from which
$$400R_1 + R_1^2 = 80\,000 + 200R_1 + 30\,000 + 100R_1$$
and
$$R_1^2 + 100R_1 - 110\,000 = 0.$$

Solving by the quadratic formula gives
$$R_1 = \frac{-100 \pm \sqrt{[100^2 - (4)(1)(-110\,000)]}}{2} = \frac{-100 \pm 670.8}{2} = \mathbf{285.4\,\Omega}$$

(neglecting the negative value).
  The iterative impedance at port 2 in Fig. 27 is shown as R_2. Hence

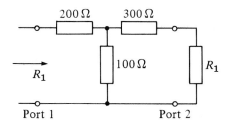

Figure 26

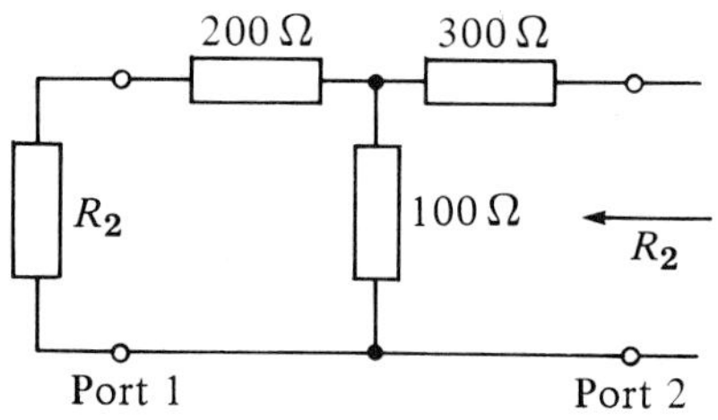

Figure 27

$$R_2 = 300 + \frac{(100)(200 + R_2)}{100 + 200 + R_2} = 300 + \frac{20\,000 + 100R_2}{300 + R_2}$$

from which

$$300R_2 + R_2^2 = 90\,000 + 300R_2 + 20\,000 + 100R_2$$

and

$$R_2^2 - 100R_2 - 110\,000 = 0.$$

Thus

$$R_2 = \frac{100 \pm \sqrt{[(-100)^2 - (4)(1)(-110\,000)]}}{2} = \frac{100 \pm 670.8}{2}$$

$$= 385.4\,\Omega$$

Thus the iterative impedances of the section shown in Fig. 24 are 285.4 Ω and 385.4 Ω.

Problem 2. An asymmetrical π-section attenuator is shown in Fig. 28. Determine for the section (a) the image impedances, and (b) the iterative impedances.

(a) The image resistance R_{0A} seen at port 1 is given by $R_{0A} = \sqrt{(R_{OC})(R_{SC})}$, where the open-circuit impedance at port 2,

$$R_{OC} = \frac{(1000)(5000)}{1000 + 5000} = 833\,\Omega$$

and the short-circuit impedance at port 2,

$$R_{SC} = \frac{(1000)(3000)}{1000 + 3000} = 750\,\Omega$$

Hence $R_{0A} = \sqrt{[(833)(750)]} = \mathbf{790\,\Omega.}$
Similarly, $R_{0B} = \sqrt{(R_{OC})(R_{SC})}$, where the open-circuit impedance at port 1,

$$R_{OC} = \frac{(2000)(4000)}{2000 + 4000} = 1333\,\Omega$$

and the short-circuit impedance at port 1,

$$R_{SC} = \frac{(2000)(3000)}{2000 + 3000} = 1200\,\Omega$$

Hence $R_{0B} = \sqrt{[(1333)(1200)]} = \mathbf{1265\,\Omega.}$
Thus the image impedances are 790 Ω and 1265 Ω.

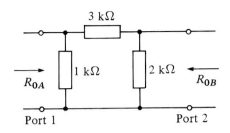

Figure 28

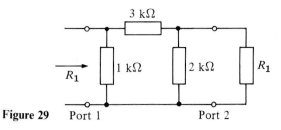

Figure 29

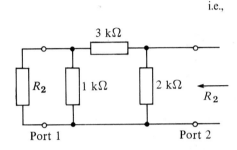

Figure 30

(b) The iterative impedance at port 1 in Fig. 29 is shown as R_1. From circuit theory,

$$R_1 = \frac{1000[3000 + (2000R_1/(2000 + R_1))]}{1000 + 3000 + (2000R_1/(2000 + R_1))}$$

i.e.,

$$R_1 = \frac{3 \times 10^6 + (2 \times 10^6 R_1/(2000 + R_1))}{4000 + (2000R_1/(2000 + R_1))}$$

$$4000R_1 + \frac{2000R_1^2}{2000 + R_1} = 3 \times 10^6 + \frac{2 \times 10^6 R_1}{2000 + R_1}$$

$$8 \times 10^6 R_1 + 4000R_1^2 + 2000R_1^2 = 6 \times 10^9 + 3 \times 10^6 R_1 + 2 \times 10^6 R_1$$

$$6000R_1^2 + 3 \times 10^6 R_1 - 6 \times 10^9 = 0$$

$$2R_1^2 + 1000R_1 - 2 \times 10^6 = 0$$

Using the quadratic formula gives

$$R_1 = \frac{-1000 \pm \sqrt{[(1000)^2 - (4)(2)(-2 \times 10^6)]}}{4}$$

$$= \frac{-1000 \pm 4123}{4} = \textbf{781 } \Omega$$

(neglecting the negative value).
The iterative impedance at port 2 in Fig. 30 is shown as R_2.

$$R_2 = \frac{2000[3000 + (1000R_2/(1000 + R_2))]}{2000 + 3000 + (1000R_2/(1000 + R_2))}$$

$$= \frac{6 \times 10^6 + (2 \times 10^6 R_2/(1000 + R_2))}{5000 + (1000R_2/(1000 + R_2))}$$

Hence

$$5000R_2 + \frac{1000R_2^2}{1000 + R_2} = 6 \times 10^6 + \frac{2 \times 10^6 R_2}{1000 + R_2}$$

$$5 \times 10^6 R_2 + 5000R_2^2 + 1000R_2^2 = 6 \times 10^9 + 6 \times 10^6 R_2 + 2 \times 10^6 R_2$$

$$6000 R_2^2 - 3 \times 10^6 R_2 - 6 \times 10^9 = 0$$

$$2R_2^2 - 1000R_2 - 2 \times 10^6 = 0$$

from which

$$R_2 = \frac{1000 \pm \sqrt{[(-1000)^2 - (4)(2)(-2 \times 10^6)]}}{4} = \frac{1000 \pm 4123}{4}$$

$$= \textbf{1281 } \Omega$$

Thus the iterative impedances of the section shown in Fig. 28 are 781 Ω and 1281 Ω.

Further problems on asymmetrical T- and π-sections may be found in section 9, problems 22 to 24, page 347.

7. The L-section attenuator

A typical L-section attenuator pad is shown in Fig. 31. Such a pad is used for matching purposes only, the design being such that the attenuation introduced is a minimum. In order to derive values for R_1 and R_2, consider the resistances seen from either end of the section.

Looking in at port 1,

$$R_{0A} = R_1 + \frac{R_2 R_{0B}}{R_2 + R_{0B}}$$

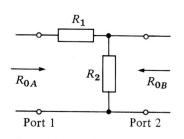

Figure 31 L-section attenuator pad

from which

$$R_{0A} R_2 + R_{0A} R_{0B} = R_1 R_2 + R_1 R_{0B} + R_2 R_{0B} \tag{22}$$

Looking in at port 2,

$$R_{0B} = \frac{R_2(R_1 + R_{0A})}{R_1 + R_{0A} + R_2}$$

from which

$$R_{0B} R_1 + R_{0A} R_{0B} + R_{0B} R_2 = R_1 R_2 + R_2 R_{0A} \tag{23}$$

Adding equations (22) and (23) gives

$$R_{0A} R_2 + 2R_{0A} R_{0B} + R_{0B} R_1 + R_{0B} R_2 = 2R_1 R_2 + R_1 R_{0B} \\ + R_2 R_{0B} + R_2 R_{0A}$$

i.e.,

$$2R_{0A} R_{0B} = 2R_1 R_2$$

and

$$R_1 = \frac{R_{0A} R_{0B}}{R_2} \tag{24}$$

Substituting this expression for R_1 into equation (22) gives

$$R_{0A} R_2 + R_{0A} R_{0B} = \left(\frac{R_{0A} R_{0B}}{R_2}\right) R_2 + \left(\frac{R_{0A} R_{0B}}{R_2}\right) R_{0B} + R_2 R_{0B}$$

i.e.,

$$R_{0A} R_2 + R_{0A} R_{0B} = R_{0A} R_{0B} + \frac{R_{0A} R_{0B}^2}{R_2} + R_2 R_{0B}$$

from which

$$R_2(R_{0A} - R_{0B}) = \frac{R_{0A} R_{0B}^2}{R_2}$$

$$R_2^2(R_{0A} - R_{0B}) = R_{0A} R_{0B}^2$$

and

$$\text{resistance, } \boldsymbol{R_2} = \sqrt{\left(\frac{R_{0A} R_{0B}^2}{R_{0A} - R_{0B}}\right)} \tag{25}$$

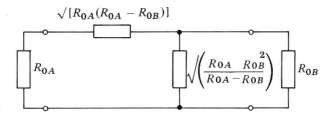

Figure 32

Thus, from equation (24),

$$R_1 = \frac{R_{0A}R_{0B}}{\sqrt{(R_{0A}R_{0B}^2/(R_{0A} - R_{0B}))}} = \frac{R_{0A}R_{0B}}{R_{0B}\sqrt{(R_{0A}/(R_{0A} - R_{0B}))}}$$

$$= \frac{R_{0A}}{\sqrt{R_{0A}}}\sqrt{(R_{0A} - R_{0B})}$$

Hence

$$\text{resistance, } R_1 = \sqrt{[R_{0A}(R_{0A} - R_{0B})]} \qquad (26)$$

Figure 32 shows an L-section attenuator pad with its resistances expressed in terms of the input and output resistances, R_{0A} and R_{0B}.

Worked problem on L-section attenuator

Problem 1. A generator having an internal resistance of $500\,\Omega$ is connected to a $100\,\Omega$ load via an impedance-matching resistance pad as shown in Fig. 33. Determine (a) the values of resistances R_1 and R_2, (b) the attenuation of the pad in decibels, and (c) its insertion loss.

(a) From equation (26), $R_1 = \sqrt{[500(500 - 100)]} = $ **447.2 Ω**

  From equation (25), $R_2 = \sqrt{\left(\dfrac{(500)(100)^2}{500 - 100}\right)} = $ **111.8 Ω**

(b) From section 3, the attenuation is given by $10\lg(P_1/P_2)$dB. Note that, for an asymmetrical section such as that shown in Fig. 33, the expression $20\lg(V_1/V_2)$ or $20\lg(I_1/I_2)$ may **not** be used for attenuation since the terminals of the pad are not matched to equal impedances. In Fig. 34,

$$\text{current } I_1 = \frac{E}{500 + 447.2 + (111.8 \times 100/(111.8 + 100))} = \frac{E}{1000}$$

and

$$\text{current } I_2 = \left(\frac{111.8}{111.8 + 100}\right)I_1 = \left(\frac{111.8}{211.8}\right)\left(\frac{E}{1000}\right) = \frac{E}{1894.5}$$

Figure 33 R_1

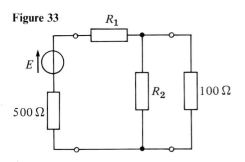

Figure 34 I_1 $R_1 = 447.2\,\Omega$ I_2

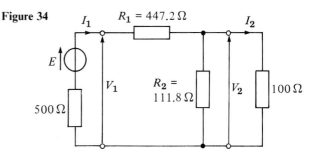

Thus

$$\text{input power, } P_1 = I_1^2(500) = \left(\frac{E}{1000}\right)^2(500)$$

and

$$\text{output power, } P_2 = I_2^2(100) = \left(\frac{E}{1894.5}\right)^2(100)$$

Hence

$$\text{attenuation} = 10\lg\frac{P_1}{P_2} = 10\lg\left\{\frac{(E/1000)^2(500)}{(E/1894.5)^2(100)}\right\}$$

$$= 10\lg\left\{\left(\frac{1894.5}{1000}\right)^2(5)\right\}\text{dB}$$

i.e., **attenuation = 12.54 dB.**

(c) Insertion loss A_L is defined as

$$\frac{\text{voltage across load when connected directly to the generator}}{\text{voltage across load when the two-port network is connected}}$$

Figure 35 shows the generator connected directly to the load.

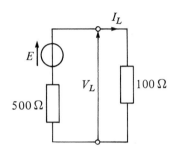

Figure 35

$$\text{Load current, } I_L = \frac{E}{500+100} = \frac{E}{600}$$

and

$$\text{voltage, } V_L = I_L(100) = \frac{E}{600}(100) = \frac{E}{6}.$$

From Fig. 34,

$$\text{voltage, } V_1 = E - I_1(500) = E - (E/1000)500 \qquad \text{from part (b)}$$

i.e.,

$$V_1 = 0.5E$$

$$\text{Voltage, } V_2 = V_1 - I_1 R_1 = 0.5E - \left(\frac{E}{1000}\right)(447.2) = 0.0528E$$

$$\textbf{insertion loss, } A_L = \frac{V_L}{V_2} = \frac{E/6}{0.0528E} = \textbf{3.157}$$

$$\textbf{In decibels, the insertion loss} = 20\lg\frac{V_L}{V_2} = 20\lg 3.157 = \textbf{9.99 dB}$$

Further problems on L-section attenuators may be found in section 9, problems 25 and 26, page 348.

8. Two-port networks in cascade

Often two-port networks are connected in cascade, i.e., the output from the first network becomes the input to the second network, and so on, as shown in Fig. 36. Thus an attenuator may consist of several cascaded sections so as to achieve a particular desired overall performance.

If the cascade is arranged so that the impedance measured at one port and the impedance with which the other port is terminated have the same value, then each section (assuming they are symmetrical) will have the same characteristic impedance Z_0 and the last network will be terminated in Z_0. Thus each network will have a matched termination and hence the

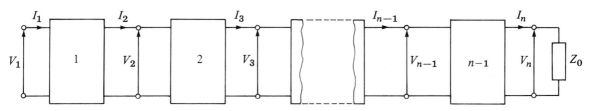

Figure 36 Two-port networks connected in cascade

attenuation in decibels of section 1 in Fig. 36 is given by $a_1 = 20 \lg (V_1/V_2)$. Similarly, the attenuation of section 2 is given by $a_2 = 20 \lg (V_2/V_3)$, and so on.

The overall attenuation is given by

$$a = 20 \lg \frac{V_1}{V_n}$$

$$= 20 \lg \left(\frac{V_1}{V_2} \times \frac{V_2}{V_3} \times \frac{V_3}{V_4} \times \cdots \times \frac{V_{n-1}}{V_n} \right)$$

$$= 20 \lg \frac{V_1}{V_2} + 20 \lg \frac{V_2}{V_3} + \cdots + 20 \lg \frac{V_{n-1}}{V_n}$$

by the laws of logarithms, i.e.,

overall attenuation, $a = a_1 + a_2 + \cdots + a_{n-1}$ \hfill (27)

Thus the overall attenuation is the sum of the attenuations (in decibels) of the matched sections.

Worked problems on cascading two-port networks

Problem 1. Five identical attenuator sections are connected in cascade. The overall attenuation is 70 dB and the voltage input to the first section is 20 mV. Determine (a) the attenuation of each individual attenuator section, (b) the voltage output of the final stage, and (c) the voltage output of the third stage.

(a) From equation (27), the overall attenuation is equal to the sum of the attenuations of the individual sections and, since in this case each section is identical, **the attenuation of each section** $= 70/5 = $ **14 dB.**
(b) If $V_1 = $ the input voltage to the first stage and $V_0 = $ the output voltage of the final stage, then the overall attenuation $= 20 \lg (V_1/V_0)$, i.e.,

$$70 = 20 \lg \left(\frac{20}{V_0} \right) \qquad \text{where } V_0 \text{ is in millivolts}$$

$$3.5 = \lg \left(\frac{20}{V_0} \right)$$

$$10^{3.5} = \frac{20}{V_0}$$

from which

output voltage of final stage, $V_0 = \dfrac{20}{10^{3.5}} = 6.32 \times 10^{-3} \, \text{mV}$

$$= 6.32 \, \mu \text{V}$$

(c) The overall attenuation of three identical stages is $3 \times 14 = 42\,\text{dB}$. Hence $42 = 20\lg(V_1/V_3)$, where V_3 is the voltage output of the third stage. Thus

$$\frac{42}{20} = \lg\left(\frac{20}{V_3}\right), \qquad 10^{42/20} = \frac{20}{V_3}$$

from which **the voltage output of the third stage**, $V_3 = 20/10^{2.1} = \mathbf{0.159\,mV}$.

Problem 2. A d.c. generator has an internal resistance of $450\,\Omega$ and supplies a $450\,\Omega$ load.
(a) Design a T-network attenuator pad having a characteristic impedance of $450\,\Omega$ which, when connected between the generator and the load, will reduce the load current to $\frac{1}{8}$ of its initial value.
(b) If two such networks as designed in (a) were connected in series between the generator and the load, determine the fraction of the initial current that would now flow in the load.
(c) Determine the attenuation in decibels given by four such sections as designed in (a).

The T-network attenuator is shown in Fig. 37 connected between the generator and the load. Since it is matching equal impedances, the network is symmetrical.

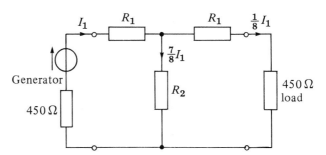

Figure 37

(a) Since the load current is to be reduced to $\frac{1}{8}$ of its initial value, the attenuation, $N = 8$. From equation (13),

$$\text{resistance}, R_1 = \frac{R_0(N-1)}{(N+1)} = 450\frac{(8-1)}{(8+1)} = \mathbf{350\,\Omega}$$

and from equation (14),

$$\text{resistance}, R_2 = R_0\left(\frac{2N}{N^2-1}\right) = 450\left(\frac{2\times 8}{8^2-1}\right) = \mathbf{114\,\Omega}$$

(b) When two such networks are connected in series, as shown in Fig. 38, current I_1 flows into the first stage and $\frac{1}{8}I_1$ flows out of the first stage into the second. Again, $\frac{1}{8}$ of this current flows out of the second stage, i.e., $\frac{1}{8} \times \frac{1}{8}I_1$, i.e., $\frac{1}{64}$ of I_1 flows into the load.
    Thus $\frac{1}{64}$ of the original current flows in the load.
(c) The attenuation of a single stage is 8. Expressed in decibels, the attenuation is $20\lg(I_1/I_2) = 20\lg 8 = 18.06\,\text{dB}$. From equation (23), the overall attenuation of four identical stages is given by $18.06 + 18.06 + 18.06 + 18.06$ i.e., **72.24 dB**.

Further problems on cascading two-port networks may be found in the following section (9), problems 27 to 29, page 348.

Figure 38

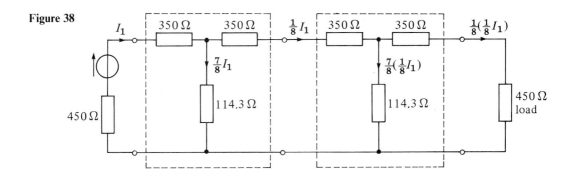

9. Further problems

Logarithmic ratios

1. The ratio of two powers is (a) 3, (b) 10, (c) 30, (d) 10000. Determine the decibel power ratio for each.

[(a) 4.77 dB (b) 10 dB (c) 14.8 dB (d) 40 dB]

2. The ratio of two powers is (a) $\frac{1}{10}$, (b) $\frac{1}{2}$, (c) $\frac{1}{40}$, (d) $\frac{1}{1000}$. Determine the decibel power ratio for each.

[(a) -10 dB (b) -3 dB (c) -16 dB (d) -30 dB]

3. An amplifier has (a) a gain of 25 dB, (b) an attenuation of 25 dB. If the input power is 12 mW, determine the output power in each case.

[(a) 3795 mW (b) 37.9 μW]

4. 7.5% of the power supplied to a cable appears at the output terminals. Determine the attenuation in decibels.

[11.25 dB]

5. The voltage output of an attenuator is 300 μV. If the input and load resistances are equal, determine the voltage input given that the logarithmic voltage ratio is (a) -4.44 dB (b) -1.87 Np.

[(a) 500 μV (b) 1.95 mV]

6. The current input of a system is 250 mA. If the current ratio of the system is (i) 15 dB, (ii) -8 dB, determine (a) the current output and (b) the current ratio expressed in nepers.

$$\begin{bmatrix} \text{(i) (a) 1.406 A} & \text{(b) 1.727 Np} \\ \text{(ii) (a) 99.53 mA} & \text{(b) } -0.921 \text{ Np} \end{bmatrix}$$

Symmetric T- *and* π-*attenuators*

7. Determine the characteristic impedances of the T-network attenuator sections shown in Fig. 39.

[(a) 26.46 Ω (b) 244.9 Ω (c) 1.342 kΩ]

Figure 39

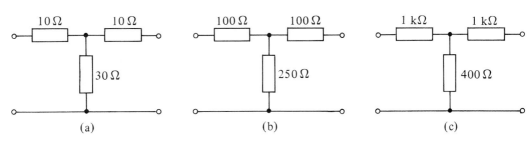

(a)   (b)   (c)

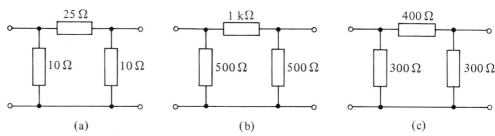

Figure 40
(a)   (b)   (c)

8. Determine the characteristic impedances of the π-network attenuator pads shown in Fig. 40.

[(a) 7.45 Ω   (b) 353.6 Ω   (c) 189.7 Ω]

9. A T-section attenuator is to provide 18 dB voltage attenuation per section and is to match a 1.5 kΩ line. Determine the resistance values necessary per section.

$[R_1 = 1165 \, \Omega, \; R_2 = 384 \, \Omega]$

10. A π- section attenuator has a series resistance of 500 Ω and shunt resistances of 2 kΩ. Determine (a) the characteristic impedance, and (b) the attenuation produced by the network.

[(a) 667 Ω   (b) 6 dB]

11. For each of the attenuator pads shown in Fig. 41 determine (a) the input resistance when the output port is open-circuited, (b) the input resistance when the output port is short-circuited, and (c) the characteristic impedance.

$$\begin{bmatrix} \text{(i) (a) } 50 \, \Omega & \text{(b) } 42 \, \Omega & \text{(c) } 45.83 \, \Omega \\ \text{(ii) (a) } 285.7 \, \Omega & \text{(b) } 240 \, \Omega & \text{(c) } 261.9 \, \Omega \end{bmatrix}$$

12. A television signal received from an aerial through a length of coaxial cable of characteristic impedance 100 Ω has to be attenuated by 15 dB before entering the receiver. If the input impedance of the receiver is also 100 Ω, design a suitable T-attenuator network to give the necessary reduction.

$[R_1 = 69.8 \, \Omega, \; R_2 = 36.7 \, \Omega]$

13. Design (a) a T-section symmetrical attenuator pad, and (b) a π-section symmetrical attenuator pad, to provide a voltage attenuation of 15 dB and having a characteristic impedance of 500 Ω.

$[\text{(a) } R_1 = 349 \, \Omega, \; R_2 = 184 \, \Omega \quad \text{(b) } R_1 = 1.36 \, \text{k}\Omega, \; R_2 = 716 \, \Omega]$

14. Determine the values of the shunt and series resistances for T-pad attenuators of characteristic impedance 400 Ω to provide the following voltage attenuations: (a) 12 dB (b) 25 dB (c) 36 dB.

$$\begin{bmatrix} \text{(a) } R_1 = 239.4 \, \Omega, \; R_2 = 214.5 \, \Omega \\ \text{(b) } R_1 = 357.4 \, \Omega, \; R_2 = 45.14 \, \Omega \\ \text{(c) } R_1 = 387.5 \, \Omega, \; R_2 = 12.68 \, \Omega \end{bmatrix}$$

15. Design a π-section symmetrical attenuator network to provide a voltage attenuation of 24 dB and having a characteristic impedance of 600 Ω.

$[R_1 = 4.736 \, \text{k}\Omega, \; R_2 = 680.8 \, \Omega]$

Figure 41

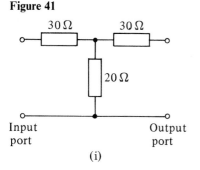

Input   Output
port   port

(i)

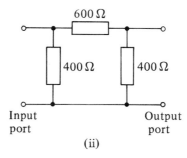

Input   Output
port   port

(ii)

16. Explain what is meant by "the characteristic impedance of an attenuator section". Determine the values of the shunt and series resistances for π-pad attenuator sections of characteristic impedance $600\,\Omega$ to give the following attenuations: (a) $8\,dB$, (b) $20\,dB$, (c) $32\,dB$.

$$\left[\begin{array}{l} \text{(a) } R_1 = 634.1\,\Omega,\ R_2 = 1393.7\,\Omega \\ \text{(b) } R_1 = 2.97\,k\Omega,\ R_2 = 733.3\,\Omega \\ \text{(c) } R_1 = 11.94\,k\Omega,\ R\ \ = 630.9\,\Omega \end{array}\right]$$

17. A battery of emf E and negligible internal resistance is connected across the input terminals of the T network shown in Fig. 42. Determine, in terms of E, the current drawn from the battery when (a) the output terminals is open-circuited, (b) the output terminals are short-circuited, (c) the network is correctly terminated. (d) For the last case, determine the attenuation of the network in decibels.

$$\left[\text{(a) } \frac{E}{320}\,A \quad \text{(b) } \frac{E}{195}\,A \quad \text{(c) } \frac{E}{249.8}\,A \quad \text{(d) } 9.09\,dB\right]$$

18. A d.c. generator has an internal resistance of $600\,\Omega$ and supplies a $600\,\Omega$ load. Design a symmetrical (a) T-network and (b) π-network attenuator pad, having a characteristic impedance of $600\,\Omega$ which when connected between the generator and load will reduce the load current to $\frac{1}{4}$ its initial value.

$$[\text{(a) } R_1 = 360\,\Omega,\ R_2 = 320\,\Omega \qquad \text{(b) } R_1 = 1125\,\Omega,\ R_2 = 1000\,\Omega]$$

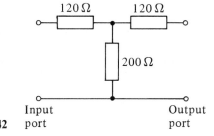

Figure 42   Input port   Output port

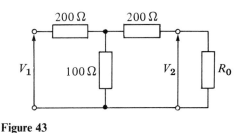

Figure 43

Insertion loss

19. The attenuator section shown in Fig. 43 feeds a matched load. Determine (a) the characteristic impedance R_0 and (b) the insertion loss.

$$[\text{(a) } 282.8\,\Omega \qquad \text{(b) } 15.31\,dB]$$

20. A 0–$10\,k\Omega$ variable resistor is connected across the output of a generator of internal resistance $500\,\Omega$. If a load of $1500\,\Omega$ is connected across the variable resistor, determine the insertion loss in decibels at a tapping of (a) $7.5\,k\Omega$ (b) $2.5\,k\Omega$.

$$[\text{(a) } 8.13\,dB \qquad \text{(b) } 17.09\,dB]$$

21. A symmetrical π attenuator pad has a series arm resistance of $800\,\Omega$ and shunt arms each of $250\,\Omega$. Determine (a) the characteristic impedance of the section, and (b) the insertion loss when feeding a matched load.

$$[\text{(a) } 196.1\,\Omega \qquad \text{(b) } 18.36\,dB]$$

Asymmetric T- and π-attenuators

22. An asymmetric section is shown in Fig. 44. Determine for the section (a) the image impedances, and (b) the iterative impedances.

$$[\text{(a) } 144.9\,\Omega,\ 241.5\,\Omega \qquad \text{(b) } 143.6\,\Omega,\ 243.6\,\Omega]$$

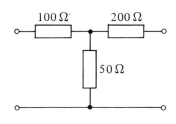

Figure 44

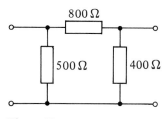

Figure 45

23. An asymmetric π-section is shown in Fig. 45. Determine for the section (a) the image impedances, and (b) the iterative impedances.
                    [(a) 329.5 Ω, 285.6 Ω   (b) 331.2 Ω, 284.2 Ω]
24. Distinguish between image and iterative impedances of a network. An asymmetric T-attenuator section has series arms of resistance 200 Ω and 400 Ω respectively, and a shunt arm of resistance 300 Ω. Determine the image and iterative impedances of the section.
                    [(a) 430.9 Ω, 603.3 Ω; 419.6 Ω, 619.6 Ω]

L-section attenuators

25. Fig. 46 shows an L-section attenuator. The resistance across the input terminals is 250 Ω and the resistance across the output terminals is 100 Ω. Determine the values R_1 and R_2.
                    $[R_1 = 193.6\,\Omega, R_2 = 129.1\,\Omega]$

Figure 46

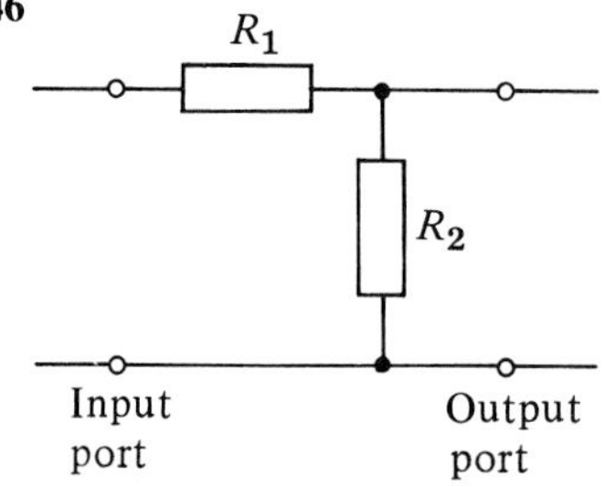

Figure 47

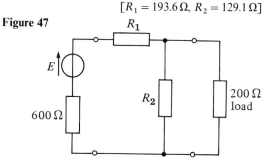

26. A generator having an internal resistance of 600 Ω is connected to a 200 Ω load via an impedance-matching resistive pad as shown in Fig. 47. Determine (a) the values of resistances R_1 and R_2, (b) the attenuation of the matching pad, and (c) its insertion loss.
                    [(a) $R_1 = 490\,\Omega, R_2 = 245\,\Omega$ (b) 9.96 dB   (c) 8.71 dB.]

Cascading two-port networks

27. The input to an attenuator is 24 V and the output is 4 V. Determine the attenuation in decibels. If five such identical attenuators are cascaded, determine the overall attenuation.
                    [15.56 dB; 77.80 dB]
28. Four identical attenuator sections are connected in cascade. The overall attenuation is 60 dB. The input to the first section is 50 mV. Determine (a) the attenuation of each section, (b) the output of the final stage, and (c) the output of the second stage.
                    [(a) 15 dB   (b) 50 μV   (c) 1.58 mV]
29. A d.c. generator has an internal resistance of 300 Ω and supplies a 300 Ω load.
    (a) Design a symmetrical T network attenuator pad having a characteristic impedance of 300 Ω which, when connected between the generator and the load, will reduce the load current to $\frac{1}{3}$ its initial value.
    (b) If two such networks as in (a) were connected in series between the generator and the load, what fraction of the initial current would the load take?
    (c) Determine the fraction of the initial current that the load would take if six such networks were cascaded between the generator and the load.
    (d) Determine the attenuation in decibels provided by five such identical stages as in (a).
                    [(a) $R_1 = 150\,\Omega, R_2 = 225\,\Omega$ (b) $\frac{1}{9}$ (c) $\frac{1}{729}$ (d) 47.71 dB]

14 Transmission lines

1. Introduction

A **transmission line** is a system of conductors connecting one point to another and along which electromagnetic energy can be sent. Thus telephone lines and power distribution lines are typical examples of transmission lines; in electronics, however, the term usually implies a line used for the transmission of radio-frequency (r.f.) energy such as that from a radio transmitter to the antenna.

An important feature of a transmission line is that it should guide energy from a source at the sending end to a load at the receiving end without loss by radiation. One form of construction often used consists of two similar conductors mounted close together at a constant separation. The two conductors form the two sides of a balanced circuit and any radiation from one of them is neutralised by that from the other. Such twin-wire lines are used for carrying high r.f. power, for example, at transmitters. The coaxial form of construction is commonly employed for low power use, one conductor being in the form of a cylinder which surrounds the other at its centre, and thus acts as a screen. Such cables are often used to couple f.m. and television receivers to their antennas.

At frequencies greater than 1000 MHz, transmission lines are usually in the form of a waveguide which may be regarded as coaxial lines without the centre conductor, the energy being launched into the guide or abstracted from it by probes or loops projecting into the guide.

2. Transmission line primary constants

Let an a.c. generator be connected to the input terminals of a pair of parallel conductors of infinite length. A sinusoidal wave will move along the line and a finite current will flow into the line. The variation of voltage with distance along the line will resemble the variation of applied voltage with time. The moving wave, sinusoidal in this case, is called a voltage **travelling wave**. As the wave moves along the line the capacitance of the line is charged up and the moving charges cause magnetic energy to be stored. Thus the propagation of such an **electromagnetic wave** constitutes a flow of energy.

After sufficient time the magnitude of the wave may be measured at any point along the line. The line does not therefore appear to the generator as an open circuit but presents a definite load Z_0. If the sending-end voltage is V_s and the sending end current is I_s then $Z_0 = V_s/I_s$. Thus all of the energy is absorbed by the line and the line behaves in a similar manner to the generator as would a single "lumped" impedance of value Z_0 connected directly across the generator terminals.

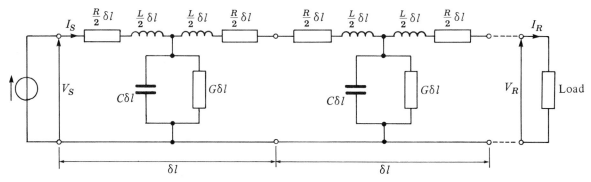

Figure 1

There are **four parameters** associated with transmission lines, these being resistance, inductance, capacitance and conductance.

(i) **Resistance R** is given by $R = \rho l/A$, where ρ is the resistivity of the conductor material, A is the cross-sectional area of each conductor and l is the length of the conductor (for a two-wire system, l represents twice the length of the line). Resistance is stated in ohms per metre length of a line and represents the imperfection of the conductor. A resistance stated in ohms per loop metre is a little more specific since it takes into consideration the fact that there are two conductors in a particular length of line.

(ii) **Inductance L** is due to the magnetic field surrounding the conductors of a transmission line when a current flows through them. The inductance of an isolated twin line is considered in section 7, chapter 12. From equation (23), page 309, the inductance L is given by

$$L = \frac{\mu_0 \mu_r}{\pi} \left\{ \frac{1}{4} + \ln \frac{D}{a} \right\} \text{ henry/metre}$$

where D is the distance between centres of the conductor and a is the radius of each conductor. In most practical lines $\mu_r = 1$. An inductance stated in henrys per loop metre takes into consideration the fact that there are two conductors in a particular length of line.

(iii) **Capacitance C** exists as a result of the electric field between conductors of a transmission line. The capacitance of an isolated twin line is considered in section 3, chapter 12. From equation (14), page 298, the capacitance between the two conductors is given by

$$C = \frac{\pi \varepsilon_0 \varepsilon_r}{\ln (D/a)} \text{ farads/metre}$$

In most practical lines $\varepsilon_r = 1$.

(iv) **Conductance G** is due to the insulation of the line allowing some current to leak from one conductor to the other. Conductance is measured in siemens per metre length of line and represents the imperfection of the insulation. Another name for conductance is leakance.

Each of the four transmission line constants, R, L, C and G, known as the **primary constants**, are uniformly distributed along the line.

From chapter 13, when a symmetrical T-network is terminated in its characteristic impedance Z_0, the input impedance of the network is also equal to Z_0. Similarly, if a number of identical T-sections are connected in cascade, the input impedance of the network will also be equal to Z_0.

A transmission line can be considered to consist of a network of a very large number of cascaded T-sections each a very short length (δl) of transmission line, as shown in Fig. 1. This is an approximation of the uniformly distributed line: the larger the number of lumped parameter sections, the nearer does it approach the true distributed nature of the line. When the generator V_s is connected, a current I_s flows which divides between that flowing through the leakage conductance G, which is lost, and that which progressively charges each capacitor C and which sets up the voltage travelling wave moving along the transmission line. The loss or attenuation in the line is caused by both the conductance G and the series resistance R.

3. Phase delay, wavelength and velocity of propagation

Each section of that shown in Fig. 1 is simply a low-pass filter possessing losses R and G. If losses are neglected, and R and G are removed, the circuit simplifies and the infinite line reduces to a repetitive T-section low-pass filter network as shown in Fig. 2. Let a generator be connected to the line as shown and let the voltage be rising to a maximum positive value just at the instant when the line is connected to it. A current I_s flows through inductance L_1 into capacitor C_1. The capacitor charges and a voltage develops across it. The voltage sends a current through inductance L_1' and L_2 into capacitor C_2. The capacitor charges and the voltage developed across it sends a current through L_2' and L_3 into C_3, and so on. Thus all capacitors will in turn charge up to the maximum input voltage. When the generator voltage falls, each capacitor is charged in turn in opposite polarity, and as before the input charge is progressively passed along to the next capacitor. In this manner voltage and current waves travel along the line together and depend on each other.

Figure 2

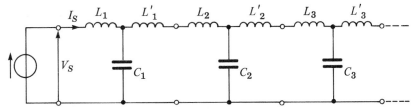

The process outlined above takes time; for example, by the time capacitor C_3 has reached its maximum voltage, the generator input may be at zero or moving towards its minimum value. There will therefore be a time, and thus a phase difference between the generator input voltage and the voltage at any point on the line.

Phase delay

Since the line shown in Fig. 2 is a ladder network of low-pass T-section filters, it may be shown that the phase delay, β, is given by

$$\beta = \omega\sqrt{(LC)} \text{ radians/metre} \qquad (1)$$

where L and C are the inductance and capacitance per metre of the line.

Wavelength

The wavelength λ on a line is the distance between a given point and the next point along the line at which the voltage is the same phase, the initial point leading the latter point by 2π radian. Since in one wavelength a phase change of 2π radians occurs, the phase change per metre is $2\pi/\lambda$. Hence, phase change per metre, $\beta = 2\pi/\lambda$

$$\text{or \textbf{wavelength}, } \lambda = \frac{2\pi}{\beta} \text{ \textbf{metres}} \tag{2}$$

Velocity of propagation

The velocity of propagation, u, is given by $u = f\lambda$, where f is the frequency and λ the wavelength. Hence

$$u = f\lambda = f\left(\frac{2\pi}{\beta}\right) = \frac{2\pi f}{\beta} = \frac{\omega}{\beta} \tag{3}$$

The velocity of propagation of free space is the same as that of light, i.e., approximately 300×10^6 m/s. The velocity of electrical energy along a line is always less than the velocity in free space. The wavelength λ of radiation in free space is given by $\lambda = c/f$, where c is the velocity of light. Since the velocity along a line is always less than c, the wavelength corresponding to any particular frequency is always shorter on the line than it would be in free space.

Worked problems on phase delay, wavelength and velocity of propagation

Problem 1. A parallel-wire air-spaced transmission line operating at 1910 Hz has a phase shift of 0.05 rad/km. Determine (a) the wavelength on the line, and (b) the speed of transmission of a signal.

(a) From equation (2), wavelength $\lambda = 2\pi/\beta = 2\pi/0.05 = $ **125.7 km.**
(b) From equation (3), speed of transmission,

$$u = f\lambda = (1910)(125.7)$$
$$= \textbf{240} \times \textbf{10}^3 \textbf{ km/s}$$
$$\text{or } \textbf{240} \times \textbf{10}^6 \textbf{ m/s}$$

Problem 2. A transmission line has an inductance of 4 mH/loop km and a capacitance of $0.004\,\mu$F/km. Determine, for a frequency of operation of 1 kHz, (a) the phase delay, (b) the wavelength on the line, and (c) the velocity of propagation in metres per second of the signal.

(a) From equation (1), phase delay,

$$\beta = \omega\sqrt{(LC)} = (2\pi 1000)\sqrt{[(4 \times 10^{-3})(0.004 \times 10^{-6})]}$$
$$= \textbf{0.025 rad/km}$$

(b) From equation (2), wavelength $\lambda = 2\pi/\beta = 2\pi/0.025 = $ **251 km.**

(c) From equation (3), velocity of propagation,

$$u = f\lambda = (1000)(251)\,\text{km/s}$$
$$\equiv \mathbf{251 \times 10^6\,m/s}$$

Further problems on phase delay, wavelength and velocity of propagation may be found in section 9, problems 1 to 3, page 376.

4. Current and voltage relationships

Figure 3 shows a voltage source V_s applied to the input terminals of an infinite line, or a line terminated in its characteristic impedance, such that a current I_s flows into the line. At a point, say, 1 km down the line let the current be I_1. The current I_1 will not have the same magnitude as I_s because of line attenuation; also I_1 will lag I_s by some angle β. The ratio I_s/I_1 is therefore a phasor quantity. Let the current a further 1 km down the line be I_2, and so on, as shown in Fig. 3. Each unit length of line can be treated as a section of a repetitive network, as explained in section 2. The attenuation is in the form of a logarithmic decay and

$$\frac{I_s}{I_1} = \frac{I_1}{I_2} = \frac{I_2}{I_3} = e^\gamma$$

where γ is the **propagation constant**. γ has no unit.

Figure 3

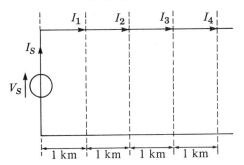

The propagation constant is a complex quantity given by $\gamma = \alpha + j\beta$, where α is the **attenuation constant**, whose unit is the neper, and β is the **phase shift coefficient**, whose unit is the radian. For n such 1 km sections, $I_s/I_R = e^{n\gamma}$, where I_R is the current at the receiving end. Hence

$$\frac{I_s}{I_R} = e^{n(\alpha + j\beta)} = e^{(n\alpha + jn\beta)} = e^{n\alpha}\angle n\beta$$

from which

$$I_R = I_s e^{-n\gamma} = I_s e^{-n\alpha}\angle -n\beta \qquad (4)$$

In equation (4), the attenuation on the line is given by $n\alpha$ nepers and the phase shift is $n\beta$ radians.

At all points along an infinite line, the ratio of voltage to current is Z_0, the characteristic impedance. Thus from equation (4) it follows that

$$\textbf{receiving end voltage, } V_R = V_s e^{-n\gamma} = V_s e^{-n\alpha}\angle -n\beta \qquad (5)$$

Z_0, γ, α and β are referred to as the **secondary line constants** or **coefficients**.

Worked problems on current and voltage relationships

Problem 1. When operating at a frequency of 2 kHz, a cable has an attenuation of 0.25 Np/km and a phase shift of 0.20 rad/km. If a 5V rms signal is applied at the sending end, determine the voltage at a point 10 km down the line, assuming that the termination is equal to the characteristic impedance of the line.

Let V_R be the voltage at a point n km from the sending end, then from equation (5),

$$V_R = V_s e^{-n\gamma}$$

i.e.,

$$V_R = V_s e^{-n\alpha} \angle -n\beta$$

Since $\alpha = 0.25$ Np/km, $\beta = 0.20$ rad/km, $V_s = 5$ V and $n = 10$ km, then

$$V_R = (5)e^{-(10)(0.25)} \angle -(10)(0.20)$$
$$= 5e^{-2.5} \angle -2.0 \text{ V}$$
$$= \mathbf{0.41} \angle -\mathbf{2.0} \text{ V or } \mathbf{0.41} \angle -\mathbf{114.6°} \text{ V}$$

Thus the voltage 10 km down the line is 0.41 V rms lagging the sending end voltage of 5 V by 2.0 rad or 114.6°.

Problem 2. A transmission line 5 km long has a characteristic impedance $800 \angle -25°$ Ω. At a particular frequency, the attenuation coefficient of the line is 0.5 Np/km and the phase shift coefficient is 0.25 rad/km. Determine the magnitude and phase of the current at the receiving end, if the sending end voltage is $2.0 \angle 0°$ V rms.

The receiving end voltage (from equation (5)) is given by

$$V_R = V_s e^{-n\gamma} = V_s e^{-n\alpha} \angle -n\beta$$
$$= (2.0 \angle 0°)e^{-(5)(0.5)} \angle -(5)(0.25)$$
$$= 2.0e^{-2.5} \angle -1.25$$
$$= 0.1642 \angle -71.62° \text{ V}$$

Receiving end current,

$$I_R = \frac{V_R}{Z_0} = \frac{0.1642 \angle -71.62°}{800 \angle -25°} = 2.05 \times 10^{-4} \angle (-71.62° - (-25°)) \text{ A}$$
$$= \mathbf{0.205} \angle -\mathbf{46.62°} \text{ mA}$$

Problem 3. The voltages at the input and at the output of a transmission line properly terminated in its characteristic impedance are 8.0 V and 2.0 V rms respectively. Determine the output voltage if the length of the line is doubled.

The receiving-end voltage V_R is given by $V_R = V_s e^{-n\gamma}$. Hence $2.0 = 8.0e^{-n\gamma}$, from which

$$e^{-n\gamma} = \frac{2.0}{8.0} = 0.25$$

If the line is doubled in length, then

$$V_R = 8.0e^{-2n\gamma} = 8.0(e^{-n\gamma})^2$$
$$= 8.0(0.25)^2 = \mathbf{0.50} \text{ V}$$

Further problems on current and voltage relationships may be found in section 9, problems 4 to 6, page 376.

5. Characteristic impedance and propagation coefficient in terms of the primary constants

Characteristic impedance

At all points along an infinite line, the ratio of voltage to current is called the characteristic impedance Z_0. The value of Z_0 is independent of the length of the line; it merely describes a property of a line that is a function of the physical construction of the line. Since a short length of line may be considered as a ladder of identical low-pass filter sections, the characteristic impedance may be determined from equation (2), chapter 13, i.e.,

$$Z_0 = \sqrt{(Z_{oc}Z_{sc})} \tag{6}$$

since the open-circuit impedance Z_{oc} and the short-circuit impedance Z_{sc} may be easily measured.

The characteristic impedance of a transmission line may also be expressed in terms of the primary constants, R, L, G and C. Measurements of the primary constants may be obtained for a particular line and manufacturers usually state them for a standard length.

Figure 4

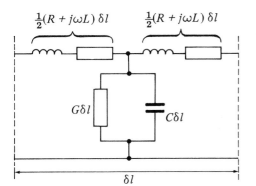

Figure 5

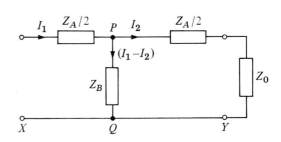

Let a very short length of line δl metres be as shown in Fig. 4 comprising a single T-section. Each series arm impedance is $Z_1 = \frac{1}{2}(R + j\omega L)\delta l$ ohms, and the shunt arm impedance is

$$Z_2 = \frac{1}{Y_2} = \frac{1}{(G + j\omega C)\delta l}$$

$\left(\text{i.e., from chapter 3, the total admittance } Y_2 \text{ is the sum of the admittance}\right.$

of the two parallel arms, i.e., in this case, the sum of

$$G\delta l \quad \text{and} \quad \left(\frac{1}{1/(j\omega C)}\right)\delta l \Bigg).$$

From equation (1), chapter 13, the characteristic impedance Z_0 of a T-section having in each series arm an impedance Z_1 and a shunt arm impedance Z_2 is given by

$$Z_0 = \sqrt{(Z_1^2 + 2Z_1Z_2)}$$

Hence the characteristic impedance of the section shown in Fig. 4 is

$$Z_0 = \sqrt{\left\{[\tfrac{1}{2}(R + j\omega L)\delta l]^2 + 2[\tfrac{1}{2}(R + j\omega L)\delta l]\left[\frac{1}{(G + j\omega C)\delta l}\right]\right\}}$$

The term Z_1^2 involves δl^2 and, since δl is a very short length of line, δl^2 is negligible. Hence

$$Z_0 = \sqrt{\left(\frac{R + j\omega L}{G + j\omega C}\right)} \text{ ohms} \tag{7}$$

If losses R and G are neglected, then

$$Z_0 = \sqrt{\left(\frac{L}{C}\right)} \text{ ohms} \tag{8}$$

Propagation coefficient

Figure 5 shows a T-section with the series arm impedances each expressed as $Z_A/2$ ohms per unit length and the shunt impedance as Z_B ohms per unit length. The p.d. between points P and Q is given by

$$V_{PQ} = (I_1 - I_2)Z_B = I_2\left(\frac{Z_A}{2} + Z_0\right)$$

i.e.,

$$I_1 Z_B - I_2 Z_B = \frac{I_2 Z_A}{2} + I_2 Z_0$$

Hence

$$I_1 Z_B = I_2\left(Z_B + \frac{Z_A}{2} + Z_0\right)$$

from which

$$\frac{I_1}{I_2} = \frac{Z_B + (Z_A/2) + Z_0}{Z_B}$$

From equation (1), chapter 13, $Z_0 = \sqrt{(Z_1^2 + 2Z_1 Z_2)}$. In Fig. 5, $Z_1 \equiv Z_A/2$ and $Z_2 \equiv Z_B$. Thus

$$Z_0 = \sqrt{\left[\left(\frac{Z_A}{2}\right)^2 + 2\left(\frac{Z_A}{2}\right)Z_B\right]} = \sqrt{\left(\frac{Z_A^2}{4} + Z_A Z_B\right)}$$

Thus

$$\frac{I_1}{I_2} = \frac{Z_B + (Z_A/2) + \sqrt{(Z_A Z_B + (Z_A^2/4))}}{Z_B}$$

$$= \frac{Z_B}{Z_B} + \frac{(Z_A/2)}{Z_B} + \frac{\sqrt{(Z_A Z_B + (Z_A^2/4))}}{Z_B}$$

$$= 1 + \frac{1}{2}\left(\frac{Z_A}{Z_B}\right) + \sqrt{\left(\frac{Z_A Z_B}{Z_B^2} + \frac{(Z_A^2/4)}{Z_B^2}\right)}$$

i.e.,

$$\frac{I_1}{I_2} = 1 + \frac{1}{2}\left(\frac{Z_A}{Z_B}\right) + \left[\frac{Z_A}{Z_B} + \frac{1}{4}\left(\frac{Z_A}{Z_B}\right)^2\right]^{1/2} \tag{9}$$

From section 4, $I_1/I_2 = e^\gamma$, where γ is the propagation coefficient. Also, from the binomial theorem.

$$(a+b)^n = a^n + na^{n-1}b + \frac{n(n-1)}{2!}a^{n-2}b^2 + \cdots$$

Thus

$$\left[\frac{Z_A}{Z_B} + \frac{1}{4}\left(\frac{Z_A}{Z_B}\right)^2\right]^{1/2} = \left(\frac{Z_A}{Z_B}\right)^{1/2} + \frac{1}{2}\left(\frac{Z_A}{Z_B}\right)^{-1/2}\frac{1}{4}\left(\frac{Z_A}{Z_B}\right)^2 + \cdots$$

Hence, from equation (9),

$$\frac{I_1}{I_2} = e^\gamma = 1 + \frac{1}{2}\left(\frac{Z_A}{Z_B}\right) + \left[\left(\frac{Z_A}{Z_B}\right)^{1/2} + \frac{1}{8}\left(\frac{Z_A}{Z_B}\right)^{3/2} + \cdots\right]$$

Rearranging gives

$$e^\gamma = 1 + \left(\frac{Z_A}{Z_B}\right)^{1/2} + \frac{1}{2}\left(\frac{Z_A}{Z_B}\right) + \frac{1}{8}\left(\frac{Z_A}{Z_B}\right)^{3/2} + \cdots$$

Let length XY in Fig. 5 be a very short length of line δl and let impedance $Z_A = Z\delta l$, where $Z = R + j\omega L$ and $Z_B = 1/(Y\delta l)$, where $Y = G + j\omega C$. Then

$$e^{\gamma\delta l} = 1 + \left(\frac{Z\delta l}{(1/Y\delta l)}\right)^{1/2} + \frac{1}{2}\left(\frac{Z\delta l}{(1/Y\delta l)}\right) + \frac{1}{8}\left(\frac{Z\delta l}{(1/Y\delta l)}\right)^{3/2} + \cdots$$

$$= 1 + (ZY\delta l^2)^{1/2} + \tfrac{1}{2}(ZY\delta l^2) + \tfrac{1}{8}(ZY\delta l^2)^{3/2} + \cdots$$

$$= 1 + (ZY)^{1/2}\delta l + \tfrac{1}{2}(ZY)(\delta l)^2 + \tfrac{1}{8}(ZY)^{3/2}(\delta l)^3 + \cdots$$

$$= 1 + (ZY)^{1/2}\delta l, \quad \text{if } (\delta l)^2, (\delta l)^3$$

and higher powers are considered as negligible.

e^x may be expressed as a series:

$$e^x = 1 + x + \frac{x^2}{2!} + \frac{x^3}{3!} + \cdots$$

Comparison with $e^{\gamma\delta l} = 1 + (ZY)^{1/2}\delta l$ shows that $\gamma\delta l = (ZY)^{1/2}\delta l$ i.e., $\gamma = \sqrt{(ZY)}$. Thus

$$\text{propagation coefficient, } \gamma = \sqrt{[(R+j\omega L)(G+j\omega C)]} \tag{10}$$

The unit of γ is $\sqrt{(\Omega)(S)}$, i.e., $\sqrt{(\Omega)(1/\Omega)}$, thus γ is dimensionless, as expected, since $I_1/I_2 = e^\gamma$, from which $\gamma = \ln(I_1/I_2)$, i.e., a ratio of two currents. For a lossless line, $R = G = 0$ and

$$\gamma = \sqrt{(j\omega L)(j\omega C)} = j\omega\sqrt{(LC)} \tag{11}$$

Equations (7) and (10) are used to determine the characteristic impedance Z_0 and propagation coefficient γ of a transmission line in terms of the primary constants R, L, G and C. When $R = G = 0$, i.e., losses are neglected, equations (8) and (11) are used to determine Z_0 and γ.

Worked problems on the characteristic impedance and the propagation coefficient in terms of the primary constants

Problem 1. At a frequency of 1.5 kHz the open-circuit impedance of a length of transmission line is $800\underline{/-50°}\ \Omega$ and the short-circuit impedance is $413\underline{/-20°}\ \Omega$. Determine the characteristic impedance of the line at this frequency.

From equation (6),

$$\text{characteristic impedance } Z_0 = \sqrt{(Z_{OC}Z_{SC})} = \sqrt{[(800\underline{/-50°})(413\underline{/-20°})]}$$
$$= \sqrt{(330400\underline{/-70°})}$$
$$= \mathbf{575\underline{/-35°}\ \Omega}$$

by de Moivre's theorem.

Problem 2. A transmission line has the following primary constants: resistance $R = 15\ \Omega/\text{loop km}$, inductance $L = 3.4\ \text{mH/loop km}$, conductance $G = 3\ \mu\text{S/km}$ and capacitance $C = 10\ \text{nF/km}$. Determine the characteristic impedance of the line when the frequency is 2 kHz.

From equation (7),

$$\text{characteristic impedance } Z_0 = \sqrt{\left(\frac{R + j\omega L}{G + j\omega C}\right)}$$

$$R + j\omega L = 15 + j(2\pi 2000)(3.4 \times 10^{-3}) = (15 + j42.73)\Omega = 45.29\underline{/70.66°}\ \Omega$$
$$G + j\omega C = 3 \times 10^{-6} + j(2\pi 2000)(10 \times 10^{-9}) = (3 + j125.66)10^{-6}\text{S}$$
$$= 125.7 \times 10^{-6}\underline{/88.63°}\ \text{S}$$

Hence

$$Z_0 = \sqrt{\left(\frac{45.29\underline{/70.66°}}{125.7 \times 10^{-6}\underline{/88.63°}}\right)} = \sqrt{(0.360 \times 10^6\underline{/-17.97°})}$$

i.e., characteristic impedance, $\mathbf{Z_0 = 600\underline{/-8.99°}\ \Omega.}$

Problem 3. A transmission line having negligible losses has primary line constants of inductance $L = 0.5\ \text{mH/loop km}$ and capacitance $C = 0.12\ \mu\text{F/km}$. Determine, at an operating frequency of 400 kHz, (a) the characteristic impedance, (b) the propagation coefficient, (c) the wavelength on the line, and (d) the velocity of propagation, in metres per second, of a signal.

(a) Since the line is lossfree, from equation (8), the characteristic impedance Z_0 is given by

$$Z_0 = \sqrt{\left(\frac{L}{C}\right)} = \sqrt{\left(\frac{0.5 \times 10^{-3}}{0.12 \times 10^{-6}}\right)} = \mathbf{64.55\ \Omega}$$

(b) From equation (11), for a lossfree line, the propagation coefficient γ is given by

$$\gamma = j\omega\sqrt{(LC)} = j(2\pi 400 \times 10^3)\sqrt{[(0.5 \times 10^{-3})(0.12 \times 10^{-6})]} = j19.47$$

i.e.,

$$\gamma = \mathbf{0 + j19.47}$$

Since $\gamma = \alpha + j\beta$, the attenuation coefficient $\alpha = 0$ and the phase-shift coefficient, $\beta = 19.47\ \text{rad/km}$.

(c) From equation (2),

$$\text{wavelength } \lambda = \frac{2\pi}{\beta} = \frac{2\pi}{19.47} = \textbf{0.323 km or 323 m}$$

(d) From equation (3), velocity of propagation $u = f\lambda = (400 \times 10^3)(323)$
$$= \textbf{129} \times \textbf{10}^6 \textbf{ m/s.}$$

Problem 4. At a frequency 1 kHz the primary constants of a transmission line are resistance $R = 25\,\Omega/\text{loop km}$, inductance $L = 5\,\text{mH/loop km}$, capacitance $C = 0.04\,\mu\text{F/km}$ and conductance $G = 80\,\mu\text{S/km}$. Determine for the line (a) the characteristic impedance, (b) the propagation coefficient, (c) the attenuation coefficient and (d) the phase-shift coefficient.

(a) From equation (7),

$$\text{characteristic impedance } Z_0 = \sqrt{\left(\frac{R + j\omega L}{G + j\omega C}\right)}$$

$$R + j\omega L = 25 + j(2\pi 1000)(5 \times 10^{-3}) = (25 + j31.42) = 40.15\underline{/\ 51.49°}\ \Omega$$
$$G + j\omega C = 80 \times 10^{-6} + j(2\pi 1000)(0.04 \times 10^{-6}) = (80 + j251.33)10^{-6}$$
$$= 263.76 \times 10^{-6}\underline{/\ 72.34°}\ \text{S}$$

Thus characteristic impedance

$$Z_0 = \sqrt{\left(\frac{40.15\underline{/\ 51.49°}}{263.76 \times 10^{-6}\underline{/\ 72.34°}}\right)} = \textbf{390.2}\underline{/\ \textbf{- 10.43° }}\ \Omega$$

(b) From equation (10), propagation coefficient

$$\gamma = \sqrt{[(R + j\omega L)(G + j\omega C)]}$$
$$= \sqrt{[(40.15\underline{/\ 51.49°})(263.76 \times 10^{-6}\underline{/\ 72.34°})]}$$
$$= \sqrt{(0.01059\underline{/\ 123.83°})}$$
$$= \textbf{0.1029}\underline{/\ \textbf{61.92°}}$$

(c) $\gamma = \alpha + j\beta = 0.1029\,(\cos 61.92° + j\sin 61.92°)$, i.e.,

$$\gamma = 0.0484 + j0.0908$$

Thus the attenuation coefficient, $\boldsymbol{\alpha} = \textbf{0.0484 nepers/km.}$
(d) The phase shift coefficient, $\boldsymbol{\beta} = \textbf{0.0908 rad/km.}$

Problem 5. An open wire line is 300 km long and is terminated in its characteristic impedance. At the sending end is a generator having an open-circuit emf of 10.0 V, an internal impedance of $(400 + j0)\,\Omega$ and a frequency of 1 kHz. If the line primary constants are $R = 8\,\Omega/\text{loop km}$, $L = 3\,\text{mH/loop km}$, $C = 7500\,\text{pF/km}$ and $G = 0.25\,\mu\text{S/km}$, determine (a) the characteristic impedance, (b) the propagation coefficient, (c) the attenuation and phase-shift coefficients, (d) the sending-end current, (e) the receiving-end current, (f) the wavelength on the line, and (g) the speed of transmission of signal.

(a) From equation (7),

$$\text{characteristic impedance, } Z_0 = \sqrt{\left(\frac{R + j\omega L}{G + j\omega C}\right)}$$

$$R + j\omega L = 8 + j(2\pi 1000)(3 \times 10^{-3}) = 8 + j6\pi = 20.48\underline{/\ 67.0°}\ \Omega$$
$$G + j\omega C = 0.25 \times 10^{-6} + j(2\pi 1000)(7500 \times 10^{-12}) = (0.25 + j47.12)10^{-6}$$
$$= 47.12 \times 10^{-6}\underline{/\ 89.70°}\ \text{S}$$

Hence characteristic impedance

$$Z_0 = \sqrt{\left(\frac{20.48 \angle 67.0°}{47.12 \times 10^{-6} \angle 89.70°}\right)} = 659.3 \angle -11.35° \ \Omega$$

(b) From equation (10),

propagation coefficient $\gamma = \sqrt{[(R + j\omega L)(G + j\omega C)]}$
$$= \sqrt{[(20.48 \angle 67.0°)(47.12 \times 10^{-6} \angle 89.70°)]}$$
$$= \mathbf{0.03106 \angle 78.35°}$$

(c) $\gamma = \alpha + j\beta = 0.03106 (\cos 78.35° + j \sin 78.35°)$
$$= 0.00627 + j0.03042$$

Hence the attenuation coefficient, $\alpha = \mathbf{0.00627\,Np/km}$ and the phase shift coefficient, $\beta = \mathbf{0.03042\,rad/km.}$

(d) With reference to Fig. 6, since the line is matched, i.e., terminated in its characteristic impedance, $V_S/I_S = Z_0$. Also

$$V_S = V_G - I_S Z_G = 10.0 - I_S(400 + j0)$$

Thus

$$I_S = \frac{V_S}{Z_0} = \frac{10.0 - 400\,I_S}{Z_0}$$

Rearranging gives $I_S Z_0 = 10.0 - 400\,I_S$, from which $I_S(Z_0 + 400) = 10.0$. Thus the sending-end current,

$$I_S = \frac{10.0}{Z_0 + 400} = \frac{10.0}{659.3 \angle -11.35° + 400} = \frac{10.0}{646.41 - j129.75 + 400}$$

$$= \frac{10.0}{1054.4 \angle -7.07°} = \mathbf{9.484 \angle 7.07°\,mA}$$

Figure 6

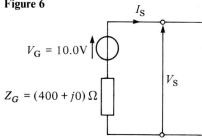

$V_G = 10.0\text{V}$

I_S

V_S

$Z_G = (400 + j0)\ \Omega$

(e) From equation (4), the receiving-end current,

$$I_R = I_S e^{-n\gamma} = I_S e^{-n\alpha} \angle -n\beta$$
$$= (9.484 \angle 7.07°)e^{-(300)(0.00627)} \angle -(300)(0.03042)$$
$$= 9.484 \angle 7.07° e^{-1.881} \angle -9.13\,\text{rad}$$
$$= 1.446 \angle -516°\,\text{mA} = \mathbf{1.446 \angle -156°\,mA}$$

(f) From equation (2),

$$\text{wavelength } \lambda = \frac{2\pi}{\beta} = \frac{2\pi}{0.03042} = \mathbf{206.5\,km}$$

(g) From equation (3),

$$\text{speed of transmission, } u = f\lambda = (1000)(206.5)$$
$$= 206.5 \times 10^3\,\text{km/s}$$
$$= \mathbf{206.5 \times 10^6\,m/s}$$

Problem 6. At a frequency of 1200 Hz the input impedance of a transmission line is $1.5\underline{/\,30°}\,\text{k}\Omega$ with the termination open-circuited and $256\underline{/\,-50°}\,\Omega$ with the termination short-circuited. If the propagation coefficient for the line is $(0.04 + j0.20)$, determine the values of the primary constants R, L, G and C.

From equation (6), characteristic impedance,

$$Z_0 = \sqrt{(Z_{oc}Z_{sc})}$$
$$= \sqrt{[(1500\underline{/\,30°})(256\underline{/\,-50°})]}$$
$$= 620\underline{/\,-10°}\,\Omega$$

The characteristic impedance is also given by

$$Z_0 = \sqrt{\left(\frac{R + j\omega L}{G + j\omega C}\right)}$$

and the propagation coefficient is given by

$$\gamma = \sqrt{[(R + j\omega L)(G + j\omega C)]}$$

Thus

$$\gamma Z_0 = \sqrt{[(R + j\omega L)(G + j\omega C)]}\sqrt{\left(\frac{R + j\omega L}{G + j\omega C}\right)} = R + j\omega L$$

$$\gamma = (0.04 + j0.20) = 0.204\underline{/\,78.69°}$$

Thus

$$\gamma Z_0 = (0.204\underline{/\,78.69°})(620\underline{/\,-10°}) = 126.5\underline{/\,68.69°} = R + j\omega L$$

i.e.,

$$R + j\omega L = (45.97 + j117.85)\Omega$$

from which **resistance $R = 45.97\,\Omega$** and $\omega L = 117.85\,\Omega$. Thus

$$\textbf{inductance, } L = \frac{117.85}{\omega} = \frac{117.85}{2\pi1200} = \textbf{0.0156 H or 15.6 mH}$$

Also

$$\frac{\gamma}{Z_0} = \frac{\sqrt{[(R + j\omega L)(G + j\omega C)]}}{\sqrt{\left(\frac{R + j\omega L}{G + j\omega C}\right)}} = G + j\omega C$$

Thus

$$\frac{\gamma}{Z_0} = \frac{0.204\underline{/\,78.69°}}{620\underline{/\,-10°}} = 3.29 \times 10^{-4}\underline{/\,88.69°} = G + j\omega C$$

i.e.,

$$G + j\omega C = 7.52 \times 10^{-6} + j3.29 \times 10^{-4}$$

from which **conductance, $G = 7.52 \times 10^{-6}\,\text{S}$ or $7.52\,\mu\text{S}$** and $\omega C = 3.29 \times 10^{-4}$. Thus

$$\textbf{capacitance, } C = \frac{3.29 \times 10^{-4}}{\omega} = \frac{3.29 \times 10^{-4}}{2\pi1200} \equiv \textbf{0.0436 }\mu\textbf{F}$$

Further problems on the characteristic impedance and the propagation coefficient in terms of the primary constants may be found in section 9, problems 7 to 12, page 376.

6. Distortion on transmission lines

If the waveform at the receiving end of a transmission line is not the same shape as the waveform at the sending end, **distortion** is said to have occurred. The three main causes of distortion on transmission lines are as follows.

(i) The characteristic impedance Z_0 of a line varies with the operating frequency, i.e., from equation (7),

$$Z_0 = \sqrt{\left(\frac{R + j\omega L}{G + j\omega C}\right)}$$

The terminating impedance of the line may not vary with frequency in the same manner.

In the above equation for Z_0, if the frequency is very low, ω is low and $Z_0 \approx \sqrt{(R/G)}$. If the frequency is very high, then $\omega L \gg R$, $\omega C \gg G$ and $Z_0 \approx \sqrt{(L/C)}$. A graph showing the variation of Z_0 with frequency f is shown in Fig. 7.

If the characteristic impedance is to be constant throughout the entire operating frequency range then the following condition is required: $\sqrt{(L/C)} = \sqrt{(R/G)}$, i.e., $L/C = R/G$, from which

$$LG = CR \tag{12}$$

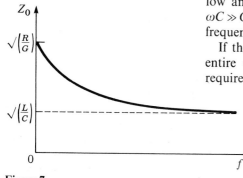

Figure 7

Thus, in a transmission line, if $LG = CR$ it is possible to provide a termination equal to the characteristic impedance Z_0 at all frequencies.

(ii) The attenuation of a line varies with the operating frequency (since $\gamma = \sqrt{[(R + j\omega L)(G + j\omega C)]}$, from equation (10)), thus waves of differing frequencies and component frequencies of complex waves are attenuated by different amounts.

From the above equation for the propagation coefficient;

$$\gamma^2 = (R + j\omega L)(G + j\omega C)$$
$$= RG + j\omega(LG + CR) - \omega^2 LC$$

If $LG = CR = x$, then $LG + CR = 2x$ and $LG + CR$ may be written as $2\sqrt{x^2}$, i.e., $LG + CR$ may be written as $2\sqrt{[(LG)(CR)]}$. Thus

$$\gamma^2 = RG + j\omega(2\sqrt{[(LG)(CR)]}) - \omega^2 LC$$
$$= [\sqrt{(RG)} + j\omega\sqrt{(LC)}]^2$$

and $\gamma = \sqrt{(RG)} + j\omega\sqrt{(LC)}$. Since $\gamma = \alpha + j\beta$,

attenuation coefficient, $\alpha = \sqrt{(RG)}$ $\tag{13}$

and

phase shift coefficient, $\beta = \omega\sqrt{(LC)}$ $\tag{14}$

Thus, in a transmission line, if $LG = CR$, $\alpha = \sqrt{(RG)}$, i.e., the attenuation coefficient is independent of frequency and all frequencies are equally attenuated.

(iii) The delay time, or the time of propagation, and thus the velocity of propagation, varies with frequency and therefore waves of different frequencies arrive at the termination with differing delays. From equation (14), the phase-shift coefficient, $\beta = \omega \sqrt{(LC)}$ when $LG = CR$.

$$\text{Velocity of propagation, } v = \frac{\omega}{\beta} = \frac{\omega}{\omega \sqrt{(LC)}} = \frac{1}{\sqrt{(LC)}} \qquad (15)$$

Thus, in a transmission line, if $LG = CR$, the velocity of propagation, and hence the time delay, is independent of frequency.

From the above it appears that the condition $LG = CR$ is appropriate for the design of a transmission line, since under this condition no distortion is introduced. This means that the signal at the receiving end is the same as the sending-end signal—except that it is reduced in amplitude and delayed by a fixed time. Also, with no distortion, the attenuation on the line is a minimum. In practice, however, $R/L \gg G/C$. The inductance is usually low and the capacitance is large and not easily reduced. Thus if the condition $LG = CR$ is to be achieved in practice, either L or G must be increased since neither C or R can really be altered. It is undesirable to increase G since the attenuation and power losses increase. Thus the inductance L is the quantity that needs to be increased and such an artificial increase in the line inductance is called **loading**. This is achieved either by inserting inductance coils at intervals along the transmission line—this being called "**lumped loading**"—or by wrapping the conductors with a high-permeability metal tape—this being called "**continuous loading**".

Problem 1. An underground cable has the following primary constants: resistance $R = 10\,\Omega/\text{loop km}$, inductance $L = 1.5\,\text{mH/loop km}$, conductance $G = 1.2\,\mu\text{S/km}$ and capacitance $C = 0.06\,\mu\text{F/km}$. Determine by how much the inductance should be increased to satisfy the condition for minimum distortion.

From equation (12), the condition for minimum distortion is given by $LG = CR$, from which

$$\text{inductance } L = \frac{CR}{G} = \frac{(0.06 \times 10^{-6})(10)}{(1.2 \times 10^{-6})} = 0.5\,\text{H or } 500\,\text{mH}$$

Thus the inductance should be increased by $(500 - 1.5)\,\text{mH}$, i.e., **498.5 mH** per loop km, for minimum distortion.

Problem 2. A cable has the following primary constants: resistance $R = 80\,\Omega/\text{loop km}$, conductance, $G = 2\,\mu\text{S/km}$, and capacitance $C = 5\,\text{nF/km}$. Determine, for minimum distortion at a frequency of 1.5 kHz (a) the value of inductance per loop kilometre required, (b) the propagation coefficient, (c) the velocity of propagation of signal and (d) the wavelength on the line.

(a) From equation (12), for minimum distortion, $LG = CR$, from which

$$\text{inductance per loop kilometre, } L = \frac{CR}{G} = \frac{(5 \times 10^{-9})(80)}{(2 \times 10^{-6})}$$

$$= 0.20\,\text{H or } 200\,\text{mH}$$

(b) From equation (13),

$$\text{attenuation coefficient, } \alpha = \sqrt{(RG)} = \sqrt{[(80)(2 \times 10^{-6})]}$$
$$= 0.0126 \, \text{Np/km}$$

and from equation (14), phase shift coefficient,

$$\beta = \omega \sqrt{(LC)} = (2\pi 1500)\sqrt{[(0.20)(5 \times 10^{-9})]} = 0.2980 \, \text{rad/km}$$

Hence the propagation coefficient, $\gamma = \alpha + j\beta = \mathbf{(0.0126 + j0.2980)}$ or $\mathbf{0.2983 \underline{/\ 87.58°}}$.

(c) From equation (15),

velocity of propagation,

$$u = \frac{1}{\sqrt{(LC)}} = \frac{1}{\sqrt{[(0.2)(5 \times 10^{-9})]}} = \mathbf{31620 \, km/s}$$

or $\mathbf{31.62 \times 10^6 \, m/s}$

(d) Wavelength, $\lambda = \dfrac{u}{f} = \dfrac{31.62 \times 10^6}{1500}\,\text{m} = \mathbf{21.08 \, km}$

Further problems on distortion on transmission lines may be found in section 9, problems 13 and 14, page 377.

7. Wave reflection and the reflection coefficient

In earlier sections of this chapter it was assumed that the transmission line had been properly terminated in its characteristic impedance or regarded as an infinite line. In practice, of course, all lines have a definite length and often the terminating impedance does not have the same value as the characteristic impedance of the line. When this is the case, the transmission line is said to have a "**mis matched load**".

The forward-travelling wave moving from the source to the load is called the **incident wave** or the sending-end wave. With a mismatched load the termination will absorb only a part of the energy of the incident wave, the remainder being forced to return back along the line toward the source. This latter wave is called the **reflected wave**.

Electrical energy is transmitted by a transmission line; when such energy arrives at a termination that has a value different from the characteristic impedance, it experiences a sudden change in the impedance of the medium. When this occurs, some reflection of incident energy occurs and the reflected energy is lost to the receiving load. (Reflections commonly occur in nature when a change of transmission medium occurs; for example, sound waves are reflected at a wall, which can produce echoes, and light rays are reflected by mirrors.)

If a transmission line is terminated in its characteristic impedance, no reflection occurs; if terminated in an open circuit or a short circuit, total reflection occurs, i.e., the whole of the incident wave reflects along the line. Between these extreme possibilities, all degrees of reflection are possible.

Open-circuited termination

If a length of transmission line is open-circuited at the termination, no current can flow in it and thus no power can be absorbed by the termination. This condition is achieved if a current is imagined to be reflected from the termination, the reflected current having the same magnitude as the incident wave but with a phase difference of 180°. Also, since no power is absorbed at the termination (it is all returned back along the line), the reflected voltage wave at the termination must be equal to the incident wave. Thus the voltage at the termination must be doubled by the open circuit. The resultant current (and voltage) at any point on the transmission line and at any instant of time is given by the sum of the currents (and voltages) due to the incident and reflected waves (see section 8).

Short-circuit termination

If the termination of a transmission line is short-circuited, the impedance is zero, and hence the voltage developed across it must be zero. As with the open-circuit condition, no power is absorbed by the termination. To obtain zero voltage at the termination, the reflected voltage wave must be equal in amplitude but opposite in phase (i.e., 180° phase difference) to the incident wave. Since no power is absorbed, the reflected current wave at the termination must be equal to the incident current wave and thus the current at the end of the line must be doubled at the short circuit. As with the open-circuited case, the resultant voltage (and current) at any point on the line and at any instant of time is given by the sum of the voltage (and currents) due to the incident and reflected waves.

Energy associated with a travelling wave

A travelling wave on a transmission line may be thought of as being made up of electric and magnetic components. Energy is stored in the magnetic field due to the current (energy $= \frac{1}{2} L I^2$ — see page 312) and energy is stored in the electric field due to the voltage (energy $= \frac{1}{2} C V^2$ — see page 300). It is the continual interchange of energy between the magnetic and electric fields, and *vice versa*, that causes the transmission of the total electromagnetic energy along the transmission line.

When a wave reaches an open-circuited termination the magnetic field collapses since the current I is zero. Energy cannot be lost, but it can change form. In this case it is converted into electrical energy, adding to that already caused by the existing electric field. The voltage at the termination consequently doubles and this increased voltage starts the movement of a reflected wave back along the line. A magnetic field will be set up by this movement and the total energy of the reflected wave will again be shared between the magnetic and electric field components.

When a wave meets a short-circuited termination, the electric field collapses and its energy changes form to the magnetic energy. This results in a doubling of the current.

Reflection coefficient

Let a generator having impedance Z_0 (this being equal to the characteristic impedance of the line) be connected to the input terminals of a transmission line which is terminated in an impedance Z_R, where $Z_0 \neq Z_R$, as shown in Fig. 8. The sending-end or incident current I_i flowing from the source generator flows along the line and, until it arrives at the termination Z_R, behaves as though the line were infinitely long or properly terminated in its characteristic impedance, Z_0.

The incident voltage V_i shown in Fig. 8 is given by

$$V_i = I_i Z_0 \qquad (12)$$

from which

$$I_i = \frac{V_i}{Z_0} \qquad (13)$$

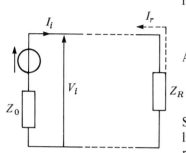

Figure 8

At the termination the conditions must be such that

$$Z_R = \frac{\text{total voltage}}{\text{total current}}$$

Since $Z_R \neq Z_0$, part of the incident wave will be reflected back along the line from the load to the source. Let the reflected voltage be V_r and the reflected current be I_r. Then

$$V_r = -I_r Z_0 \qquad (14)$$

from which

$$I_r = -\frac{V_r}{Z_0} \qquad (15)$$

(Note the minus sign, since the reflected voltage and current waveforms travel in the opposite direction to the incident waveforms.)

Thus, at the termination,

$$Z_R = \frac{\text{total voltage}}{\text{total current}} = \frac{V_i + V_r}{I_i + I_r}$$

$$= \frac{I_i Z_0 - I_r Z_0}{I_i + I_r} \quad \text{from equation (12) and (14)}$$

i.e.,

$$Z_R = \frac{Z_0(I_i - I_r)}{(I_i + I_r)}$$

Hence

$$Z_R(I_i + I_r) = Z_0(I_i - I_r)$$
$$I_r(Z_0 + Z_R) = I_i(Z_0 - Z_R)$$

from which

$$\frac{I_r}{I_i} = \frac{Z_0 - Z_R}{Z_0 + Z_R}$$

The ratio of the reflected current to the incident current is called the **reflection coefficient** and is often given the symbol ρ, i.e.,

$$\frac{I_r}{I_i} = \rho = \frac{Z_0 - R_R}{Z_0 + Z_R} \tag{16}$$

By similar reasoning to above an expression for the ratio of the reflected to the incident voltage may be obtained. From above,

$$Z_R = \frac{V_i + V_r}{I_i + I_r} = \frac{V_i + V_r}{(V_i/Z_0) - (V_r/Z_0)} \quad \text{from equations (13) and (15),}$$

i.e.,

$$Z_R = \frac{V_i + V_r}{(V_i - V_r)/Z_0}$$

Hence

$$\frac{Z_R}{Z_0}(V_i - V_r) = V_i + V_r$$

$$V_i\left(\frac{Z_R}{Z_0} - 1\right) = V_r\left(1 + \frac{Z_R}{Z_0}\right)$$

$$V_i\left(\frac{Z_R - Z_0}{Z_0}\right) = V_r\left(\frac{Z_0 + Z_R}{Z_0}\right)$$

from which,

$$\frac{V_r}{V_i} = \frac{Z_R - Z_0}{Z_0 + Z_R} = -\left(\frac{Z_0 - Z_R}{Z_0 + Z_R}\right) \tag{17}$$

Hence

$$\frac{V_r}{V_i} = -\frac{I_r}{I_i} = -\rho \tag{18}$$

Thus the ratio of the reflected to the incident voltage has the same magnitude as the ratio of reflected to incident current, but is of opposite sign.

From equations (16) and (17) it is seen that when $Z_R = Z_0$, $\rho = 0$ and there is no reflection.

Worked problems on the reflection coefficient

Problem 1. A cable which has a characteristic impedance of $75\,\Omega$ is terminated in a $250\,\Omega$ resistive load. Assuming that the cable has negligible losses and the voltage measured across the terminating load is $10\,\text{V}$, calculate the value of (a) the reflection coefficient for the line, (b) the incident current, (c) the incident voltage, (d) the reflected current, and (e) the reflected voltage.

(a) From equation (16),

$$\textbf{reflection coefficient, } \rho = \frac{Z_0 - Z_R}{Z_0 + Z_R} = \frac{75 - 250}{75 + 250} = \frac{-175}{325} = \textbf{-0.538}$$

(b) The circuit diagram is shown in Fig. 9. Current flowing in the terminating load,

$$I_R = \frac{V_R}{Z_R} = \frac{10}{250} = 0.04\,\text{A}$$

Figure 9

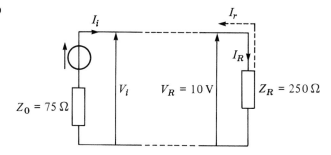

However, current $I_R = I_i + I_r$. From equation (16), $I_r = \rho I_i$. Thus

$$I_R = I_i + \rho I_i = I_i(1 + \rho)$$

from which

$$\textbf{incident current, } I_i = \frac{I_R}{(1 + \rho)} = \frac{0.04}{1 + (-0.538)} = \textbf{0.0866 A} \text{ or}$$

$$\textbf{86.6 mA}$$

(c) From equation (12), **incident voltage**, $V_i = I_i Z_0 = (0.0866)(75) = \textbf{6.50 V}$.
(d) Since $I_R = I_i + I_r$,

$$\textbf{reflected current, } I_r = I_R - I_i = 0.04 - 0.0866 = -\textbf{0.0466 A} \text{ or}$$

$$-\textbf{46.6 mA}$$

(e) From equation (14),

$$\textbf{reflected voltage, } V_r = -I_r Z_0 = -(-0.0466)(75)$$

$$= \textbf{3.50 V}$$

Problem 2. A long transmission line has a characteristic impedance of $(500 - j40)\,\Omega$ and is terminated in an impedance of (a) $(500 + j40)\,\Omega$ and (b) $(600 + j20)\,\Omega$. Determine the magnitude of the reflection coefficient in each case.

(a) From equation (16), reflection coefficient,

$$\rho = \frac{Z_0 - Z_R}{Z_0 + Z_R}$$

When $Z_0 = (500 - j40)\,\Omega$ and $Z_R = (500 + j40)\,\Omega$,

$$\rho = \frac{(500 - j40) - (500 + j40)}{(500 - j40) + (500 + j40)} = \frac{-j80}{1000} = -j0.08$$

Hence the magnitude of the reflection coefficient, $|\rho| = \textbf{0.08}$.
(b) When $Z_0 = (500 - j40)\,\Omega$ and $Z_R = (600 + j20)\,\Omega$,

$$\rho = \frac{(500 - j40) - (600 + j20)}{(500 - j40) + (600 + j20)} = \frac{-100 - j60}{1100 - j20} = \frac{116.62\angle -149.04°}{1100.18\angle -1.04°}$$

$$= 0.106\angle -148°$$

Hence the magnitude of the reflection coefficient, $|\rho| = \textbf{0.106}$.

Problem 3. A lossfree transmission line has a characteristic impedance of $500\angle 0°\,\Omega$ and is connected to an aerial of impedance $(320 + j240)\,\Omega$. Determine

(a) the magnitude of the ratio of the reflected to the incident voltage wave, and
(b) the incident voltage if the reflected voltage is $20\underline{/\,35°}$ V.

(a) From equation (17), the ratio of the reflected to the incident voltage is given by

$$\frac{V_r}{V_i} = \frac{Z_R - Z_0}{Z_0 + Z_R}$$

where Z_0 is the characteristic impedance $500\underline{/\,0°}\,\Omega$ and Z_R is the terminating impedance $(320 + j240)\,\Omega$. Thus

$$\frac{V_r}{V_i} = \frac{(320 + j240) - 500\underline{/\,0°}}{500\underline{/\,0°} + (320 + j240)} = \frac{-180 + j240}{820 + j240} = \frac{300\underline{/\,126.87°}}{854.4\underline{/\,16.31°}}$$

$$= 0.351\underline{/\,110.56°}$$

Hence the magnitude of the ratio $V_r:V_i$ **is 0.351**.

(b) Since $V_r/V_i = 0.351\underline{/\,110.59°}$,

$$\text{incident voltage, } V_i = \frac{V_r}{0.351\underline{/\,110.56°}}$$

Thus, when $V_r = 20\underline{/\,35°}$ V,

$$V_i = \frac{20\underline{/\,35°}}{0.351\underline{/\,110.56°}} = \textbf{57.0}\underline{/\,\textbf{-75.56°}}\textbf{ V}$$

Further problems on the reflection coefficient may be found in section 9, problems 15 to 18, page 377.

---

8. Standing waves and the standing wave ratio

Consider a lossfree transmission line **open-circuited** at its termination. An incident current waveform is completely reflected at the termination, and, as stated in section 7, the reflected current is of the same magnitude as the incident current but is 180° out of phase. Figure 10(a) shows the incident and reflected current waveforms drawn separately (shown as I_i moving to the right and I_r moving to the left respectively) at a time $t = 0$, with $I_i = 0$ and decreasing at the termination.

The resultant of the two waves is obtained by adding them at intervals. In this case the resultant is seen to be zero. Figures 10(b) and (c) show the incident and reflected waves drawn separately as times $t = T/8$ seconds and $t = T/4$, where T is the periodic time of the signal. Again, the resultant is obtained by adding the incident and reflected waveforms at intervals. Figures 10(d) to (h) show the incident and reflected current waveforms plotted on the same axis, together with their resultant waveform, at times $t = 3T/8$ to $t = 7T/8$ at intervals of $T/8$.

If the resultant waveforms shown in Figs 10(a) to (g) are superimposed one upon the other, Fig. 11 results. (Note that the scale has been doubled for clarity.) The waveforms show clearly that waveform (a) moves to (b) after $T/8$, then to (c) after a further period of $T/8$, then to (d), (e), (f), (g) and (h) at intervals of $T/8$. It is noted that at any particular point the current varies sinusoidally with time, but the amplitude of oscillation is different at different points on the line.

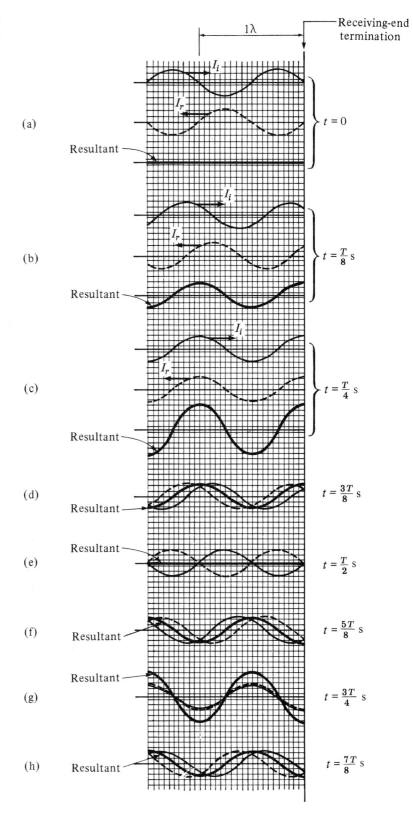

Figure 10 Current waveforms on an open-circuited transmission line

Figure 11

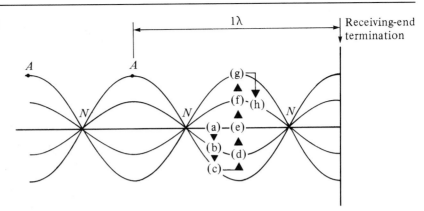

Whenever two waves of the same frequency and amplitude travelling in opposite directions are superimposed on each other as above, interference takes place between the two waves and a **standing** or **stationary wave** is produced. The points at which the current is always zero are called **nodes** (labelled N in Fig. 11). The standing wave does not progress to the left or right and the nodes do not oscillate. Those points on the wave that undergo maximum disturbance are called **antinodes** (labelled A in Fig. 11). The distance between adjacent nodes or adjacent antinodes is $\lambda/2$, where λ is the wavelength. A standing wave is therefore seen to be a periodic variation in the vertical plane taking place on the transmission line without travel in either direction.

The resultant of the incident and reflected voltage for the open-circuit termination may be deduced in a similar manner to that for current. However, as stated in section 7, when the incident voltage wave reaches the termination it is reflected without phase change. Figure 12 shows the resultant waveforms of incident and reflected voltages at intervals of $t = T/8$. Figure 13 shows all the resultant waveforms of Fig. 12(a) to (h) superimposed: again, standing waves are seen to result. Nodes (labelled N) and antinodes (labelled A) are shown in Fig. 13 and, in comparison with the current waves, are seen to occur $90°$ out of phase.

If the transmission line is short-circuited at the termination, it is the incident voltage that is reflected without phase change and the incident current that is reflected with a phase change of $180°$. Thus the diagrams shown in Figs 10 and 11 representing current at an open-circuited termination may be used to represent voltage conditions at a short-circuited termination and the diagrams shown in Figs 12 and 13 representing voltage at an opencircuited termination may be used to represent current conditions at a short-circuited termination.

Figure 14 shows the rms current and voltage waveforms plotted on the same axis against distance for the case of total reflection, deduced from Figs 11 and 13. The rms values are equal to the amplitudes of the waveforms shown in Figs 11 and 13, except that they are each divided by $\sqrt{2}$ (since, for a sine wave, rms value $= (1/\sqrt{2}) \times$ maximum value). With total reflection, the standing-wave patterns of rms voltage and current consist of a succession of positive sine waves with the voltage node located at the

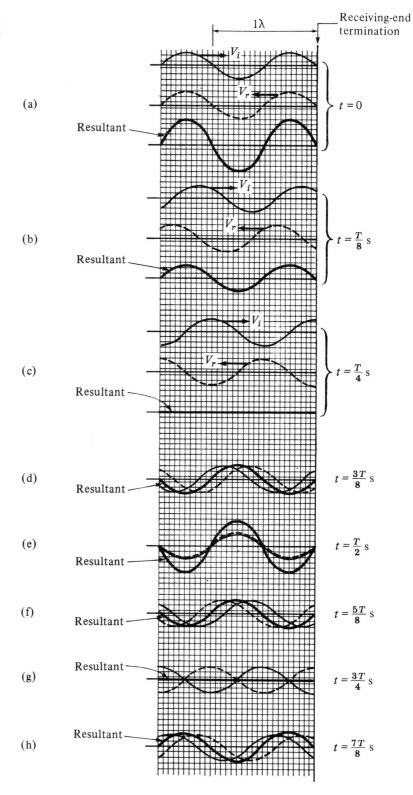

Figure 12 Voltage waveforms on an open-circuited transmission line

(a)

Resultant

$t = 0$

(b)

Resultant

$t = \frac{T}{8}$ s

(c)

Resultant

$t = \frac{T}{4}$ s

(d)

Resultant

$t = \frac{3T}{8}$ s

(e)

Resultant

$t = \frac{T}{2}$ s

(f)

Resultant

$t = \frac{5T}{8}$ s

(g)

Resultant

$t = \frac{3T}{4}$ s

(h)

Resultant

$t = \frac{7T}{8}$ s

1λ

Receiving-end termination

V_i

V_r

V_i

V_r

V_i

V_r

Figure 13

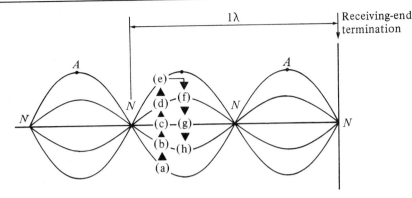

current antinode and the current node located at the voltage antinode. The termination is a current nodal point. The rms values of current and voltage may be recorded on a suitable rms instrument moving along the line. Such measurements of the maximum and minimum voltage and current can provide a reasonably accurate indication of the wavelength, and also provide information regarding the amount of reflected energy relative to the incident energy that is absorbed at the termination, as shown below.

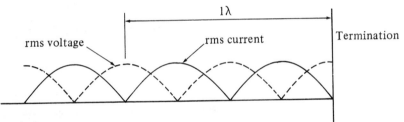

Figure 14

Standing-wave ratio

Let the incident current flowing from the source of a mismatched low-loss transmission line be I_i and the current reflected at the termination be I_r. If I_{MAX} is the sum of the incident and reflected current, and I_{MIN} is their difference, then the **standing-wave ratio (SWR)** on the line is defined as

$$SWR = \frac{I_{MAX}}{I_{MIN}} = \frac{I_i + I_r}{I_i - I_r} \qquad (19)$$

Hence

$$SWR(I_i - I_r) = I_i + I_r$$
$$I_i(SWR - 1) = I_r(SWR + 1)$$

i.e.,

$$\frac{I_r}{I_i} = \frac{SWR - 1}{SWR + 1} \qquad (20)$$

The power absorbed in the termination $P_t = I_i^2 Z_0$ and the reflected power, $P_r = I_r^2 Z_0$. Thus

$$\frac{P_r}{P_t} = \frac{I_r^2 Z_0}{I_i^2 Z_0} = \left(\frac{I_r}{I_i}\right)^2$$

Hence, from equation (20),

$$\frac{P_r}{P_t} = \left(\frac{SWR-1}{SWR+1}\right)^2 \tag{21}$$

Thus the ratio of the reflected to the transmitted power may be calculated directly from the standing-wave ratio, which may be calculated from measurements of I_{MAX} and I_{MIN}. When a transmission line is properly terminated there is no reflection, i.e., $I_r = 0$, and from equation (19) the standing wave ratio is 1. From equation (21), when $SWR = 1$, $P_r = 0$, i.e., there is no reflected power. In practice, the standing-wave ratio is kept as close to unity as possible.

From equation (16), the reflection coefficient, $\rho = I_r/I_i$. Thus, from equation (20),

$$|\rho| = \frac{SWR-1}{SWR+1}$$

Rearranging gives

$$|\rho|(SWR+1) = (SWR-1)$$
$$|\rho|SWR + |\rho| = SWR - 1$$
$$1 + |\rho| = SWR(1 - |\rho|)$$

from which

$$SWR = \frac{1+|\rho|}{1-|\rho|} \tag{22}$$

Equation (22) gives an expression for the standing-wave ratio in terms of the magnitude of the reflection coefficient.

Worked problems on the standing-wave ratio

Problem 1. A transmission line has a characteristic impedance of $600 \angle 0° \, \Omega$ and negligible loss. If the terminating impedance of the line is $(400 + j250)\Omega$, determine (a) the reflection coefficient and (b) the standing-wave ratio.

(a) From equation (16),

$$\text{reflection coefficient, } \rho = \frac{Z_0 - Z_R}{Z_0 + Z_R} = \frac{600 \angle 0° - (400 + j250)}{600 \angle 0° + (400 + j250)}$$

$$= \frac{200 - j250}{1000 + j250} = \frac{320.16 \angle -51.34°}{1030.78 \angle 14.04°}$$

Hence

$$\rho = 0.3106 \angle -65.38°$$

(b) From above, $|\rho| = 0.3106$. Thus from equation (22),

$$SWR = \frac{1+|\rho|}{1-|\rho|} = \frac{1+0.3106}{1-0.3106} = 1.901$$

Problem 2. A low-loss transmission line has a mismatched load such that the reflection coefficient at the termination is $0.2 \angle -120°$. The characteristic impedance of the line is $80\,\Omega$. Calculate (a) the standing-wave ratio, (b) the load impedance, and (c) the incident current flowing if the reflected current is $10\,\text{mA}$.

(a) From equation (22),

$$SWR = \frac{1+|\rho|}{1-|\rho|} = \frac{1+0.2}{1-0.2} = \frac{1.2}{0.8}$$
$$= 1.5$$

(b) From equation (16),

$$\text{reflection coefficient, } \rho = \frac{Z_0 - Z_R}{Z_0 + Z_R}$$

Rearranging gives $\rho(Z_0 + Z_R) = Z_0 - Z_R$, from which $Z_R(\rho + 1) = Z_0(1 - \rho)$ and

$$\frac{Z_R}{Z_0} = \frac{1-\rho}{1+\rho} = \frac{1-0.2\underline{/-120°}}{1+0.2\underline{/-120°}}$$

$$= \frac{1-(-0.10-j0.173)}{1+(-0.10-j0.173)} = \frac{1.10+j0.173}{0.90-j0.173}$$

$$= \frac{1.1135\underline{/\,8.94°}}{0.9165\underline{/-10.80°}} = 1.215\underline{/\,19.82°}$$

Hence

$$\text{load impedance } Z_R = Z_0(1.215\underline{/\,19.82°})$$
$$= (80)(1.215\underline{/\,19.82°})$$
$$= 97.2\underline{/\,19.82°}\,\Omega \text{ or}$$
$$(91.4 + j33.0)\,\Omega$$

(c) From equation (20),

$$\frac{I_r}{I_i} = \frac{SWR-1}{SWR+1}$$

Hence

$$\frac{10}{I_i} = \frac{1.5-1}{1.5+1} = \frac{0.5}{2.5} = 0.2$$

Thus the **incident current, I_i** $= 10/0.2 = $ **50 mA**.

Problem 3. The standing-wave ratio on a mismatched line is calculated as 1.60. If the incident power arriving at the termination is $200\,\text{mW}$, determine the value of the reflected power.

From equation (21),

$$\frac{P_r}{P_t} = \left(\frac{SWR-1}{SWR+1}\right)^2 = \left(\frac{1.60-1}{1.60+1}\right)^2 = \left(\frac{0.60}{2.60}\right)^2 = 0.0533$$

Hence the **reflected power, P_r** $= 0.0533 P_t = (0.0533)(200) = $ **10.66 mW**.

Further problems on the standing wave ratio may be found in the following section 9, problems 19 to 23, page 378.

9. Further problems

Phase delay, wavelength and velocity of propagation

1. A parallel-wire air-spaced line has a phase-shift of 0.03 rad/km. Determine (a) the wavelength on the line, and (b) the speed of transmission of a signal of frequency 1.2 kHz.

$$[(a)\ 209.4\ \text{km} \qquad (b)\ 251.3 \times 10^6\ \text{m/s}]$$

2. A transmission line has an inductance of $5\ \mu\text{H/m}$ and a capacitance of $3.49\ \text{pF/m}$. Determine, for an operating frequency of 5 kHz, (a) the phase delay, (b) the wavelength on the line and (c) the velocity of propagation of the signal in metres per second.

$$[(a)\ 0.131\ \text{rad/km} \qquad (b)\ 48\ \text{km} \qquad (c)\ 240 \times 10^6\ \text{m/s}]$$

3. An air-spaced transmission line has a capacitance of $6.0\ \text{pF/m}$ and the velocity of propagation of a signal is $225 \times 10^6\ \text{m/s}$. If the operating frequency is 20 kHz, determine (a) the inductance per metre, (b) the phase delay, and (c) the wavelength on the line.

$$[(a)\ 3.29\ \mu\text{H/m} \qquad (b)\ 0.558 \times 10^{-3}\ \text{rad/m} \qquad (c)\ 11.25\ \text{km}]$$

Current and voltage relationships

4. When the working frequency of a cable is 1.35 kHz, its attenuation is 0.40 Np/km and its phase-shift is 0.25 rad/km. The sending-end voltage and current are 8.0 V rms and 10.0 mA rms. Determine the voltage and current at a point 25 km down the line, assuming that the termination is equal to the characteristic impedance of the line.

$$\left[\begin{array}{l} V_R = 0.363 \underline{/-6.25}\ \text{mV or } 0.363 \underline{/\ 1.90°}\ \text{mV} \\ I_R = 0.454 \underline{/-6.25}\ \mu\text{A or } 0.454 \underline{/\ 1.90°}\ \mu\text{A} \end{array} \right]$$

5. A transmission line 8 km long has a characteristic impedance $600 \underline{/-30°}\ \Omega$. At a particular frequency the attenuation coefficient of the line is 0.4 Np/km and the phase-shift coefficient is 0.20 rad/km. Determine the magnitude and phase of the current at the receiving end if the sending-end voltage is $5 \underline{/\ 0°}$ V rms.

$$[0.340 \underline{/-61.67}\ \text{mA}]$$

6. The voltages at the input and at the output of a transmission line properly terminated in its characteristic impedance are 10 V and 4 V rms respectively. Determine the output voltage if the length of the line is trebled.

$$[0.64\ \text{V}]$$

Characteristic impedance and propagation constant

7. At a frequency of 800 Hz, the open-circuit impedance of a length of transmission line is measured as $500 \underline{/-35°}\ \Omega$ and the short-circuit impedance as $300 \underline{/-15°}\ \Omega$. Determine the characteristic impedance of the line at this frequency.

$$[387.3 \underline{/-25°}\ \Omega]$$

8. (a) Derive an expression for the characteristic impedance of a transmission line in terms of the primary constants.

   (b) A transmission line has the following primary constants per loop kilometre run: $R = 12\ \Omega$, $L = 3\ \text{mH}$, $G = 4\ \mu\text{S}$ and $C = 0.02\ \mu\text{F}$. Determine the characteristic of the line when the frequency is 750 Hz.

$$[(b)\ 443.3 \underline{/-18.95°}\ \Omega]$$

9. A transmission line having negligible losses has primary constants: inductance $L = 1.0\ \text{mH/loop km}$ and capacitance $C = 0.20\ \mu\text{F/km}$. Determine, at an operating frequency of 50 kHz, (a) the characteristic impedance, (b) the

propagation coefficient, (c) the attenuation and phase-shift coefficients, (d) the wavelength on the line, and (e) the velocity of propagation of signal in metres per second.

[(a) 70.71 Ω (b) $j4.443$ (c) 0; 4.443 rad/km   (d) 1.414 km
(e) 70.7 × 10^6 m/s]

10. At a frequency of 5 kHz the primary constants of a transmission line are: resistance $R = 12\,\Omega$/loop km, inductance $L = 0.50$ mH/loop km, capacitance $C = 0.01\ \mu$F/km and $G = 60\ \mu$S/km. Determine for the line (a) the characteristic impedance, (b) the propagation coefficient, (c) the attenuation coefficient, and (d) the phase-shift coefficient.

$$\begin{bmatrix} \text{(a) } 248.6 \underline{/ -13.28°}\ \Omega & \text{(b) } 0.0795 \underline{/ 65.90°} \\ \text{(c) } 0.0325\ \text{Np/km} & \text{(d) } 0.0726\ \text{rad/km} \end{bmatrix}$$

11. A transmission line is 50 km in length and is terminated in its characteristic impedance. At the sending end a signal emanates from a generator which has an open-circuit emf of 20.0 V, an internal impedance of $(250 + j0)\,\Omega$ at a frequency of 1592 Hz. If the line primary constants are $R = 30\,\Omega$/loop km, $L = 4.0$ mH/loop km, $G = 5.0\ \mu$S/km, and $C = 0.01\ \mu$F/km, determine (a) the value of the characteristic impedance, (b) the propagation coefficient, (c) the attenuation and phase-shift coefficients, (d) the sending-end current, (e) the receiving-end current, (f) the wavelength on the line and (g) the speed of transmission of a signal, in metres per second.

$$\begin{bmatrix} \text{(a) } 706.7 \underline{/ -17°}\ \Omega & \text{(b) } 0.0708 \underline{/ 70.14°} \\ \text{(c) } 0.024\ \text{Np/km; } 0.067\ \text{rad/km} \\ \text{(d) } 21.1 \underline{/ 12.58°}\ \text{mA} & \text{(e) } 6.32 \underline{/ -178.21°}\ \text{mA} \\ \text{(f) } 94.34\ \text{km} & \text{(g) } 150.2 × 10^6\ \text{m/s} \end{bmatrix}$$

12. The input impedance of a transmission line is $700 \underline{/ 40°}\ \Omega$ with the termination open-circuited and $212 \underline{/ -54°}\ \Omega$ with the termination shortcircuited, each being measured at a frequency of 1 kHz. If the propagation coefficient for the line is $(0.02 + j0.15)$, determine the values of the primary constants.

[$R = 14.68\,\Omega,$ $L = 8.98$ mH, $G = 4.04\ \mu$S, $C = 62.50$ nF]

Distortion on transmission lines

13. A cable has the following primary constants: resistance $R = 90\,\Omega$/loop km, inductance $L = 2.0$ mH/loop km, capacitance $C = 0.05\ \mu$F/km and conductance $G = 3.0\ \mu$S/km. Determine the value to which the inductance should be increased to satisfy the condition for minimum distortion.

[1.5 H]

14. A condition of minimum distortion is required for a cable. Its primary constants are: $R = 40\,\Omega$/loop km, $L = 2.0$ mH/loop km, $G = 2.0\ \mu$S/km and $C = 0.08\ \mu$F/km. At a frequency of 100 Hz determine (a) the increase in inductance required, (b) the propagation coefficient, (c) the speed of signal transmission and (d) the wavelength on the line.

[(a) 1.598   (b) $(8.94 + j225)10^{-3}$ (c) 2.795 × 10^6 m/s   (d) 27.95 km]

Reflection coefficient

15. A coaxial line has a characteristic impedance of 100 Ω and is terminated in a 400 Ω resistive load. The voltage measured across the termination is 15 V. The cable is assumed to have negligible losses. Calculate for the line the values of (a) the reflection coefficient, (b) the incident current, (c) the incident voltage,

(d) the reflected current, and (e) the reflected voltage.

[(a) -0.60 (b) $93.75 \, \text{mA}$ (c) $9.375 \, \text{V}$ (d) $-56.25 \, \text{mA}$ (e) $5.625 \, \text{V}$]

16. (a) Explain what happens when a wave on a transmission line meets (i) an open-circuited termination, (ii) a short-circuited termination.

(b) A long transmission line has a characteristic impedance of $(400 - j50) \, \Omega$ and is terminated in an impedance of (i) $(400 + j50) \, \Omega$, (ii) $(500 + j60) \, \Omega$ and (iii) $400 \underline{/\ 0°} \, \Omega$. Determine the magnitude of the reflection coefficient in each case.

[(b)(i) 0.125 (ii) 0.165 (iii) 0.062]

17. Derive an expression for the reflection coefficient of a mismatched transmission line if the characteristic impedance is Z_0 and the load impedance is Z_R. Show that the ratio of the reflected to the incident voltage has the same magnitude as the ratio of the reflected to the incident current, but is opposite in size.

18. A transmission line which is loss-free has a characteristic impedance of $600 \underline{/\ 0°} \, \Omega$ and is connected to a load of impedance $(400 + j300) \, \Omega$. Determine (a) the magnitude of the reflection coefficient and (b) the magnitude of the sending-end voltage if the reflected voltage is $14.60 \, \text{V}$.

[(a) 0.345 (b) $42.32 \, \text{V}$]

Standing-wave ratio

19. A transmission line has a characteristic impedance of $500 \underline{/\ 0°} \, \Omega$ and negligible loss. If the terminating impedance of the line is $(320 + j200) \, \Omega$ determine (a) the reflection coefficient and (b) the standing-wave ratio.

[(a) $0.319 \underline{/ -61.72°}$ (b) 1.937]

20. A low-loss transmission line has a mismatched load such that the reflection coefficient at the termination is $0.5 \underline{/ -135°}$. The characteristic impedance of the line is $60 \, \Omega$. Calculate (a) the standing-wave ratio, (b) the load impedance, and (c) the incident current flowing if the reflected current is $25 \, \text{mA}$.

[(a) 3 (b) $113.93 \underline{/\ 43.32°} \, \Omega$ (c) $50 \, \text{mA}$]

21. (a) With reference to a transmission line, explain the meaning of the standing-wave ratio.

(b) The standing-wave ratio on a mismatched line is calculated as 2.20. If the incident power arriving at the termination is $100 \, \text{mW}$, determine the value of the reflected power.

[$14.06 \, \text{mW}$]

22. The termination of a coaxial cable may be represented as a $150 \, \Omega$ resistance in series with a $0.20 \, \mu\text{H}$ inductance. If the characteristic impedance of the line is $100 \underline{/\ 0°} \, \Omega$ and the operating frequency is $80 \, \text{MHz}$, determine (a) the reflection coefficient and (b) the standing-wave ratio.

[(a) $0.417 \underline{/ -138.35°}$ (b) 2.43]

23. A cable has a characteristic impedance of $70 \underline{/\ 0°} \, \Omega$. The cable is terminated by an impedance of $60 \underline{/\ 30°} \, \Omega$. Determine the ratio of the maximum to minimum current along the line.

[1.77]

Index